MULTILINGUAL DICTIONARY OF FISHING GEAR

ARTES DE PESCA
FISKEREDSKABER
FISCHFANGGERÄT
ΑΛΙΕΥΤΙΚΑ ΕΡΓΑΛΕΙΑ
FISHING GEAR
ENGINS DE PÊCHE
ATTREZZI DA PESCA
VISTUIG
ARTES DE PESCA

2nd Edition

Commission of the European Communities

MULTILINGUAL
DICTIONARY
OF FISHING GEAR

ARTES DE PESCA
FISKEREDSKABER
FISCHFANGGERÄT
ΑΛΙΕΥΤΙΚΑ ΕΡΓΑΛΕΙΑ
ENGINS DE PÊCHE
ATTREZZI DA PESCA
VISTUIG
ARTES DE PESCA

2nd Edition

Fishing News Books

Office for Official Publications
of the European Communities

Published by Fishing News Books
A division of Blackwell Scientific Publications Ltd
Osney Mead, Oxford OX2 0EL
and
Office for Official Publications of the European Communities
2 rue Mercier, L-2985 Luxembourg

Printed in Belgium

British Library
Cataloguing in Publication Data
A catalogue record for this book
is available from the British Library
ISBN 0-85238-192-1

Office for Official Publications of the European Communities
2 rue Mercier, L-2985 Luxembourg
ISBN 92-826-4380-8

1992 — XX, 333 pp. — 17.6 × 25 cm

EUR 14426

First edition published under the title
Glossarium — Fishing gear
by the Commission of the European Communities,
Directorate-General
Telecommunications, Information Industries and Innovation
L-2920 Luxembourg, 1987

Second edition published jointly by
Fishing News Books and the
Office for Official Publications of the European Communities, 1992

Preface

Fishing, one of man's oldest occupations, is an economic sector which has continued to grow in importance in the European Community, especially since the accession of Spain and Portugal.

The fishing industry possesses such a richness and diversity of vocabulary that confusion frequently results, and politicians, specialists and decision-makers, as well as translators and interpreters, find themselves confronted with difficulties arising from the lack of a reliable, standardized terminology. Whether it be fixing catch quotas, agreeing on permitted fishing techniques or establishing measures for the conservation of fish stocks, it is essential to speak a 'common' language.

Directorates-General V (Employment, Industrial Relations and Social Affairs) and XIV (Fisheries) are well aware of the problems and, in an effort to surmount them, asked the Terminology Unit of the Translation Service to compile glossaries of specialized terminology covering various aspects of the fishing industry. The aim of the project was to achieve a certain degree of standardization in the use of terms, taking into account the richness of the language in this particular field, without however adopting the innumerable regional or local variations.

This is a considerably enlarged and updated edition of the *Fishing gear* glossary published by the Publications Office in 1987, and is one of a series of terminological works devoted to fishing, the others being *Fishing vessels and safety on board* and *Aquatic animals and plants*.

The use of these glossaries of fishing terminology, complete with illustrations, will henceforth enable those involved in the fishing industry, politicians, translators and interpreters, and even the general public, to employ a correct 'common' language.

Manuel Marín
Vice-President
Commission of the
European Communities

Prefacio

La pesca, uno de los oficios más antiguos que se conocen, es un sector de actividad cuya relevancia ha ido aumentando continuamente dentro de las Comunidades Europeas, sobre todo tras la adhesión de España y Portugal.

La terminología propia del sector pesquero presenta una riqueza y variedad tales que frecuentemente se presta a confusión.

La comunicación a nivel europeo entre políticos, especialistas y profesionales, sin olvidar el trabajo de traductores e intérpretes, tropieza a menudo con dificultades surgidas de la falta de una terminología fiable y armonizada dentro del sector en cuestión.

Para fijar las cuotas de captura, para llegar a un acuerdo sobre las técnicas de pesca permitidas o para discutir las medidas relativas a la conservación de los recursos pesqueros, es indispensable que se utilice un lenguaje común.

Conscientes de estos problemas, y con ánimo de solventarlos, las Direcciones Generales V «Empleo, relaciones laborales y asuntos sociales» y XIV «Pesca» han solicitado el concurso de la unidad de terminología del Servicio de Traducción para confeccionar una terminología especializada que abarcase los diferentes aspectos del mundo de la pesca.

Dos líneas directrices han orientado este trabajo: por un lado, lograr una cierta unificación en el empleo de los términos y, por el otro, tener en cuenta la riqueza expresiva propia del sector, sin incluir, no obstante, múltiples variantes regionales o locales.

Este glosario de las *artes de pesca* es una nueva edición, considerablemente ampliada y actualizada, de la publicada por la Oficina de Publicaciones bajo el mismo título en 1987. Forma parte de una serie de obras terminológicas consagradas a la pesca *(Buques pesqueros y seguridad a bordo, Animales y plantas acuáticos).*

La terminología del sector de la pesca, presentada en forma de glosario y conteniendo numerosas ilustraciones, se pone a disposición de un vasto público, entre el que se encuentran especialistas, profesionales, políticos, traductores e intérpretes, que, a partir de ahora, dispondrán de un instrumento apropiado de trabajo que les facilitará la comunicación a nivel europeo.

Manuel Marín
Vicepresidente
Comisión de las
Comunidades Europeas

Forord

Fiskeriet, som er et af verdens ældste erhverv, er en sektor, der har fået stadig større betydning i Det Europæiske Fællesskab, især efter Spaniens og Portugals tiltrædelse.

Det ordforråd, der eksisterer på fiskeriområdet, er så rigt og varieret, at det ikke sjældent giver anledning til forvirring.

Kommunikationen på europæisk plan mellem politikere, fagfolk og beslutningstagere hæmmes ofte, og oversætteres og tolkes arbejde vanskeliggøres, fordi der ikke findes pålidelig og harmoniseret terminologi på området.

Når der skal fastsættes fangstkvoter, indgås aftaler om tilladelige fangstmetoder eller drøftes foranstaltninger til bevarelse af fiskebestanden, er det væsentligt, at der tales et fælles sprog.

I erkendelse af disse problemer og til afhjælpning af dem har Generaldirektorat V »Beskæftigelse, Arbejdsmarkedsrelationer og Sociale Anliggender« og XIV »Fiskeri« anmodet Oversættelsestjenestens terminologiafdeling om at sammenstille en fagterminologi, der dækker de forskellige aspekter inden for fiskerisektoren.

Målsætningen for dette arbejde har dels været at nå frem til en vis harmonisering af sprogbrugen, og dels at tage højde for den eksisterende ordrigdom på området, uden dog at medtage de mange regionale og lokale varianter.

Dette glossar med titlen *Fiskeredskaber* er en udvidet og ajourført udgave af det glossar, som Publikationskontoret udgav i 1987 under samme titel. Det indgår i en serie terminologiske værker om fiskeri (*Fiskeskibe og Sikkerhed om Bord, Vanddyr og -planter*).

Med disse glossarer, der er forsynet med illustrationer, bliver terminologien på fiskeriområdet således tilgængelig for et bredt publikum, herunder fiskerisagkyndige, politikere, oversættere og tolke, som fremover vil have et værktøj, der kan fremme brugen af et fælles sprog.

Manuel Marín
Næstformand
Kommissionen for de
Europæiske Fælleskaber

Vorwort

Die Fischerei, eines der ältesten Handwerke der Welt, ist ein wichtiges Tätigkeitsfeld der Europäischen Gemeinschaften und hat im Laufe der Zeit, insbesondere durch den Beitritt Spaniens und Portugals, noch an Bedeutung gewonnen.

Die für diesen Bereich typische Terminologie weist eine derartige Vielfalt an sprachlichen Varianten auf, daß sie allzuoft verwirrend wirkt.

Die Kommunikation auf europäischer Ebene zwischen Politikern, Fachleuten und Entscheidungsträgern, ohne dabei die Arbeit von Übersetzern und Dolmetschern zu vergessen, wird oft durch das Fehlen einer harmonisierten Fachterminologie erschwert.

Für die Festlegung von Fangquoten, die Vereinbarung zulässiger Fangtechniken oder die Erörterung von Maßnahmen zur Erhaltung der Fischbestände ist es unerläßlich, daß eine allen verständliche „gemeinsame" Sprache gesprochen wird.

Angesichts der Wichtigkeit einer sprachlichen Harmonisierung und in dem Bestreben, die terminologischen Probleme zu lösen, haben die Verantwortlichen der Generaldirektion V, „Beschäftigung, Arbeitsbeziehungen und soziale Angelegenheiten", und der Generaldirektion XIV, „Fischerei", das Referat Terminologie des Übersetzungsdienstes beauftragt die für die verschiedenen Gebiete der Fischerei maßgebliche Fachterminologie zusammenzustellen.

Die Arbeiten waren von dem zweifachen Bemühen getragen, einerseits eine gewisse sprachliche Vereinheitlichung zu erreichen und andererseits dem erwähnten Ausdrucksreichtum Rechnung zu tragen, ohne jedoch alle regionalen bzw. lokalen Varianten zu berücksichtigen.

Das vorliegende Glossar *Fischfanggerät* ist die erweiterte und verbesserte Auflage eines unter gleichem Titel vom Amt für amtliche Veröffentlichungen der Europäischen Gemeinschaften 1987 herausgegebenen Werkes. Es ist der erste Band einer Reihe terminologischer Werke zum Thema Fischerei (die weiteren Bände: *Fischereifahrzeuge und Sicherheit an Bord, Wassertiere und -pflanzen*).

Mit diesen durch Zeichnungen anschaulich gestalteten Glossaren wird die Fischereiterminologie einem breiten Publikum zugänglich gemacht. Sachverständigen und in der Fischerei Tätigen sowie Politikern, Übersetzern und Dolmetschern ist somit ein Werkzeug an die Hand gegeben, um die Verständigung auf europäischer Ebene zu erleichtern.

Manuel Marín
Vizepräsident
Kommission der
Europäischen Gemeinschaften

Πρόλογος

Η αλιεία, που αποτελεί μία από τις αρχαιότερες ανθρώπινες δραστηριότητες, είναι ένας τομέας του οποίου η σημασία αυξάνεται συνεχώς στα πλαίσια τη Ευρωπαϊκής Κοινότητας και ιδίως μετά την ένταξη της Ισπανίας και της Πορτογαλίας.

Το λεξιλόγιο που αναφέρεται στον αλιευτικό τομέα, παρουσιάζει τόσο πλούτο και τόση ποικιλία που συχνά προκαλεί σύγχυση.

Η επικοινωνία σε ευρωπαϊκό επίπεδο μεταξύ πολιτικών, ειδικών, ατόμων που λαμβάνουν αποφάσεις και επαγγελματιών, χωρίς να υπολογίζεται η εργασία των μεταφραστών και των διερμηνέων, προσκρούει συχνά σε δυσκολίες οφειλόμενες στην έλλειψη αξιόπιστης και εναρμονισμένης ορολογίας στον αλιευτικό τομέα.

Είτε πρόκειται για τον καθορισμό ποσοστώσεων αλιευμάτων, είτε για να συμφωνηθούν οι παραδεκτές τεχνικές αλιείας είτε, τέλος, για τη διατύπωση των μέτρων διατήρησης των πόρων, είναι απαραίτητο να ομιλούμε την ίδια γλώσσα.

Γνωρίζοντας τα προβλήματα αυτά και μεριμνώντας για την επίλυσή τους, οι υπεύθυνοι των Γενικών Διευθύνσεων, V «Απασχόληση, εργασιακές σχέσεις και κοινωνικές υποθέσεις» και XIV «Αλιεία», απευθύνθηκαν στη Μονάδα Ορολογίας της Μεταφραστικής Υπηρεσίας για τον καθορισμό εξειδικευμένης ορολογίας που να καλύπτει τις διάφορες πτυχές της αλιείας.

Οι εργασίες αυτές απέβλεπαν σε δύο στόχους· αφενός να επιτευχθεί εναρμόνιση της χρήσης των όρων και αφετέρου να ληφθεί υπόψη όλος ο γλωσσικός πλούτος που αφορά τον ειδικό αυτό τομέα, χωρίς φυσικά προσκόλληση στις πολυπληθείς περιφερειακές και τοπικές παραλλαγές.

Το παρόν γλωσσάριο των «Αλιευτικών εργαλείων» είναι μια σημαντικά εκτενέστερη και ενημερωμένη έκδοση του ομώνυμου γλωσσαρίου που είχε δημοσιεύσει η Υπηρεσία Επισήμων Εκδόσεων το 1987. Αποτελεί μέρος μιας σειράς ορολογικών έργων αφιερωμένων στην αλιεία («*Αλιευτικά σκάφη και ασφάλεια πάνω σ' αυτά*», «*Υδρόβια ζώα και φυτά*»).

Η ορολογία του αλιευτικού τομέα που παρουσιάζεται με τη μορφή γλωσσαρίου και έχει εμπλουτισθεί με εικονογράφηση, τίθεται στη διάθεση του κοινού, των ειδικών της αλιείας, των επαγγελματιών, των πολιτικών, των μεταφραστών ή των διερμηνέων οι οποίοι, στο εξής, θα διαθέτουν ένα εργαλείο που θα διευκολύνει την επικοινωνία σε ευρωπαϊκό επίπεδο.

Manuel Marín
Αντιπρόεδρος
Επιτροπή των
Ευρωπαϊκών Κοινοτήτων

Préface

La pêche, qui constitue l'une des activités humaines les plus anciennes, est un secteur dont l'importance n'a fait que croître au sein de la Communauté européenne, en particulier depuis l'adhésion de l'Espagne et du Portugal.

Le vocabulaire propre au domaine halieutique présente une telle richesse, une telle variété, qu'il n'est pas rare qu'il prête à confusion.

La communication au niveau européen entre hommes politiques, spécialistes, décideurs et professionnels, sans considérer le travail des traducteurs et des interprètes, se heurte souvent à des difficultés dues à l'absence d'une terminologie fiable et harmonisée dans le secteur considéré.

Qu'il s'agisse de fixer des quotas de capture, de convenir des techniques admissibles de pêche ou de mettre au point des mesures de conservation des ressources, il est indispensable de parler un langage commun.

Conscients de ces problèmes et soucieux d'y pallier, les responsables des directions générales V (Emploi, relations industrielles et affaires sociales) et XIV (Pêche) ont sollicité l'unité «terminologie» du Service de traduction afin de mettre en place une terminologie spécialisée couvrant les différents aspects de la pêche.

Ces travaux ont été guidés par le double souci de parvenir à une certaine harmonisation dans l'emploi des termes et de prendre en compte toute la richesse linguistique propre à ce domaine, sans pour autant s'attacher aux multiples variantes régionales ou locales.

Le présent glossaire *Engins de pêche* est une édition considérablement élargie et actualisée de celui publié par l'Office des publications sous le même titre en 1987. Il fait partie d'une série d'ouvrages terminologiques consacrés à la pêche (*Bateaux de pêche et sécurité à bord, Animaux et plantes aquatiques*).

La terminologie du domaine halieutique, présentée sous forme de glossaire et enrichie d'illustrations, est ainsi mise à la disposition d'un large public, spécialistes de la pêche, professionnels, hommes politiques, traducteurs ou interprètes, qui, désormais, disposeront d'un instrument de travail destiné à faciliter la communication au niveau européen.

Manuel Marín
Vice-président
Commission des
Communautés européennes

Prefazione

La pesca, che rappresenta una delle attività umane più antiche, è un settore la cui importanza non ha fatto che accrescersi nell'ambito della Comunità europea, soprattutto dopo l'adesione della Spagna e del Portogallo.

Il vocabolario specifico dell'alieutica presenta una tale ricchezza e una tale varietà che non di rado si presta a confusione.

La comunicazione a livello europeo tra uomini, specialisti, uomini di decisione e professionisti, senza contare il lavoro dei traduttori e degli interpreti, incontra spesso delle difficoltà dovute all'assenza di una terminologia certa e armonizzata nel settore in questione.

Che si tratti di fissare le quote di cattura, di convenire le tecniche ammissibili di pesca o di mettere a punto le misure di conservazione delle risorse, si rivela indispensabile parlare un linguaggio comune.

Coscienti di tali problemi e ansiosi di risolverli, i responsabili delle direzioni generali V «Impiego, relazioni industriali e affari sociali» e XIV «Pesca» hanno sollecitato l'unità di Terminologia del servizio di traduzione a predisporre una terminologia specializzata comprendente i differenti aspetti della pesca.

Detti lavori sono stati guidati dal duplice intento di pervenire ad una certa armonizzazione nell'uso dei termini, senza peraltro trascurare tutta la ricchezza linguistica specifica del settore, sganciandosi tuttavia dalle molteplici varianti regionali o locali.

Il presente glossario *Attrezzi da pesca* è un'edizione considerevolmente ampliata e attualizzata di quella pubblicata dall'Ufficio delle pubblicazioni col medesimo titolo nel 1987. Esso fa parte di una serie di lavori terminologici consacrati alla pesca (*Navi da pesca e sicurezza a bordo, Animali e piante acquatiche*).

La terminologia alieutica, presentata sotto forma di glossario e corredata di illustrazioni, viene in tal modo messa a disposizione di un vasto pubblico: specialisti della pesca, professionisti, uomini politici, traduttori e interpreti che disporranno finalmente di uno strumento di lavoro inteso a facilitare la comunicazione a livello europeo.

Manuel Marín
Vicepresidente
Commissione delle
Comunità europee

Voorwoord

De visserij, een van de oudste menselijke bezigheden, is een sector waarvan de omvang binnen de Europese Gemeenschap voortdurend is toegenomen, vooral sinds de toetreding van Spanje en Portugal.

De woordenschat op het gebied van de visvangst is zo rijk en gevarieerd, dat dit niet zelden tot verwarring leidt.

De communicatie op Europees niveau tussen politici, specialisten, beleidmakers en beroepsvissers, om maar te zwijgen van het werk van vertalers en tolken, stuit vaak op problemen door gebrek aan betrouwbare en geharmoniseerde terminologie op dit gebied.

Of het nu gaat om het vaststellen van vangstquota, het bepalen van toelaatbare visserijmethoden of het ontwikkelen van maatregelen ter bescherming van de visbestanden, het is van groot belang dat iedereen dezelfde vaktaal hanteert.

De beleidmakers van de Directoraten-generaal V „Werkgelegenheid, industriële betrekkingen en sociale zaken" en XIV „Visserij" zijn zich van deze problemen bewust en willen er iets aan doen. Daarom hebben zij de afdeling Terminologie van de vertaaldienst verzocht, een gespecialiseerde terminologielijst op te stellen, waarin de verschillende aspecten van de visserij aan de orde komen.

Deze werkzaamheden zijn opgezet met de tweeledige bedoeling een zekere mate van harmonisatie te bewerkstelligen en tevens rekening te houden met de taalrijkdom op dit gebied. Hierbij zijn echter de talrijke regionale of lokale varianten buiten beschouwing gelaten.

Dit glossarium *Vistuig* is een sterk uitgebreide en bijgewerkte versie van het werk dat in 1987 onder dezelfde titel werd uitgegeven door het Publikatiebureau. Het maakt deel uit van een serie terminologische publikaties over de visserij (*Vissersvaartuigen en veiligheid aan boord Waterdieren en waterplanten*).

De terminologie op het gebied van de visvangst komt zo, in de vorm van een glossarium en voorzien van illustraties, binnen het bereik van een breed publiek van visserij-specialisten, beroepsvissers, politici, vertalers en tolken, die daarmee nu beschikken over een instrument dat het gebruik van een gemeenschappelijke vaktaal op Europees niveau mogelijk maakt.

Manuel Marín
Vice-voorzitter
Commissie van de
Europese Gemeenschappen

Prefácio

A pesca, que é uma das mais antigas actividades humanas, é um sector de crescente importância na Comunidade Europeia, principalmente desde a adesão de Espanha e de Portugal.

O vocabulário próprio do domínio haliêutico caracteriza-se por uma considerável riqueza e variedade, o que o torna passível de confusões.

A comunicação a nível europeu entre políticos, especialistas, decisores e profissionais, para não referir o trabalho dos tradutores e intérpretes, depara-se frequentemente com dificuldades devidas à inexistência de terminologia fiável e harmonizada neste sector.

Para a fixação das quotas de pesca, celebração de acordos em matéria de técnicas admissíveis de captura ou discussão de medidas relativas à conservação dos recursos, é essencial que se disponha de uma linguagem comum.

Por estes motivos, as direcções-gerais V, «Emprego, Relações Industriais e Assuntos Sociais», e XIV, «Pesca», solicitaram à Unidade de Terminologia do Serviço de Tradução que elaborasse um trabalho terminológico especializado abrangendo os diferentes aspectos da pesca.

O nosso trabalho reflecte a dupla preocupação de alcançar uma certa uniformização linguística, sem com isso desprezar a riqueza lexical do sector; contudo, nem todas as variantes regionais e locais foram tomadas em consideração.

O presente glossário é uma edição consideravelmente alargada e actualizada daquela que o Serviço das Publicações editou em 1987 com o mesmo título, e faz parte de uma série sobre o tema *«Pescas»* (*«Embarcações de pesca e Segurança a bordo» e «Animais e plantas aquáticas»*).

A terminologia do domínio haliêutico, apresentada sob a forma de glossário e enriquecida com ilustrações, é colocada à disposição de um vasto público em que se incluem especialistas de pesca, profissionais, políticos, tradutores ou intérpretes, que passarão a dispor de um instrumento de trabalho destinado a facilitar a comunicação a nível europeu.

Manuel Marín
Vice-presidente
Comissão das
Comunidades Europeias

Contents

List of collaborators
Lista de colaboradores
Medarbejderliste
Mitarbeiterverzeichnis
Κατάλογος συνεργατών
Liste des collaborateurs
Elenco dei collaboratori
Lijst van de medewerkers
Lista dos colaboradores

COMMISSION OF THE EUROPEAN COMMUNITIES
COMISIÓN DE LAS COMUNIDADES EUROPEAS
KOMMISSIONEN FOR DE EUROPÆISKE FÆLLESSKABER
KOMMISSION DER EUROPÄISCHEN GEMEINSCHAFTEN
ΕΠΙΤΡΟΠΗ ΤΩΝ ΕΥΡΩΠΑΪΚΩΝ ΚΟΙΝΟΤΗΤΩΝ
COMMISSION DES COMMUNAUTÉS EUROPÉENNES
COMMISSIONE DELLE COMUNITÀ EUROPEE
COMMISSIE VAN DE EUROPESE GEMEENSCHAPPEN
COMISSÃO DAS COMUNIDADES EUROPEIAS

TERMINOLOGY

IRMGARD FIAMOZZI
Coordination

STEVE BREDDY
LISE JENSEN
CLAUDE MARCHAND
ARGHIRI VERNARDAKI
HUBERT WELLENSTEIN

JON ETXABE
HENDRIK KOCKAERT
Probationary staff

LORETO GANZINI
External staff

WORD-PROCESSING

MYRIAM BIRCHEN
ALEXANDRE KAMANIS
INGE PHILP
KALLIOPI SAMARA
JOKE SMITH

MEMBER STATES
ESTADOS MIEMBROS
MEDLEMSSTATER
MITGLIEDSTAATEN
ΚΡΑΤΗ ΜΕΛΗ
PAYS MEMBRES
PAESI MEMBRI
LID STATEN
ESTADOS-MEMBROS

We thank the following people for their valuable contributions

MANUEL BARATA METELO
  INIP, Instituto Nacional de Investigação das Pescas, Lisboa
DR. HANSJÜRGEN BOHL
  Bundesforschungsanstalt für Fischerei, Hamburg
PATRICK DORVAL
  Université de Bretagne Occidentale, Lorient
DAVID BRAGANÇA GIL
  INIP, Instituto Nacional de Investigação das Pescas, Lisboa
MARIO FERRETTI
  ICRAM, Istituto centrale per la ricerca scientifica e tecnologica applicata al mare, Roma
R.S.T. FERRO
  Department of Agriculture and Fisheries for Scotland, Marine Laboratory, Aberdeen
VASILIS FILOPOULOS
  ELOT, Hellenic Organization for Standardization
DIMITRI GEORGOPOULOS
  Centre national de recherche maritime, Athènes
EBERHARD GRÜNEWALD
  Bundesministerium für Ernährung, Landwirtschaft und Forsten, Hamburg
ULRIK JES HANSEN
  DIFTA, Dansk Institut for Fiskeriteknologi og Akvakultur, Hirtshals
MARIA KARKANI
  Centre national de recherche maritime, Athènes
JAIME LEGARRA
  Instituto Politécnico Marítimo Pesquero de Pasajes, San Sebastián
ALBERTO MACHADO LEITE
  Direcção-Geral das Pescas, Lisboa
JOÃO CARLOS MENEZES
  INIP, Instituto Nacional de Investigação das Pescas, Lisboa
CLAUDE NEDELEC
  IFREMER, Institut français de recherche pour l'exploitation de la mer, Brest
ALEX OTAEGUI
  Departamento de Agricultura y Pesca, Vitoria
APOSTOLOS PAPANIKOLAOU
  National Technical University, Dept. of Naval Architecture and Marine Engineering,
  Athens
GUNNAR PETERSEN
  Dansk Fiskeriteknologisk Institut, Hirtshals
BOB VAN MARLEN
  RIVO, Rijksinstituut voor Visserijonderzoek, IJmuiden
JOÃO ALFREDO VIEGAS
  INIP, Instituto Nacional de Investigação das Pescas, Lisboa

Acknowledgements

We would like to thank the FAO, Fishing News Books (Oxford), and all others who have authorized the reproduction of illustrations from their publications.

Special thanks also to our colleagues Mr Luc Dutailly (Health and Safety Directorate, DG V) and Mr Hubert Onidi (Directorate-General XIV, Fisheries) for their encouragement and support in this work.

Corpus

3001

ES red

Trama de hilo, cordel, torzal, etc., anudado o trenzado formando malla, que se usa para pescar.

DA net; fiskenet

Fiskeredskaber der hovedsagelig består af net. Net som anvendes i forbindelse med fiskeredskaber.

DE Netz; Fischnetz

Aus Natur- oder Kunststoffgarn geknüpfte, auf die jeweilige Fischereiart abgestimmte Konstruktion.

GR δίχτυ· αλιευτικό δίχτυ

Αλιευτικό εργαλείο που αποτελείται κυρίως από δικτύωμα

EN net; fishing net

A fishing implement comprised mainly of netting. An open-work fabric forming meshes of suitable size for catching fish.

FR filet; filet de pêche

Réseau de mailles plus ou moins larges constituées de fils de diamètres et matériaux variables selon le type de l'engin de pêche.
Engin de pêche fabriqué à partir de nappes de filet.

IT rete; rete da pesca

Attrezzo da pesca costruito mediante pezze di rete.
Termine che può indicare sia la pezza di rete sia gli attrezzi da pesca formati da pezze di rete.

NL net; visnet

Samenstel van aan elkaar geknoopte of anderszins gebonden garens voor het vangen van vis.

PT rede

Trama constituída por um tecido resultante do entrelaçamento de cabos ou fios, de fibras naturais ou artificiais, e que se usa na pesca.

3002

ES red de arrastre; red de bou; trol

Red remolcada que comprende un cuerpo en forma de cono, cerrado por un copo o saco, que se ensancha en la boca mediante alas. Puede ser remolcada por una o dos embarcaciones y, según el tipo, se utilizan en el fondo o a profundidad media (pelágica).
En algunos casos, como en la pesca de arrastre de camarones o peces planos, se puede armar el barco con tangones especiales para arrastrar dos (o hasta cuatro) redes al mismo tiempo (aparejo doble).

DA trawl; trawlnet

Konisk net der består af forpart, krop og pose, samles af to eller 4 (sjældent flere) paneler: over-, under- og sidepaneler, der er sømmet sammen i siden. Slæbes af et eller to fartøjer på bunden eller oppe i vandet.
Ordet trawl kan være såvel fælles- som intetkøn.

DE Schleppnetz; Trawlnetz

Hauptfanggerät der Hochseefischerei. Es besteht aus mehreren gleichen oder ungleichen Netzblättern, die an den Laschen (2-Laschennetze, 4-Laschennetze) miteinander verbunden werden; ihr Querschnitt ist elliptisch oder kreisförmig. Das Netzmaul wird durch die Maulleinen gebildet. Zur Aufnahme des Fangs dient der aus dickeren oder doppelten Netzfäden gefertigte Steert.

GR τράτα· δίχτυ τράτας

Ρυμουλκούμενο δίχτυ αποτελούμενο από ένα κύριο τμήμα κωνικού σχήματος, κλεισμένου από μία θήκη ή σάκο και το οποίο από το άνοιγμα και μετά προεκτείνεται σε φτερά. Μπορεί να ρυμουλκηθεί από ένα ή δύο σκάφη και ανάλογα με τον τύπο της τράτας, χρησιμοποιείται στο βυθό ή τα πελαγικά νερά
Σε ορισμένες περιπτώσεις, όπως η αλιεία γαρίδας ή πλατύψαρων με τράτα, το σκάφος μπορεί να εξοπλισθεί ειδικά με προώστες για να ρυμουλκεί δύο (ή ακόμη και τέσσερις) τράτες συγχρόνως (διπλός εξαρτισμός)

EN trawl; trawl net

Towed net consisting of a cone-shaped body, closed by a bag or codend and extended at the opening by wings. It can be towed by one or two boats and, according to the type, is used on the bottom or in midwater (pelagic).
In certain cases, as in trawling for shrimp or flatfish, the trawler can be specially rigged with outriggers to tow two (or even four) trawls at the same time (double rigging).

3002

FR chalut

Filet remorqué constitué d'un corps de forme conique, fermé par une poche et prolongé à l'ouverture par des ailes. Il peut être traîné par un ou deux bateaux et, selon le type, fonctionner au fond ou entre deux eaux (pélagique).

Dans certains cas, comme pour le chalutage de la crevette ou des poissons plats, le chalutier peut être gréé spécialement de tangons pour remorquer deux (ou même quatre) chaluts simultanément (gréement double).

IT rete da traino

Rete da pesca che viene trainata in mare al fine di catturare, nel suo progressivo avanzamento, organismi marini.

NL trawl; trawlnet; sleepnet; treil

Net met afzonderlijk schrik- en vanggedeelte, dat mechanisch of met handkracht door het water wordt gesleept om de vis actief te vangen.

PT rede de arrastar

Tipo de arte de pesca rebocada que é constituída por um corpo de forma aproximadamente cónica, fechado por um saco e prolongado por asas até à boca (abertura). Pode ser rebocada por uma ou duas embarcações e, de acordo com o respectivo tipo, pode funcionar no fundo ou entre duas águas.

Em certos casos, como por exemplo no arrasto de camarão ou de peixes chatos, a embarcação (arrastão) pode ser especialmente armada com retrancas para poder rebocar duas (ou mesmo quatro) redes de arrasto em simultâneo (armamento duplo).

3003

ES red de tiro danesa; red danesa

Red que cierra en forma de red de arrastre (con bandas y saco), que normalmente se cala desde una embarcación y se puede maniobrar desde la misma embarcación.

DA snurrevod

Konisk net (ligner trawl) ofte med lange arme. Lange vodtove lægges ud på havbunden sammen med voddet, så de dækker et stort areal. Når de hales sammen med voddet, skræmmer tovene fiskene ind i voddets bane.

DE Snurrewade; dänische Ringwade; dänisches Wadennetz; Grundzugnetz

Zugnetz der Küstenfischerei.

Nach Setzen einer Ankerboje werden die Leinen im ca. rechten Winkel ausgefahren. Das anschließende Netz wird mit dem zweiten Leinenpaar im Halbkreis zur Ankerboje zurückgebracht. Der Kutter wird an der Ankerboje festgemacht, und das Netz wird mit möglichst konstanter Leinengeschwindigkeit gehievt. Die von den Leinen ausgehende Scheuchwirkung ist ein entscheidender Faktor für die Fängigkeit. Energiesparende Fischereimethode.

GR δανέζικη τράτα· δανέζα· σχοινότρατα· δανέζικος γρίπος· δανέζικη σαγήνη

Γρίπος με ενσωματωμένο ένα χωνοειδές δίχτυ (με φτερά και σάκο) και πολύ μακριά σχοινιά απλωμένα στον πυθμένα, που σύρονται από ένα σκάφος στην ανοιχτή θάλασσα

EN Danish seine

A seine incorporating a funnel-shaped net (with wings and codend) and very long ropes set out on the sea bed and hauled to a vessel in the open sea.

FR senne danoise; senne de fond

La conception de ce filet, formé de deux ailes, d'un corps et d'une poche, rappelle en bien des points celle des chaluts. Manœuvrée à partir d'un bateau, cette senne est généralement utilisée sur le fond où elle est halée par deux cordages, habituellement très longs, mis à l'eau de manière à assurer le plus grand rabattement possible du poisson vers l'ouverture du filet.

3003

| IT | sciabica danese

| NL | Deens zegennet; snurrevod; snurrevaad;
Deens seinenet; Deense zegen; snurrevaat

Vistuig bestaande uit een zakvormig net, voorzien
van vlerken met daaraan de vislijnen, dat vanuit een
zekere lokatie bij het vangen door één voor anker
liggend schip wordt ingehaald.

| PT | rede de cerco dinamarquesa

Rede envolvente arrastante incorporando uma
secção com a forma de bolsa e constituída por asas,
boca e saco, dotada de compridos cabos de alagem
(cordas) e que é manobrada por uma só
embarcação.

3003

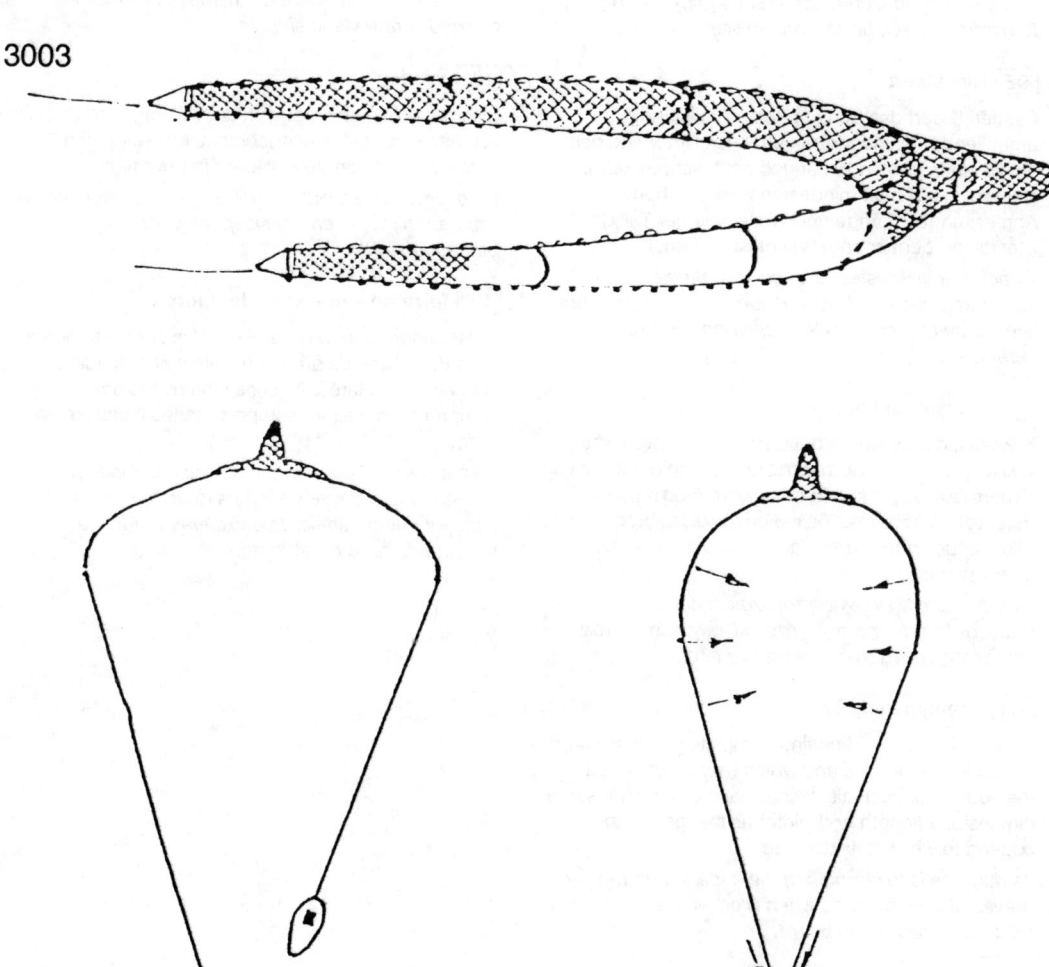

3004

ES | **Cubierta de refuerzo; forro de refuerzo**

Paño o pieza de red de forma cilíndrica que rodea completamente el copo del arte de arrastre, y que se fija al mismo, a determinados intervalos. Tiene, al menos, las mismas dimensiones (longitud y anchura) que la parte del copo a la que se haya fijado.

Está destinada a reforzar el copo de la red de arrastre con el fin de evitar que éste reviente cuando esté muy cargado de pescado y cuando se vire la red a bordo.

DA | **forstærkningspose**

Et cylinderformet stykke net, der fuldstændigt omkranser trawlens fangstpose, og som kan være fastgjort til fangstposen flere steder.

Forstærkningsposen forstærker fangstposen og forhindrer, at den brister, når trawlen hales ind.

DE | **Hievsteert**

Ein den Steert des Schleppnetzes vollständig umhüllendes zylinderförmiges Stück Netzwerk, das am Steert in Abständen angebracht werden kann. Der Hievsteert muß mindestens die gleichen Abmessungen (Länge und Weite) wie der Teil des Steerts, an dem er angebracht ist, haben.

Zweck eines Hievsteerts ist es, den Steert zu verstärken und ein Platzen des mit Fischen gefüllten Steerts beim Einholen des Schleppnetzes zu verhindern.

GR | **ενισχυτική θήκη**

Κυλινδρικό κομμάτι δικτυώματος που περιβάλλει τελείως το σάκο μιας τράτας και το οποίο μπορεί να είναι προσαρτημένο στο σάκο κατά διαστήματα. Έχει τουλάχιστον τις ίδιες διαστάσεις (μήκος, πλάτος) με το τμήμα του σάκου στο οποίο είναι προσαρτημένο

Σκοπό έχει την ενίσχυση του σάκου και την προστασία του από ρήξη όταν είναι γεμάτος ψάρια και όταν η τράτα ανασύρεται στο πλοίο

EN | **strengthening bag**

A cylindrical piece of netting completely surrounding the codend of a trawl and which may be attached to the codend at intervals. It shall have at least the same dimensions (length and width) as that part of the codend to which it is attached.

Its purpose is to strengthen the codend and to prevent it from bursting when filled with fish and when the trawl is hauled on board.

FR | **fourreau de renforcement**

Nappe ou pièce de filet de forme cylindrique entourant complètement le cul d'un chalut et fixée au cul du chalut à certains intervalles. Elle a au moins les mêmes dimensions (longueur et largeur) que la partie du cul du chalut à laquelle elle est fixée.

Destiné à renforcer le cul du chalut afin d'éviter que celui-ci n'éclate lorsqu'il est rempli abondamment de poisson et lors du virage du chalut à bord.

IT | **fodera di rinforzo**

Pezza di rete di forma cilindrica che avvolge completamente il sacco della rete da traino e può essere attaccata ad esso di tratto in tratto.

Serve a rafforzare il sacco della rete da traino e ad evitarne la rottura quando è riempita di pesce e quando la rete viene salpata.

NL | **overkuil**

Cilindervormig stuk netwerk dat volledig rond de kuil van een sleepnet is aangebracht, en dat op een aantal punten aan de kuil kan zijn bevestigd.

De overkuil verstevigt de kuil, zodat deze niet barst wanneer hij vol is en het sleepnet wordt binnengehaald.

PT | **forra do saco, saco de reforço**

Peça cilíndrica de rede que envolve completamente o saco das redes de arrasto e a ele é apontoada a intervalos regulares. As suas dimensões em comprimento e altura são pelo menos iguais às do saco.

Tem por finalidade reforçar o saco das redes de arrasto evitando que se rompa quando se encontra completamente cheio de peixe e/ou durante a viragem da rede para bordo.

3005

ES manga doble; copo dividido en dos compartimientos iguales

Destinado a reducir el riesgo de pérdida total de las capturas en caso de pesca sobre un fondo accidentado y a que el pescado no sufra tanto en los grandes lances.

DA buksepose

Særlig type trawlpose, der er delt i to adskilte poser ved siden af hinanden.

Tjener flere formål: bevirker at fangsten får mindre trykbelastning under indhalingen om bord på fartøjet, mindsker risikoen for, at hele fangsten mistes, hvis trawlen rives under fiskeri på hård bund.

DE Hosensteert; Hosen-Steert

In Längsrichtung geteilter Steert.

Zweck eines Hosen-Steerts ist es, die Gefahr eines totalen Verlustes der Fänge beim Fischen auf rauhem Meeresboden zu verringern.
In der großen und kleinen Hochseefischerei nicht mehr gebräuchlich.

GR σάκος-παντελόνι

Σκοπός του σάκου-παντελόνι είναι να μειώνει τον κίνδυνο ολικής απώλειας των αλιευμάτων κατά την αλιεία σε ανώμαλους βυθούς

EN double codend; double codends; trouser codend; trouser codend

Two codends in the form of a Siamese pair, joined together at the leading ends only.
Its purpose is to reduce the risk of a total loss of catches when fishing on rough grounds.

FR cul pantalon

Cul comportant deux poches séparées par une couture longitudinale destiné à réduire le risque de perte totale des captures en cas de pêche sur fond accidenté.

IT sacco a pantalone

Serve a ridurre il rischio di una perdita totale delle catture durante la pesca su fondali accidentati.

NL tweelingkuil

Samenstel van twee onderling verbonden kuilen.
Het gebruik van een tweelingkuil voorkomt het verlies van de totale vangst bij vissen op ruwe bodem (indien het net beschadigd wordt).

PT saco duplo; saco duplo geminado

Saco de rede de arrasto constituído por dois sacos gémeos separados por uma costura longitudinal e que se destina a reduzir o risco de perda total de capturas em caso de pesca em fundo acidentado.

3005

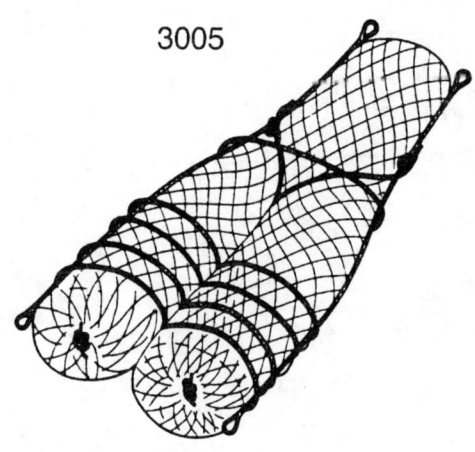

3006

ES barreta; cabo de refuerzo

Cualquier cabo distinto de una relinga de costado fijado en cualquier punto del arte de arrastre.

Destinada a reforzar la red de arrastre o a evitar que las piedras o cualquier resto llegue al copo.

DA forstærkningstov

Alle tove, undtagen tæller og stropper, fastgjort hvor som helst til trawlen.

Tjener forskellige formål: egentlig forstærkning af redskabet, overfører belastningen fra nettet til torvet, forhindrer sten og andre genstande i at nå ned til fangstposen.

DE Verstärkungstau

Jedes andere Tau als eine Laschverstärkung, das an irgendeinem Teil des Schleppnetzes angebracht ist.

Zweck des Verstärkungstaues ist es, das Schleppnetz zu verstärken.

GR ενισχυτικό σχοινί· σχοινί ενίσχυσης

Οποιοδήποτε ενισχυτικό σχοινί, εκτός από γραντί, προσαρτημένο σ'οποιοδήποτε τμήμα της τράτας

Σκοπός του είναι να ενισχύσει την τράτα ή/και να εμποδίσει πέτρες και άχρηστα υλικά να φθάσουν στο σάκο

EN strengthening rope

Any reinforcing rope, other than mounting ropes, attached to any part of the trawl.

Its purpose is to strengthen the trawl, or to prevent stones and debris from reaching the codend.

FR barrette

Tout cordage, autre que les ralingues de côté ou d'ouverture, fixé à tout endroit du chalut et destiné à renforcer le chalut ou à éviter que des pierres ou des débris n'atteignent le cul.

IT corda di rinforzo

Qualsiasi corda, diversa da una lima o ralinga, attaccata a qualsiasi parte della rete da traino.

NL verstevigingslijn

Elke lijn, met uitzondering van een naadlijn, die aan een deel van het sleepnet is bevestigd.

Wordt gebruikt om het sleepnet te versterken of om stenen of ander afval buiten de kuil te houden.

PT cabo de reforço

Qualquer cabo, com excepção dos cabos de porfio, ligado a qualquer parte da rede.

Destinado a reforçar as redes de arrasto ou a evitar que pedras e outros detritos cheguem ao saco.

3007

ES red cribadora; red seleccionadora; red selectiva

Pieza de paño, cuyo tamaño de malla debe ser, por lo menos, el doble del saco.
Su fin es capturar pescado, gambas u otras especies selectivamente.

DA sorteringspanel; selektionspanel; sinet

Et ekstra stykke net, der fastgøres inden i trawlen.
Tjener til at lede arter af fisk, krebsdyr m.m. med forskellig adfærd til forskellige dele af trawlen (eller ud af trawlen). Udformning og placering tilpasses de bestemte fiskerier.

DE Siebnetz

Stück Netzwerk, dessen Maschenöffnung mindestens doppelt so groß sein muß wie die des Steerts.
Zweck eines Siebnetzes ist es, Fische, Garnelen und andere Arten getrennt zu fangen.

GR δίχτυ επιλογής· δίχτυ κοσκινίσματος· δίχτυ-κόσκινο

Κομμάτι δικτυώματος με μέγεθος ματιών που πρέπει να είναι τουλάχιστον διπλάσιο από το μέγεθος ματιών του σάκου
Σκοπός του είναι να συλλαμβάνει ψάρια, γαρίδες ή άλλα είδη, εκλεκτικά

EN sieve netting; separator panel

A piece of netting with a mesh size which must be at least twice the mesh size of the codend.
Its purpose is to catch fish, shrimps, or other species selectively.

FR nappe de sélectivité; filet tamiseur

Pièce de filet présentant un maillage adapté à l'espèce à séparer, fixée à l'intérieur du chalut dans le corps ou dans les ailes. Destinée à assurer la capture sélective de poissons, de crevettes ou d'autres espèces.
Son maillage est en général au moins double de celui de la poche.

IT pezza selettiva

Pezza di rete avente maglie di dimensione almeno doppia di quella delle maglie del sacco della rete.
Serve per la pesca selettiva di pesci, gamberetti od altre specie.

NL zeeflap

Stuk netwerk waarvan de mazen ten minste tweemaal zo groot zijn als de mazen van de kuil.
Wordt gebruikt voor het selectief vangen van vis, garnaal e.d. .

PT pano de rede selectivo; forra interior

Pano de rede com malhagem pelo menos dupla da utilizada no saco.
Destinado a assegurar a captura selectiva de peixes, camarões ou outras espécies.

3008

ES dimensión efectiva de la malla; malla efectiva

DA indvendig maskestørrelse

Den indvendige åbning af maskerne i et trawlnet,
som målt ved den autoriserede målemetode. Dette
mål anvendes ved angivelse af et trawlredskabs
mindste tilladte maskestørrelse.
Se tillige 3242 og 3315.

DE wirkliche Maschenöffnung

GR πραγματικό μέγεθος ματιού

EN actual mesh size

Opening of mesh determined by an authorized
testing procedure.

FR maillage effectif

IT dimensione effettiva delle maglie

NL feitelijke maaswijdte

Afmeting van binnenkantknoop tot binnenkantknoop
bij gestrekt net.

PT malhagem efectiva

3009

ES rebenque del copo; sereta; sireta

Cabo que permite cerrar la parte posterior del copo
del arte de arrastre y/o las cubiertas de refuerzo, sea
mediante un nudo que se pueda alargar fácilmente,
sea mediante un dispositivo mecánico.

DA bindestrik; snørebånd

Et tov, hvormed den bageste del af fangstposen og/
eller forstærkningsposerne kan snøres sammen
enten ved hjælp af en letopbindelig knude eller en
mekanisk anordning.

DE Steertleine; Codleine

Leine, die es ermöglicht, das hintere Ende des
Steerts und/oder der Hievsteerte entweder mittels
eines leicht lösbaren Knotens oder einer
mechanischen Vorrichtung abzubinden.

GR γάιδαρος· γαϊδουρόσχοινο· σχοινί δεσίματος
του σάκου

Σχοινί που χρησιμεύει στο κλείσιμο του πίσω μέρους
του σάκου και/ή των ενισχυτικών θηκών με τη
βοήθεια ενός κόμπου που μπορεί εύκολα να
χαλαρώνει ή ενός μηχανικού μέσου

EN codline; cod line; cod-lashing; codend
lashing; codend tier

A rope making it possible to close the rear of the
codend and/or strengthening bags by means of
either a knot which can be easily loosened or a
mechanical device.

3009

FR | **raban de cul**

Cordage permettant de fermer la partie arrière du cul du chalut et/ou des fourreaux de renforcement, soit par un nœud facilement largable, soit par un dispositif mécanique.

IT | **sagola di chiusura**

Corda che permette di chiudere l'estremità posteriore del sacco della rete e/o delle fodere di rinforzo per mezzo di un nodo facilmente allentabile o di un congegno meccanico.

NL | **pooklijn**

Lijn waarmee het uiteinde van de kuil en/of de overkuil d.m.v. een (gemakkelijk los te maken) knoop of mechanisch kan worden gesloten.

PT | **estropo do cu do saco; cabo do cu do saco**

Cabo utilizado para fechar o cu do saco das redes de arrasto e/ou a respectiva forra através de um nó especial que facilmente se desfaz ou de um dispositivo mecânico.

3009

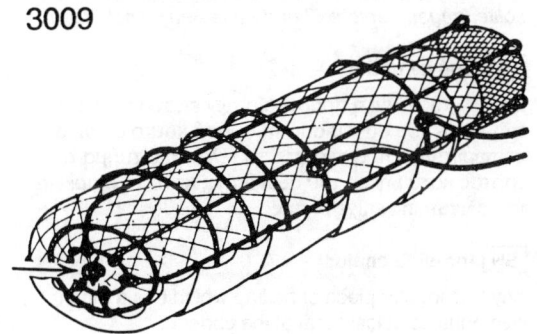

3010

ES | **faja de protección; cintura de protección**

Pequeño paño de red cilíndrico, que tiene la misma circunferencia que el copo de la red de arrastre o, eventualmente, de la red de refuerzo, que rodea el copo del arte de arrastre o las cubiertas de refuerzo situadas en el punto de fijación del estrobo para izar.

Destinado a evitar que el estrobo de izar corte el paño del saco.

DA | **løftestropslidgarn; løftestropbeskyttelsesnet**

Et kort cylinderformet stykke net, der har samme omkreds som trawlposen eller eventuelle forstærkningsposer, og som er placeret, hvor løftestroppen er fastgjort.

Dette net hindrer løftestroppen i at skære fangstposen over.

DE | **Scheuerschutzmanschette**

Kurzes zylinderförmiges Stück Netzwerk gleichen Umfangs wie der Steert oder die Hievsteerte, sofern solche vorhanden sind. Die Scheuerschutzmanschette umgibt den Steert oder die Hievsteerte an den Anbringungsstellen des Teilstropps.

Zweck einer Scheuerschutzmanschette ist es zu verhindern, daß der Teilstropp das Netzwerk des Steerts beschädigt.

GR | **προστατευτική ζώνη**

Μικρό κυλινδρικό κομμάτι δικτυώματος που έχει την ίδια περιφέρεια με το σάκο ή την ενισχυτική θήκη, αν υπάρχει, και το οποίο περιβάλλει το σάκο ή την ενισχυτική θήκη στα σημεία προσάρτησης του κοψαδούρου

Σκοπός της είναι να εμποδίσει τον κοψαδούρο να κόψει το δικτύωμα του σάκου

EN | **chafing; protection piece**

A short cylindrical piece of netting with the same circumference as the codend or strengthening bags, if any, and which surrounds the codend or the strengthening bags at the points of attachment of the lifting strap.

Its purpose is to prevent the lifting strap from cutting the netting of the codend.

3010

FR ceinture de protection

Petite nappe de filet cylindrique, présentant la même circonférence que le cul du chalut ou les fourreaux de renforcement éventuels, entourant le cul du chalut ou les fourreaux de renforcement au point de fixation de l'erse de levage.

Destinée à éviter que l'erse de levage ne coupe le filet du cul du chalut.

IT fascia di protezione

Pezza di rete corta, di forma cilindrica, avente la stessa circonferenza del sacco della rete o della eventuale fodera di rinforzo, che circonda il sacco della rete o la fodera di rinforzo nel punto di attacco dello strozzatoio.

Serve a impedire che lo strozzatoio tagli la rete del sacco.

NL beschermband

Kort cilindervormig stuk netwerk waarvan de omtrek even groot is als die van de kuil of, indien deze aanwezig is, van de overkuil, en dat rond de kuil of de overkuil is aangebracht op de plaats waar de verdeelstrop is bevestigd.

De beschermband zorgt ervoor dat de verdeelstrop het netwerk van de kuil niet doorsnijdt.

PT cinta de protecção

Pano de rede cilíndrico de fio forte que pode ser substituído e que se fixa exteriormente ao saco das redes de arrasto e cuja finalidade é reforçar o saco nos pontos de fixação dos estropos ou forcas.

3011

ES cobertura; parpalla superior

Pieza rectangular de red con una malla por lo menos igual a la del copo de la red de arrastre.

Destinada a proteger el paño superior o las bandas laterales del copo de la red de arrastre contra el desgaste, en el caso de que la parte final de la red de arrastre basculase lateralmente durante las operaciones de pesca.

DA oversideslidgarn

Rektangulært stykke net, der monteres således, at det dækker oversiden af trawlposen.

Oversideslidgarnet beskytter fangstposens overside og sider mod slid og iturivning, hvis trawlens bagende skulle begynde at rotere under fiskeriet.

DE Oberseiten-Scheuerschutz

Rechteckiges Stück Netzwerk, dessen Maschenöffnung mindestens derjenigen des Steerts entspricht.

Zweck eines Oberseiten-Scheuerschutzes ist es, das Oberblatt oder die Seitenblätter des Steerts für den Fall, daß sich das hintere Ende des Schleppnetzes bei dem Fangvorgang um seine Achse verdrehen sollte, gegen Verschleiß und Zerreißen zu schützen.

GR επάνω ποδιά

Σκοπός της είναι η προστασία των επάνω ή πλαϊνών τμημάτων του σάκου από τη φθορά και το σχίσιμο σε περίπτωση που στριφτεί το τελευταίο τμήμα της τράτας κατά μήκος του άξονά της, κατά τη διάρκεια της αλιευτικής επιχείρησης

EN top-side chafer

Any rectangular piece of netting which has a mesh size equal to at least that of the codend.

Its purpose is to protect the top or side panels of the codend from wear and tear should the rear end of the trawl twist along its axis during fishing operations.

3011

FR couverture; tablier de dessus

Toute pièce rectangulaire de filet ayant un maillage au moins égal à celui du cul du chalut.

Surtout destiné à protéger la face supérieure ou les faces latérales du cul du chalut contre l'usure, au cas où la partie terminale du chalut viendrait à basculer latéralement lors des opérations de pêche.

IT foderone superiore

Qualsiasi pezza di rete rettangolare, la cui dimensione delle maglie è almeno pari a quella delle maglie del sacco.

Serve a proteggere dall'usura il cielo o le parti laterali del sacco della rete da traino nel caso in cui la parte posteriore della rete dovesse attorcigliarsi sul proprio asse durante le operazioni di pesca.

NL sleeplap bovenzijde

Rechthoekig stuk netwerk waarvan de mazen ten minste even wijd zijn als die van de kuil.

De sleeplap bovenzijde zorgt ervoor dat de bovenkant en de zijkanten van de kuil tegen slijtage en beschadiging worden beschermd wanneer het achternet bij het vissen om zijn as draait.

PT forra superior; cobertura

Qualquer peça rectangular de rede, que tenha uma malhagem pelo menos igual à da cuada.

Tem por finalidade proteger a face superior e/ou as faces laterais do saco das redes de arrasto do desgaste no caso da parte terminal da rede oscilar lateralmente durante as operações de pesca. Desconhece-se o seu uso em Portugal.

3012

ES parpalla inferior

Puede estar formada por cualquier pieza de tela, red o cualquier otro material.

Está destinada a proteger del desgaste la parte inferior de la red de arrastre.

DA undersideslidgarn

Består af et stykke lærred, hud, kraftigt net eller lignende materiale, der monteres på undersiden af trawlposen.

Undersideslidgarn beskytter trawlens underside mod slid og iturivning.

DE Unterseiten-Scheuerschutz

Kann aus einem Stück Segeltuch, Netzwerk oder irgendeinem anderen Material bestehen.

Zweck eines Unterseiten-Scheuerschutzes ist es, die Unterseite eines Schleppnetzes gegen Verschleiß und Zerreißen zu schützen.

GR κάτω ποδιά

Μπορεί να είναι κατασκευασμένη από οποιοδήποτε κομμάτι καραβόπανου, δικτυώματος ή άλλου υλικού

Σκοπός της είναι η προστασία του κατωτέρου τμήματος της τράτας από τη φθορά και το σχίσιμο

EN bottom-side chafer

May be formed of any piece of canvas, netting, or any other material.

Its purpose is to protect the underside of the trawl from wear and tear.

3012

FR **tablier de dessous**

Peut être constitué par des pièces en toile, filet ou tout autre matériau.
Destiné à protéger la partie inférieure du chalut contre l'usure.

IT **foderone inferiore**

Può essere formato da un telo, da una rete o da qualsiasi altro materiale.
Serve soprattutto a proteggere dall'usura la parte inferiore della rete da traino.

NL **sleeplap onderzijde**

Stuk netwerk met voorgeschreven maaswijdte, zeildoek of enig ander materiaal, op voorgeschreven wijze bevestigd aan de onderkant van de kuil.
De sleeplap onderzijde zorgt ervoor, dat de onderkant van het sleepnet wordt beschermd tegen slijtage en beschadiging.

PT **avental; forra inferior**

Pode ser constituída por qualquer peça em tela, rede ou qualquer outro material.
Tem por finalidade proteger a face inferior do saco das redes de arrasto contra o desgaste.

3012

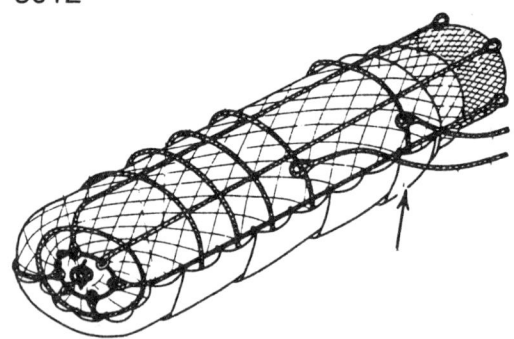

3013

ES **malla mínima**

DA **mindste maskestørrelse**

Den mindste indvendige maskestørrelse, der kan tillades i et fiskeredskab. Angives i fiskerilovgivningens tekniske bevaringsforanstaltninger.

DE **Mindestmaschenöffnung**

GR **ελάχιστο μέγεθος ματιού**

EN **minimum mesh size**

The smallest allowable size of mesh in a codend as determined by an authorized testing procedure.

FR **maillage minimal**

IT **dimensione minima della maglia**

NL **minimum maaswijdte**

Wettelijk voorgeschreven minimale maaswijdte zoals die tijdens het vissen mag worden gebruikt; wordt per vissoort, vismethode en gebied vastgesteld.

PT **malhagem mínima**

3014

ES trampa

Pieza de red de una malla al menos igual a la del copo del arte de arrastre, fijada en el interior de un arte de arrastre, de manera que permita pasar al pescado de la parte anterior de la red de arrastre hacia atrás, limitando las posibilidades de retorno.

DA lås; stopgarn; flapper; kalv

Et stykke net, der mindst skal have samme maskestørrelse som fangstposen, og som anbringes inde i trawlen på en sådan måde, at fiskene kan passere fra trawlens forende til trawlens bagende, men som begrænser deres muligheder for at undslippe.

DE Flapper

Stück Netzwerk, dessen Maschenöffnung mindestens derjenigen des Steerts entspricht und das im Inneren eines Schleppnetzes so angebracht ist, daß die Fänge vom vorderen in den hinteren Teil des Schleppnetzes gelangen können, die Möglichkeit ihrer Rückkehr aber eingeschränkt wird.

GR φλάππα

Κομμάτι δικτυώματος με μέγεθος ματιού τουλάχιστον όσο του σάκου, συνδεδεμένο στο εσωτερικό μιας τράτας, έτσι ώστε να επιτρέπει στα αλιεύματα να περάσουν από το μπροστινό στο πίσω μέρος της τράτας, αλλά περιορίζοντας τη δυνατότητα επιστροφής τους

EN flapper

A piece of netting with a mesh size at least equal to that of the codend, fastened inside a trawl in such a way that it allows catches to pass from the front to the rear of the trawl but limits their possibility of return.

FR tambour; voile

Pièce de filet présentant un maillage au moins égal à celui du cul du chalut, fixée à l'intérieur d'un chalut, de manière à permettre au poisson de passer de la partie antérieure du chalut vers l'arrière, tout en limitant les possibilités de retour.

IT enca

Pezza di rete provvista di maglie aventi almeno la stessa dimensione di quelle del sacco, fissata all'interno della rete da traino in modo da consentire alle catture di passare dalla parte anteriore a quella posteriore della rete, limitandone al tempo stesso le possibilità di ritorno.

NL keel; flap; enkel

Stuk netwerk waarvan de mazen ten minste even groot zijn als die van de kuil, dat zodanig binnen het sleepnet is aangebracht dat de vis van het voorste gedeelte naar het achterste gedeelte van het sleepnet wordt doorgelaten, en dat terugzwemmen verhindert.

PT língua

Pano de rede com malhagem igual ou superior à do saco das redes de arrasto, fixada verticalmente no seu interior e de molde a permitir a passagem dos peixes da sua parte anterior para a posterior e limitando o respectivo retorno.

3014

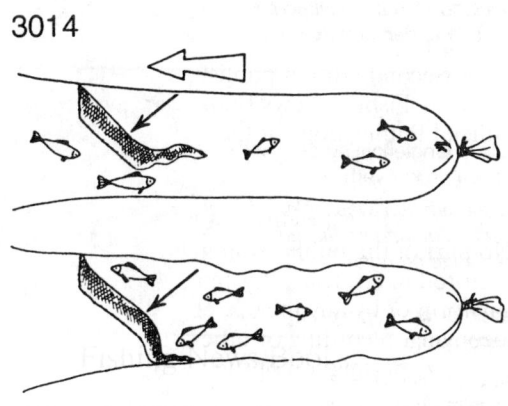

3015

ES costura mediana de un copo dividido en dos compartimientos

Las mallas podrán unirse mediante un trenzado de manera que formen dos compartimientos uniendo longitudinalmente las mitades superiores e inferiores de un copo de arte de arrastre.

DA midtersnøring af bukseposer

Langsgående samling af trawlposens over- og underside.

DE Mittellasche eines Hosensteerts

Zur Bildung eines Hosensteerts werden die Maschen im Steert in der Weise zusammengelascht, daß Ober- und Unterhälften eines Steerts in Längsrichtung verbunden werden.

GR ενδιάμεση ραφή ενός σάκου-παντελόνι

Τα μάτια μπορούν να συρραφούν έτσι ώστε να σχηματίσουν ένα σάκο-παντελόνι ενώνοντας κατά μήκος και στη μέση το επάνω και κάτω μέρος ενός σάκου

EN median lacing of a trouser codend

Meshes are laced together in order to build a trouser codend by joining lengthwise the upper and lower halves of a codend.

FR couture médiane d'un cul pantalon

Couture longitudinale qui assemble dans leur milieu la face supérieure et la face inférieure du cul de chalut et délimite ainsi deux compartiments séparés, de part et d'autre de la poche.

IT cucitura lungo la linea mediana di un sacco a pantalone

Le maglie del sacco di una rete da traino possono essere cucite assieme, allacciando longitudinalmente la metà superiore e quella inferiore, per formare un sacco a pantalone.

NL overlangse naad

Bij een tweelingkuil, naad die in lengterichting de bovenhelft en de onderhelft van de kuil verbindt.

Mazen mogen worden genaaid om een tweelingkuil te vormen door de bovenhelft en de onderhelft van een kuil in lengterichting samen te brengen.

PT porfio mediano do saco duplo geminado; costura mediana de um saco duplo

Fiada ou fiadas de malhas porfiadas mediana e longitudinalmente de tal maneira que unem as metades superiores e inferiores do saco das redes de arrasto formando um saco duplo geminado.

3016

ES copo de arte de arrastre; copo

Parte más trasera de la red de arrastre que tiene, sea una forma cilíndrica, es decir, la misma circunferencia en un extremo y en otro, sea una forma de embudo.

El copo del arte de arrastre comprende el copo propiamente dicho y la manga.

DA fangstpose og forlængelsesstykke; pose

Den bagerste del af trawlen (eller voddet) til opsamling af fangsten. Kan enten være cylindrisk overalt eller svagt konisk. Visse poser har den bagerste del udformet (forstærket) specielt til et løft, hvori fangsten løftes ind over siden på fartøjet.

I lovmæssig forstand omfatter posen: fangstposen, hvoraf løftet udgør en del, og et eventuelt forlængelsesstykke.

DE Steert

Der hinterste Teil des Schleppnetzes, der entweder zylinderförmig ist, d. h. überall den gleichen Umfang hat, oder sich nach hinten verjüngt.

Der Steert umfaßt den eigentlichen Steert und den Tunnel.

GR σάκος

Το τελευταίο τμήμα της τράτας, που έχει είτε κυλινδρικό σχήμα (δηλ. έχει την ίδια περίμετρο από το ένα άκρο στο άλλο) είτε κωνικό σχήμα

Ο σάκος περιλαμβάνει τον κυρίως σάκο και το κομμάτι επιμήκυνσης

EN codend

The rearmost part of the trawl, having either a cylindrical shape, i.e. the same circumference throughout, or a tapering shape.

The codend includes the codend itself and the lengthening piece.

3016

FR **cul de chalut; cul; poche**

Partie la plus en arrière du chalut qui présente soit
une forme cylindrique, c'est-à-dire la même
circonférence d'un bout à l'autre, soit une forme en
entonnoir, où s'accumule le poisson.

*La poche comporte le cul de chalut proprement dit et
la rallonge.*

IT **sacco della rete da traino; sacco; coda;
manica**

L'estrema parte posteriore della rete da traino, avente
forma cilindrica, ossia la stessa circonferenza in ogni
sua parte, o forma affusolata.

*Il sacco della rete da traino può comprendere il sacco
stricto sensu e l'eventuale avansacco.*

NL **kuil; staart**

Achterste deel van een sleepnet, cilindervormig of
taps.

*De kuil bestaat uit de kuil in enge zin en de tunnel.
Het begrip „kuil" staat ook voor „trawl".*

PT **saco da rede de arrasto**

Secção terminal das redes de arrasto onde se
acumulam os peixes capturados.

3016

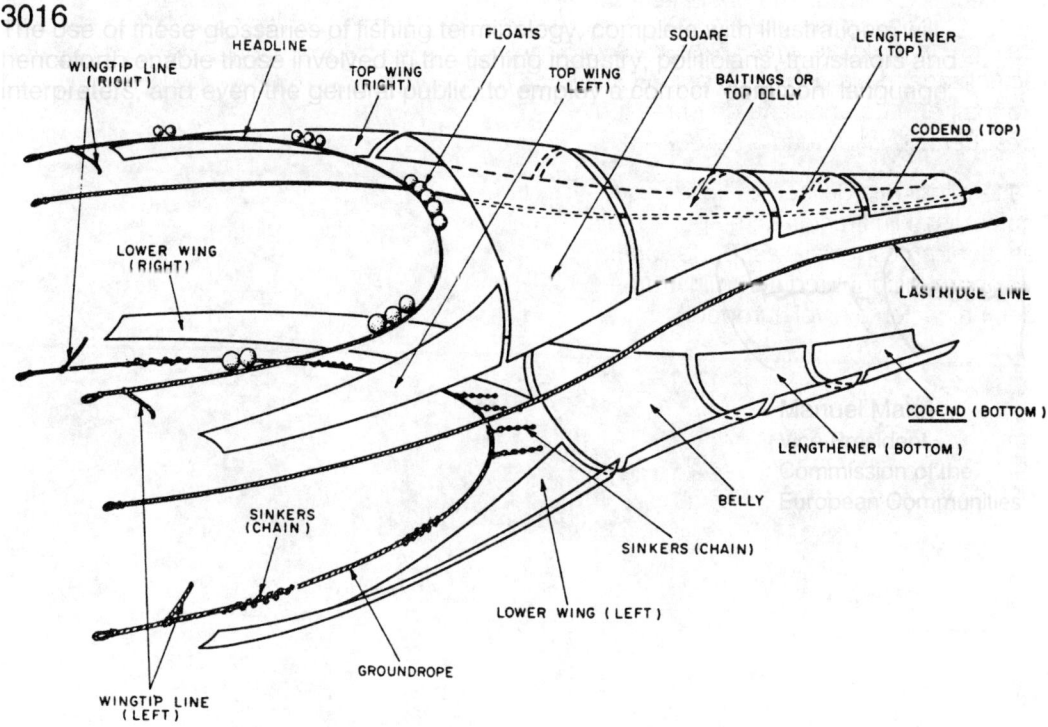

3017

ES | paño de red; paño; pieza de red

Conjunto de mallas del mismo tamaño.

DA | netstykke

DE | Stück Netzwerk

GR | κομμάτι δικτυώματος

EN | piece of netting

A section of netting consisting of a uniform size mesh.

FR | pièce de filet

Pièce constitutive d'un filet de pêche, de forme rectangulaire, trapézoïdale ou triangulaire, et qui comporte en général un maillage constant.

IT | pezza di rete

Insieme di maglie in forma e dimensioni qualsiasi ottenuto da un filo, monofilo, filato e da uno o due sistemi di fili, monofili, filati intrecciati o annodati oppure con altri metodi (per esempio per stampaggio o taglio di materiali in film o per estrusione).

NL | stuk netwerk

Samenstel van door mazen gevormd netwerk van willekeurige vorm.

PT | pano de rede; peça de rede; bocado de rede

Peça de rede com forma normalmente rectangular, triangular ou trapezoidal caracterizada pela respectiva malhagem e pelo número de malhas em altura e largura(s).

3017

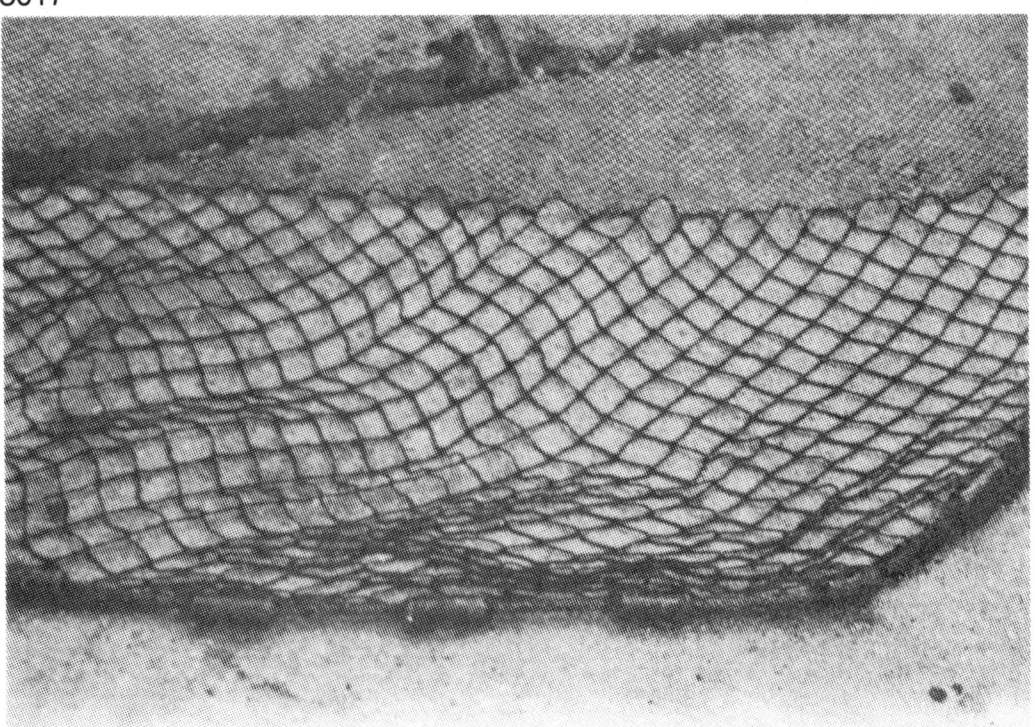

3018

ES copo de la red de arrastre propiamento dicho

Uno o varios paños (piezas de red), de la misma malla, unidos uno a otro lateralmente en el eje de la red de arrastre por un trenzado al cual podrá fijarse igualmente una relinga de costado.

DA pose i snæver forstand; fangstpose i snæver forstand

Bagerste del af trawlen efter kroppen eller et evt. forlængelsesstykke. Består af to eller flere netstykker med samme maskestørrelse.

DE eigentlicher Steert

Besteht aus einem oder mehreren Netzblättern (Stücken Netzwerk) gleicher Maschengröße, die an ihren Seiten in der Achse des Schleppnetzes durch eine Lasche, an der auch eine Laschverstärkung angebracht sein kann, miteinander verbunden sind.

GR κυρίως σάκος

Δικτυωτή θήκη κατασκευασμένη από ένα ή περισσότερα κομμάτια δικτυώματος με το ίδιο μέγεθος ματιού, συνδεδεμένα το ένα με το άλλο κατά μήκος των πλευρών τους και του άξονα της τράτας, με ραφή στην οποία μπορεί επίσης να είναι συνδεδεμένο ένα πλαϊνό σχοινί

EN codend

Netting bag made up of one or more panels (pieces of netting) of the same mesh size attached to one anothor along thoir cidos in the axis of the trawl by a seam where a side rope may also be attached.

FR cul du chalut

Partie postérieure du chalut, en forme de poche composée d'une ou de plusieurs pièces de filet du même maillage, reliées latéralement l'une à l'autre par une couture à laquelle une ralingue de côté peut également être fixée.

En français, par définition, la poche comprend le cul de chalut (partie terminale) et la rallonge (partie intermédiaire avec le corps du chalut).

IT sacco della rete da traino stricto sensu

Una o più pezze di rete, con maglie della stessa dimensione, attaccate l'una all'altra lungo i bordi nel senso della rete per mezzo di una cucitura, alla quale può essere attaccata anche una ralinga.

NL kuil in enge zin

Die delen van het net die dezelfde maaswijdte hebben en zijdelings in de richting van de lengteas van het sleepnet met elkaar zijn verbonden door een naad, waaraan eventueel ook een naadlijn bevestigd is.

Gedeelte van de kuil tussen de verdeelstrop en de pooklijn.

PT cu do saco; cuada stricto sensu

Secção terminal das redes de arrasto formada por um ou mais panos de rede com a mesma malhagem ligados lateral e longitudinalmente através de um porfio onde se liga um cabo de porfio.

3019

ES cara de la red; lado de la red

Cada una de las partes superior, inferior o lateral de un arte que puede estar formada por un paño o por varios de diferente tamaño de mallas unidos entre sí.

DA panel; netpanel

Del af et fiskenet, ofte samlet af flere netstykker, der mere eller mindre udgør en enhed, f.eks. en overside (= overpanel).

DE Netzblatt

Zugeschnittenes Netzteil als Ersatz für auszuwechselnde Teile.

GR φύλλο δικτυώματος

Τμήμα δικτυώματος που συχνά περιλαμβάνει δύο ή περισσότερα κομμάτια ενωμένα μεταξύ τους

EN panel; net panel

Sheet of netting often comprising two or more sections joined together.

FR face de filet

Partie d'un filet formée de plusieurs pièces assemblées.

IT parte della rete; faccia della rete

Insieme di pezze di rete di maglie diverse unite tramite giunzione.
Rete da traino.

NL paneel; perk

PT face de uma rede

Secção de uma rede constituída por vários panos de rede.
Redes de arrasto.

3020

ES fuerza de sustentación vertical

DA opdrift

Hydrostatisk eller hydrodynamisk kraft rettet lodret opad.

DE Auftrieb

GR άνωση

EN lift

Hydrodynamic or hydrostatic force, directed vertically upwards.

FR portance; poussée verticale

Force composante de la résistance hydrodynamique, perpendiculaire à la direction du mouvement.
La portance d'un plateau élévateur est dirigée verticalement vers le haut. Dans le cas des panneaux de chalut, la portance, appelée aussi poussée, s'exerce transversalement et assure l'écartement.

IT portanza

NL lift

De krachtscomponent werkend op een lichaam geplaatst in een stromend medium, loodrecht op de stromingsrichting.

PT força de sustentação; força de elevação; força ascensional; força de afastamento

Força componente de resistência hidrodinâmica dirigida verticalmente para cima.
No caso das portas de arrasto a força exerce-se transversalmente e assegura o afastamento das portas.

3021

ES costura de unión

Hileras de mallas que pueden unirse mediante un trenzado de forma que sirva de refuerzo a la red.

DA forstærkningssøm; blind søm

Sammenføjningen mellem to netstykker udført ved sammensnøring af flere rækker af masker for at forstærke nettet.

DE Lasche

Maschenreihen, die zur Verstärkung des Netzwerks zusammengenäht werden können.

GR ενισχυτική ραφή· ραφή

Είναι κατασκευασμένη από σειρές ματιών τα οποία μπορεί να είναι συρραμένα με σκοπό την ενίσχυση του δικτυώματος

EN strengthening lacing; seam

Rows of meshes which may be laced together in order to strengthen the netting.

FR couture d'assemblage

Assemblage de deux pièces de filet, formé par la couture de plusieurs rangées de mailles ensemble, de façon à renforcer le filet.

IT cucitura di rinforzo

Unisce due pezze di rete ed è costituita da file di maglie cucite assieme per rafforzare la rete.

NL verstevigingsnaad

Verscheidene rijen mazen die samengeregen zijn om de naad te verstevigen.

PT porfio reforçado; costura de reunião

Conjunto de fiadas longitudinais de malhas apanhadas em conjunto de maneira a reforçar a rede.

3022

ES relinga de costado; costadillo

Cabo longitudinal colocado a lo largo del empalme entre dos paños en la dirección del eje de la red de arrastre.

DA sømforstærkningstov; sømline

Et tov, der er føjet til maskerne i sømmen for at forstærke denne.

DE Laschverstärkung; Bortenleine; Nahtleine; Laschleine

Tau, das entlang der Verbindung zwischen zwei Netzblättern parallel zur Schleppnetzachse verläuft.
Zur Verstärkung der seitlichen Nähte.

GR σχοινί ραφής· πλευρική νεύρωση· γραντί ραφής

Σχοινί που εκτείνεται κατά μήκος της ένωσης δύο κομματιών δικτυώματος κατά τη διεύθυνση του άξονα της τράτας

EN lacing rope; boltrope; selvage rope

A rope running lengthwise along the join between two pieces of netting in the direction of the axis of the trawl.
Load-bearing rope fixed to length of lestridge.

FR ralingue de côté; ralingue latérale; ailière

Cordage longitudinal courant le long de la couture entre deux nappes dans la direction de l'axe du chalut.
Cordage de renfort.

IT ralinga laterale

Corda che corre longitudinalmente alla rete ed è unita a quest'ultima lungo la cucitura laterale nella direzione dell'asse della rete.

NL naadlijn; naadtouw

Lijn die loopt in de richting van de lengteas van het net langs de samenvoegingsnaad tussen twee netdelen.
Touw of lijn ter versteviging van de naden.

PT cabo de porfio

Cabo longitudinal disposto ao longo de um porfio e ao qual se apontoa ou porfia.
Cabo de reforço.

3023

ES flotador

Elemento flotante utilizado para dar al arte de arrastre una fuerza de sustentación vertical o para indicarle la posición, o ambas cosas.

DA flåd; kugle

Et flydende legeme, der enten bruges til at give opdrift eller til at markere redskabets position eller begge dele.

DE Schwimmer; Auftriebskörper

Tragfähiger Körper, der verwendet wird, um dem Netz Auftrieb zu geben, seine Position zu markieren oder beides zu bewirken.

GR πλωτήρας· σημαδούρα· φελλός

Επιπλέουσα συσκευή που χρησιμεύει στο να προσθέτει άνωση ή να επισημαίνει τη θέση ενός διχτυού ή και τα δύο

EN float

A buoyant unit used to give lift or to mark the position of a net, or both.

FR flotteur

Élément flottant utilisé pour donner une portance au filet ou pour en indiquer la position, ou les deux à la fois.

IT galleggiante

Dispositivo che serve ad impartire una forza di sollevamento alla rete o a segnalare la posizione di una rete o entrambi.

NL drijver; drijfkurk; vlotter

Drijvend deel gebruikt om onderdelen van een net aan een opdrijvende kracht te onderwerpen of om de positie van onderdelen van een vistuig aan te geven.

PT flutuador; bóia

Corpo flutuante utilizado para dar elevação vertical a uma rede de pesca, para indicar a sua posição, ou simultaneamente, para ambas as situações.

3024

ES manga

Se compone de uno o varios paños situados exactamente delante del copo propiamente dicho.

DA forlængelsesstykke; mellemstykke

Cylindrisk eller svagt konisk del af en trawl, der findes mellem fangstposen og kroppen af trawlen. Findes ikke på alle trawl. Består af to eller flere netstykker.

DE Tunnel

Zylindrisches Netzteil zwischen Belly und Steert. Der Tunnel verlängert den vorderen Teil des Steerts, damit größere Fänge aufgenommen werden können.
Besteht aus einem oder mehreren Netzblättern, die sich unmittelbar vor dem eigentlichen Steert befinden.

GR κομμάτι επιμήκυνσης· κόψη· κόψες

Τραπεζοειδές τμήμα δικτυώματος το οποίο είναι είτε τοποθετημένο μεταξύ κοιλιάς-σάκου και γούλας-σάκου, είτε συνδεδεμένο στο άκρο του σάκου για να αυξηθεί το μήκος τους

EN lengthener; lengthening piece

Untapered section of netting either inserted between the belly and batings and the codend or attached to the end of the codend to increase its length.
Is made of one or more panels located just in front of the codend.

FR rallonge; «gorget»

Partie du filet, de forme généralement cylindrique, composée d'une ou de plusieurs nappes situées juste en avant du cul du chalut.

IT avansacco; gola del sacco

È costituita da una o più pezze di rete situate immediatamente davanti al sacco stricto sensu.

NL tunnel; blinde lap

Gedeelte van een net bestaande uit één of meer panelen, dat zich vóór de kuil bevindt.

PT boca

Zona das redes de arrasto constituída por um ou mais panos de rede situados justamente antes do saco.

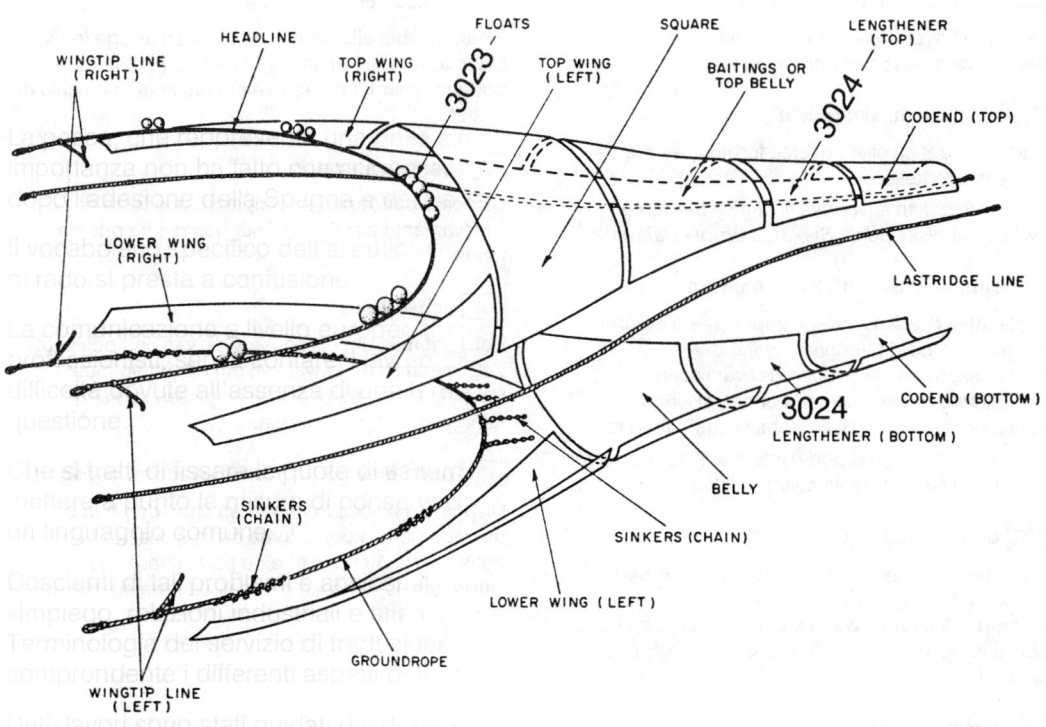

WINGTIP LINE (RIGHT)

HEADLINE

TOP WING (RIGHT)

FLOATS

3023

TOP WING (LEFT)

SQUARE

BAITINGS OR TOP BELLY

LENGTHENER (TOP)

3024

CODEND (TOP)

LOWER WING (RIGHT)

LASTRIDGE LINE

CODEND (BOTTOM)

3024
LENGTHENER (BOTTOM)

BELLY

SINKERS (CHAIN)

SINKERS (CHAIN)

LOWER WING (LEFT)

GROUNDROPE

WINGTIP LINE (LEFT)

3025

ES elevador; puerta elevadora

Elemento utilizado para dar una fuerza de sustentación vertical al arte de arrastre.

DA skærebræt; skæreplan

Plade anbragt på eller i direkte forbindelse med trawlens overtælle.
En eller flere kan anvendes til at give trawlen en hydrodynamisk opdrift. Sjældent brugt i Danmark.

DE Höhenscherbrett; Kopfscherbrett

Bestandteil des Einschiffschleppnetzes, Teil vom Vorgeschirr, das in Verbindung mit den Auftriebskörpern die Bildung der vertikalen Netzöffnung unterstützt und bei pelagischen Schleppnetzen außerdem Stabilitätsaufgaben hat.
Es ist etwa 1 qm groß und besteht aus ebenen oder gewölbten Holz- oder Aluminiumplatten.

GR αετός· υδραετός

Διατμητικό σύστημα συναρμολογημένο σε ένα ψευδές επάνω γραντί, για να ανυψώνει το πραγματικό επάνω γραντί και/ή να φοβίζει τα ψάρια οδηγώντας τα προς τα κάτω μέσα στο στόμιο του διχτυού

EN kite

A shearing device mounted on a false headline to lift the true headline and/or to scare fish downward into the mouth of the net.

FR plateau élévateur; cerf-volant

Plateau habituellement en bois ou en alliage léger, placé au-dessus de la corde de dos et dont la portance permet d'augmenter l'ouverture verticale du chalut.

IT panello elevatore

Dispositivo utilizzato per impartire una forza di sollevamento alla lima dei sugheri di una rete da traino.

NL scheerbord

Bord gebruikt om op onderdelen van een gesleept vistuig een lift op te wekken om dit vistuig meer opening te geven, of om de jaaglijnen van een haringpatent op te houden.

PT porta elevatória; porta elevadora; papagaio

Dispositivo colocado num falso cabo de pana e utilizado para imprimir força ascensional ao cabo de pana propriamente dito e/ou para afundar os cardumes na direcção da boca da rede de arrasto.

3025

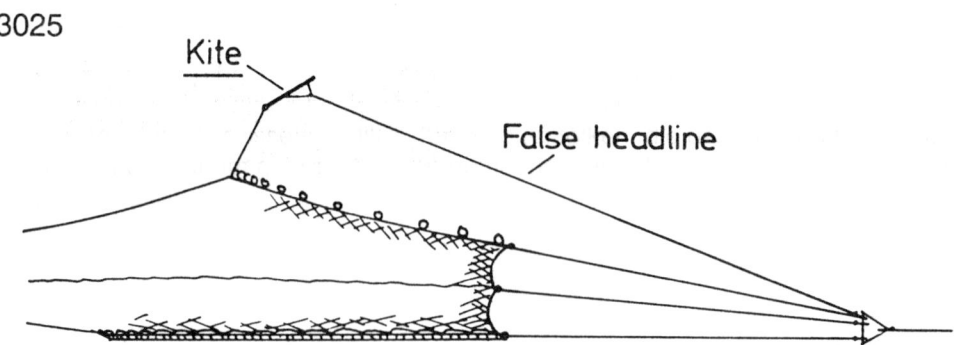

Kite

False headline

3026

ES descarte

DA kasseret fangst; »discard«; udsmid

Den del af fangsten, der ikke ilandbringes:
undermålere eller værdiløs fangst.

DE Gammel

GR σκάρτο αλίευμα

EN discarded catch

That part of the catch in the codend which is not
retained for sale.

FR poissons rejetés

IT scarto

NL ongewenste bijvangst

Gedeelte van de vangst die niet marktwaardig is en
die niet wordt aangeland.
Het kan omvatten:
- te kleine vis van een marktwaardige soort;
- vis, schaal- of schelpdieren zonder handelswaarde.

PT desperdício; lixo

Parte da captura que é rejeitada para o mar, dado
não ser vendável.

3027

ES estrobo para izar posterior

DA bagerste løftestrop; takkelstrop

Strop, der omkranser fangstposen, og som løftet
hales om bord med.

DE hinterer Hebestropp

GR οπίσθιος κοψαδούρος

EN rear lifting strap

Lifting strap nearest to the codline.

FR erse de levage arrière

IT strozzatoio posteriore

Ultimo strozzatoio.

NL achterste verdeelstrop

Verdeelstrop die het dichtst bij de pooklijn is
aangebracht.

PT estropo do laracho

Cabo que serve para estrangular o saco das redes
de arrasto e no qual se engata o gancho do aparelho
de força utilizado para virar o saco da rede.
Está ligado ao laracho.

3028

ES estrobo circular

Uno de los cabos en forma de anillo que se fijan y que rodean transversalmente al copo de arte de arrastre o a la cubierta del refuerzo, a intervalos regulares.

Tiene por objeto limitar la extensión del diámetro del copo del arte de arrastre.

DA rundstrop

Tove, der omkranser posen (og evt. beskyttelses-pose) med regelmæssige mellemrum, og som er fastgjort til denne. Benyttes for at forstærke posen og for at forhindre den i at udvide sig for meget under indhalingen.

DE Rundstropp

Eines der in regelmäßigen Abständen ringförmig um den Steert oder den Hievsteert gelegten und an diesem befestigten Taue.

Zweck eines Rundstropps ist es, eine Ausdehnung des Steertdurchmessers zu begrenzen.

GR κυκλική ζώνη· κυκλική νεύρωση

Δακτυλιοειδές σχοινί που περιβάλλει το σάκο ή την ενισχυτική θήκη κατά κανονικά διαστήματα και που είναι συνδεδεμένο σ'αυτή

Σκοπός της είναι η παρεμπόδιση της διαστολής της διαμέτρου του σάκου

EN round strap

One of the ring-shaped ropes which encircle the codend or the strengthening bag at regular intervals and which are attached to it.

Its purpose is to limit the amount of stretching of the diameter of the codend.

FR erse circulaire

Cordage en forme d'anneau encerclant transversalement le cul du chalut ou le fourreau de renforcement à intervalles réguliers et fixé à celui-ci.

A pour but de limiter l'extension du diamètre du cul du chalut.

IT cinta di rinforzo

Una delle corde a forma di anello che cingono a intervalli regolari il sacco della rete o la fodera di rinforzo a cui sono attaccate.

Serve a limitare l'estensione del diametro della rete.

NL verstevigingsstrop

Strop die op regelmatige afstanden rond de kuil of de overkuil is aangebracht en daaraan bevestigd is.

Wordt gebruikt om uitrekking van de kuil in dwarsrichting te voorkomen.

PT estropo; forca

Cabo com a forma de anel que rodeia transversalmente o saco das redes de arrasto ou a respectiva forra de reforço e a ele apontoado.

O seu número é variável e tem por finalidade limitar o aumento do diâmetro do saco quando cheio de peixe.

3029

ES estrobo de atrás

El estrobo circular situado más atrás.

DA bærering; bagerste rundstrop

Speciel kort rundstrop, monteret i kort afstand fra fangstposens sidste masker.
Stroppen har til formål at sikre, at posen får et tragtformet udløb under tømning.

DE hinterer Stropp

Hinterster Rundstropp.

GR οπίσθια ζώνη· οπίσθια νεύρωση

Η τελευταία κυκλική ζώνη

EN back strap

Rearmost round strap.

FR erse arrière

Erse circulaire située la plus en arrière.

IT cinta posteriore

Ultima cinta di rinforzo.

NL klein stropje

Achterste verstevigingsstrop.

PT estropo posterior

Último cabo com a forma de anel que rodeia transversalmente o saco das redes de arrasto.

3030

ES estrobo para izar; estrobo de la las

Cabo o cable que rodea la circunferencia del copo del arte de arrastre o de la eventual cubierta de refuerzo y que se fija a esta última mediante argollas o anillas. Podrán utilizarse varios estrobos para izar.
Un estrobo para izar está destinado a estrechar la última parte del copo de la red de arrastre con el fin de facilitar su embarque a bordo.

DA løftestrop; takkelstrop; delestrop

Strop af tov eller wire, der løst omkranser trawlposen (og evt. beskyttelsespose). For at kunne afsnøre posen går stroppen gennem ringe eller løkker fæstnet til posen.
Anvendes til at afsnøre den del af fangsten, der ligger i løftet, og løfte den om bord.

DE Teilstropp; Schleppnetzteiler; Teiler

Ein mit Hilfe von Schlaufen und Ringen befestigtes Tau oder Drahtseil, das den Umfang des Steerts oder etwaiger Hievsteerte lose umschließt. Es darf jederzeit mehr als ein Teilstropp verwendet werden.
Zweck eines Teilstropps ist es, im Hinblick auf ein einfacheres Abladen an Bord ein Abschließen des hinteren Teils des Steerts zu ermöglichen.

GR κοψαδούρος

Κομμάτι σχοινιού ή σύρματος που περιβάλλει χαλαρά την περιφέρεια του σάκου ή της ενισχυτικής θήκης, αν υπάρχει, και συνδέεται σ'αυτή κατά διαστήματα με θηλιός ή δακτυλίους. Περισσότεροι του ενός κοψαδούροι μπορούν να χρησιμοποιούνται οποτεδήποτε
Σκοπός του είναι να κλείσει η τελευταία περιοχή του σάκκου για να διευκολυνθεί η φόρτωσή του στο πλοίο

EN lifting strap; lifting strop; halving becket

A piece of rope or wire loosely encircling the circumference of the codend or the strengthening bag, if any, and attached to it by means of loops or rings. More than one lifting strap may be used at any time.
Its purpose is to make it possible to close off the rear section of the codend in order to facilitate its loading aboard.

3030

FR **erse de levage; erse de cul; coupe-cul; étrangloir**

Cordage ou câble entourant la circonférence du cul du chalut ou de l'éventuel fourreau de renforcement et fixé à ce dernier par des boucles ou des anneaux. On peut utiliser plusieurs erses de levage.

Destiné à étrangler la dernière partie du cul du chalut afin de faciliter son embarquement à bord.

IT **strozzatoio**

Pezzo di corda o di fune metallica che corre lungo la circonferenza del sacco della rete o dell'eventuale fodera di rinforzo, attaccata ad essa mediante nodi o anelli. Si può utilizzare in qualsiasi momento più di uno strozzatoio.

Serve a serrare la sezione posteriore del sacco della rete in modo da agevolarne il salpamento.

NL **verdeelstrop; kuilstrop**

Strop van touw of staaldraad die door lussen of ringen los om de kuil of de overkuil is aangebracht.

Er mag meer dan één verdeelstrop tegelijk worden aangebracht. Wordt gebruikt om het achterste deel van de kuil af te snoeren zodat de inhoud van de kuil gemakkelijker aan boord kan worden gebracht.

PT **estropo do saco; forca do saco; laracho**

Cabo que rodeia o saco das redes de arrasto ou a respectiva forra de reforço e a ele fixado por laços ou anés. Podem ser utilizados vários estropos do saco.

Destinado a estrangular secções do saco a fim de facilitar a respectiva alagem para bordo.

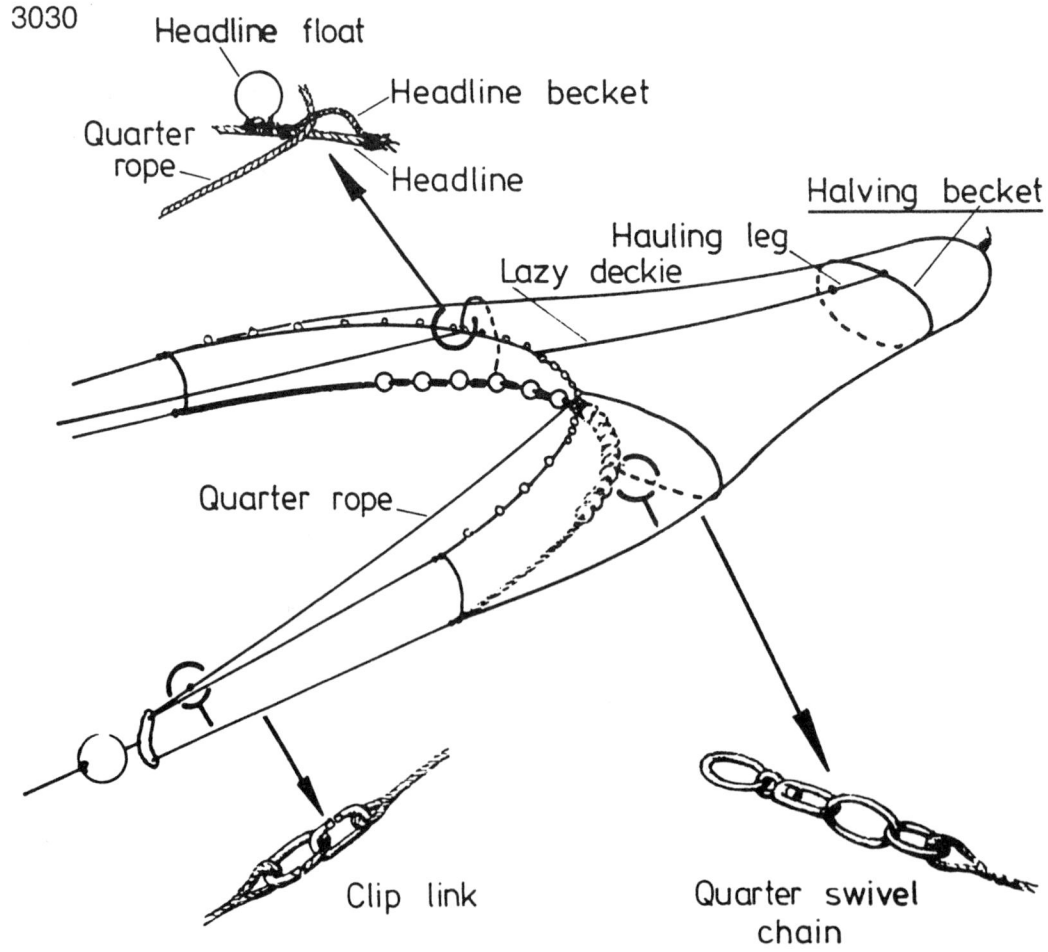

3030

Headline float

Headline becket

Quarter rope

Headline

Halving becket

Hauling leg

Lazy deckie

Quarter rope

Clip link

Quarter swivel chain

3031

| ES | saltada

Especie de trasmallos que pueden flotar horizontalmente y se mantienen tirantes mediante gruesas y secas cañas en sentido transversal. Se dejan flotando alrededor de la paradera, para que los peces retenidos por ésta vayan a enredarse en los trasmallos que suelen emplearse en albuferas.

| DA | veranda-net

Passivt fiskeredskab, type af fiskegård til springende fisk.
Anvendes ikke i Danmark.

| DE | Verandanetz

Kreisförmiges, in einem Sektor offenes, auf der Oberfläche mittels Schwimmleine und an in den Boden getriebenen Pfählen gehaltenes Netz, dessen beschwerter Untersimm bis zum Grund reicht. Der durch den offenen Sektor einkommende Fisch (Sprungfisch) versucht, das Hindernis (Netzwand) zu überwinden, und fällt in die ringförmig an der Oberfläche ausgebreitete Netzmatte.

| GR | καλαμωτή

Δίχτυα για ψάρια που πηδούν έξω από το νερό. Στο κάθετο δίχτυ και στο επάνω σχοινί των φελλών, εφαρμόζεται ένα άλλο οριζόντιο δίχτυ με καλάμια που επιπλέει στην επιφάνεια

| EN | verandah net; veranda net

Barrier netting and collecting bags made from a single piece of netting. The net forms a vertical barrier which encourages the fish to jump, and also an almost horizontal apron or verandah onto which the fish fall. The barrier can be set in a straight line or in a more or less open circle formation. In the latter case, guiding nets are used to direct the fish, usually mullet, into the catching enclosure.

| FR | sautade; cannasse; filet véranda

Longue nappe montée sur claies de roseaux, flottant horizontalement à la surface et complétée par une nappe verticale immergée; le poisson saute et se trouve capturé dans la nappe horizontale.
Filet employé pour la capture des muges ou mulets.

| IT | saltarello

Rete da posta circuitante formata da una rete verticale più una orizzontale.
Serve per catturare i pesci che potrebbero sfuggire all'accerchiamento con salti fuori dell'acqua.

| NL | veranda-net

Wordt in Nederland niet gebruikt.

| PT | salto; parreira

Arte de pesca constituída por uma rede vertical que funciona como barreira e à qual se liga, na tralha da cortiça, uma outra rede sustentada horizontalmente à superfície por meio de varas.
Destinada à captura de peixes saltadores (por exemplo, tainhas) e de peixes planadores (por exemplo, peixe voador) que ao encontrarem a rede vertical tentam passar por cima dela.

3031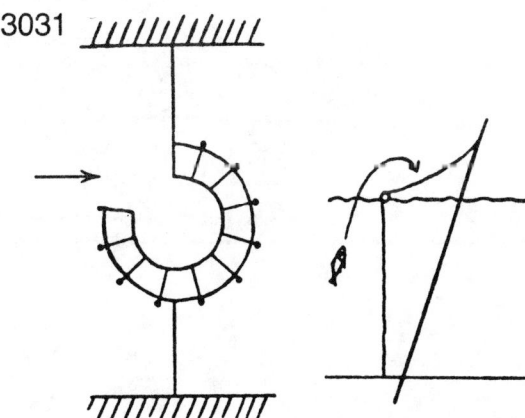

27

3032

ES trampa aérea

Los peces que saltan (por ejemplo, las lisas) y los peces planeadores (peces voladores) pueden ser capturados en la superficie con nasas, balsas, embarcaciones y redes ('saltadas'). Algunas veces se hostiga a los peces para que salten fuera del agua.

DA

Passivt fiskeredskab til fangst af springende fisk. *Anvendes ikke i Danmark.*

DE Sprungfischreuse

GR αεροπαγίδα

Ψάρια που πηδάνε (π.χ. κέφαλοι) και που ολισθαίνουν (χελιδονόψαρα) μπορούν να συλληφθούν στην επιφάνεια, σε κουτιά, σχεδίες, βάρκες και δίχτυα (καλαμωτές). Μερικές φορές εκφοβίζουμε το ψάρι προκειμένου να το αναγκάσουμε να πηδήξει έξω από το νερό

EN aerial trap

Jumping fish (e.g. mullets) and gliding fish (flying fish) can be caught on the surface in boxes, rafts, boats and nets ('veranda nets'). The fish are sometimes frightened to get them to jump out of the water.

FR piège aérien

Les poissons sauteurs (ex: mulets) ou planeurs (poissons volants ou exocets) peuvent être capturés à la surface dans des caisses, nattes ou bateaux, ainsi que dans des filets (appelés «sautades», «cannasses» ou «vérandas»). On effraie parfois le poisson pour l'inciter à sauter hors de l'eau.

IT trappola di superficie

Trappola galleggiante che cattura i pesci saltatori o volanti che nei loro salti o voli vi cadono dentro.

NL luchtnet

Vistuig voor het vangen van boven het water uitspringende vis. *Wordt in Nederland niet gebruikt.*

PT armadilha aérea

Os peixes saltadores (por exemplo, tainhas) e os peixes planadores (por exemplo, peixe voador) podem ser capturados à superfície através de caixas, jangadas, balsas, pequenas embarcações ou redes (saltos). Por vezes tenta-se assustar os peixes para que saltem da água.

3032

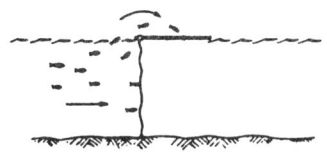

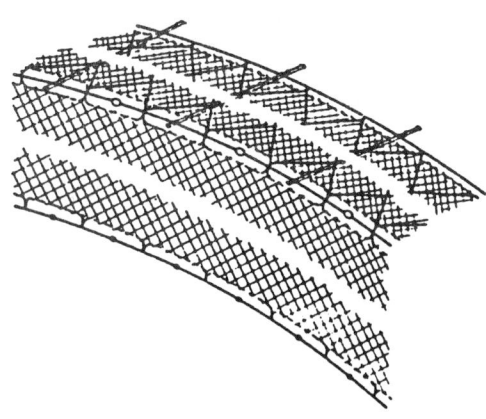

3033

ES **corral**

Arte de pesca fijo y de trampa formado por un conjunto de redes que se mantiene en sentido vertical, sujeta a estacas clavadas en el fondo o sobre pilares de mampostería. Forman un recinto en el que se introducen los peces durante la pleamar quedando apresados durante la bajamar.

DA **fiskegård**

Type af bundgarn.

Fiskegård er et gammelt dansk udtryk, der næsten ikke bruges mere.

DE **Fischzaun; Buhnen; Lenken; Gaarden; Argen**

Auf den Watten übliches stationäres Fischereigerät. Aus Reisig, Buschwerk oder Schilf hergestellte Zäune, mit denen alle im Wattenmeer vorkommenden Fische, auch Krabben, gefangen werden. Die fest in den Boden gesteckten Wände in Trichterform ergeben aneinandergereiht eine im Zickzack verlaufende Wand, an der Flut- und Ebbereusen angeordnet werden.

Regionale Bedeutung.

GR **φράχτης**

EN **corral; fish corral**

An assembly of nets supported on poles fixed in the sea bed. Fish are swept into the net at high tide and are prevented from escaping at low tide because of the V-shaped entrances.

FR **bas-parcs; filets fixes**

Engin à poste fixe constitué de filets ou autres matériaux, dont l'implantation sur l'estran en forme de demi-cercle ou de V permet de retenir les poissons à marée basse.

IT **serraglia; serragia**

Insieme di reti sostenute da paletti fissati nel suolo che incanalano il pesce verso camere di raccolta con ingressi tronco conici.

NL **weer; doolhof; rietpark; waard; vloeiwaard; sero**

Vistuig bestaande uit vleugels van twijghout in V-vorm, waarachter een fuik of een vangkamer is opgesteld. Wordt grotendeels op bij eb droogvallende platen gebruikt: bij vloed wordt de vis naar de vleugels geleid en bij eb door de vleugels opgevangen en naar de punt van het vistuig geleid.

In gebruik in Indonesië.

PT **barreira; barragem; estacada**

Arte de pesca constituída por diversos materiais (estacas, canas, ramos de árvore, panos de rede, etc.), geralmente construída em zonas de maré. Difere dos tapa-esteiros porque, quando a maré baixa, permite que os peixes não emalhados ou enredados possam passar livremente por baixo dela.

3033

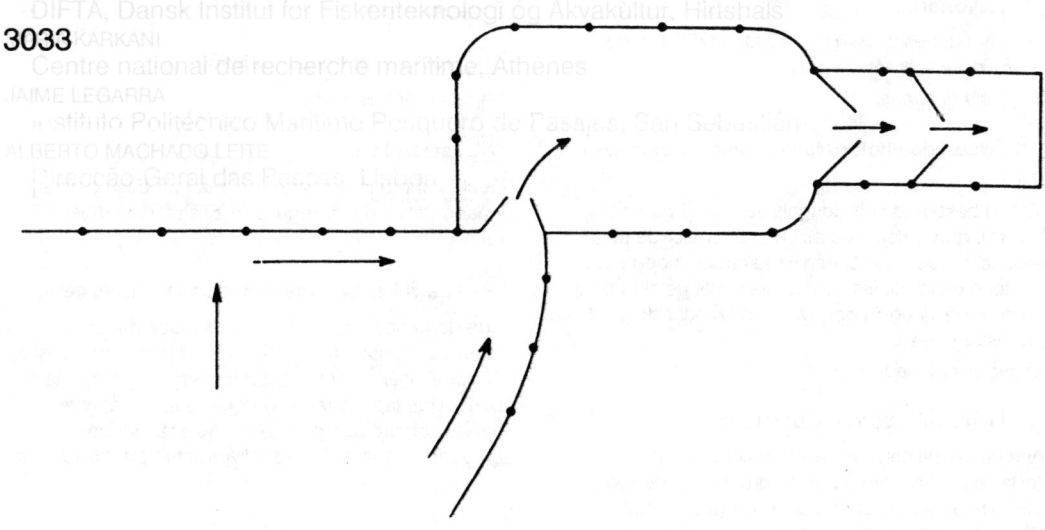

3034

ES encañizada

Corral o cerca de cañas que se forma para pescar en albuferas y otros sitios de poco fondo.

DA fiskegård; bundgarn

Passivt fiskeredskab. Fiskene eller andet bytte bliver af netrader ledt til en fiskegård, bundgarnshoved, hvorfra flugt ikke eller næsten ikke er mulig. Der anvendes forskellige anordninger til at forhindre dette.
Fiskegård er et gammelt dansk udtryk, der næsten ikke bruges mere.

DE Fischwehr; Fischzaun

Feste Fischfalle.

GR ιχθυοφραγμός
παγίδα ψαριών

EN fish weir; weir; barricade

Traplike arrangement of fences in which the fish or other prey enter a catching chamber from which escape is difficult, especially when the way out of the trap is secured by a non-return device.

FR bordigue; bourdigue

Installation fixe, constituée de clayonnages en métal ou autres matériaux, placés dans le chenal de communication d'une lagune avec la mer constituant un piège permettant de retenir ou capturer les poissons à leur entrée dans la lagune ou à leur sortie vers la mer.

IT lavoriero

Impianto di pesca fisso che sfrutta la discesa del pesce dalle valli verso il mare.
Trappola di canne.

NL weer; doolhof; rietpark; waard; vloeiwaard; sero

Vistuig bestaande uit vleugels van twijghout in V-vorm, grotendeels op bij eb droogvallende platen, waarachter een fuik of een vangkamer is opgesteld; bij vloed wordt de vis naar de vleugels geleid en bij eb door de vleugels opgevangen en naar de punt van het vistuig geleid.
In gebruik in Indonesië.

PT estacada; barreira; barragem

Arte de pesca constituída por diversos materiais (estacas, canas, ramos de árvore, panos de rede, etc.), geralmente construída em zonas de maré. Difere dos tapa-esteiros porque, quando a maré baixa, permite que os peixes não emalhados ou enredados possam passar livremente por baixo dela.

3035

ES red fija; arte fija; red barrera

Arte que permanece en la misma posición durante el tiempo que está calada. Generalmente tiene forma rectangular y es largada hasta el fondo con lastre en su base y flotadores en la parte superior por lo que adopta la forma de una valla-cortina contra la que inciden los peces para quedar atrapados en ella debido a su elaboración en malla.

DA faststående redskab

Samlebetegnelse for redskaber, der er fæstnet ved ankre, pæle eller lignende.

DE Standnetz; Stellnetz

Aneinandergereihte Netze werden quer zum Strom verankert. Bei Hochwasser werden Stellnetze in Strandnähe oder in Prielen gesetzt. Mit dem ablaufenden Wasser treiben die Fische, die sich in der flachen Uferzone aufhalten, in die Netze.

GR φράχτης· στάσιμο δίχτυ· δίχτυ καρτέρι

Γενικός όρος για κάθε απλό δίχτυ όταν διατηρείται σε σταθερή θέση, κατά τη διάρκεια της αλιείας, με άγκυρες, βαρίδια και/ή πασάλους

EN fence; set net; fixed net

General term for any simple net when it is held in fishing trim by anchors, sinkers and/or stakes.

FR parc; filet fixe

Nom générique désignant un filet à poste fixe. Il s'agit en général de filets maillants, montés sur des pieux plantés sur l'estran. Les poissons sont démaillés à marée basse.

IT rete fissa; impianto fisso

Impianto di pesca fisso.

NL staand net

Rechtstandig in het water neerhangend fijn netwerk, waarin de vis die er tegenaan zwemt, vast moet raken.

PT barreira; barragem; estacada; tapa-esteiros

Arte de pesca constituída por diversos materiais (estacas, canas, ramos de árvore, ligados por panos de rede), geralmente construída em zonas de maré. Difere dos tapa-esteiros porque, quando a maré baixa, permite que os peixes não emalhados ou enredados possam passar livremente por baixo dela.

3036

ES | barrera

Arte hecha de diversos materiales (estacas, ramas, juncos, redes, etc.) que se construye, por regla general, en las zonas intermareales. Difiere de las redes de enmalle fijas en que, en estas últimas, al bajar la marea, los peces que no han quedado enmallados o enredados pueden escapar por debajo de la relinga inferior.

DA | fiskespærring

Fiskefælde til brug i tidevandszoner.

DE | Fangbau

Fällt unter den Oberbegriff Fischfalle. Kategorie der Fischereigeräte für die sogenannte stille Fischerei im Flachwasser und Tidenbereich.

GR | φράγμα

Το εργαλείο αυτού του τύπου, κατασκευασμένο από διάφορα υλικά (πασσάλους, βραχίονες, σπάρτινα σχοινιά-βουρλιές, δικτύωμα, κλπ.), χρησιμοποιείται συνήθως σε περιοχές με παλίρροια. Διαφέρει συνήθως από τα σταθερά απλάδια τα οποία, όταν τα νερά αποσύρονται, μπορεί να αφήσουν τα ψάρια να περάσουν ελεύθερα κάτω από το κάτω γραντί

EN | barrier

Gear, made from various materials (stakes, branches, reeds, netting, etc.), usually constructed in tidal waters. Differs from the fixed gillnets which, when the tide ebbs, may eventually allow the fish not entangled or gilled to pass freely underneath their bottom line.

FR | barrage

Fabriqué en matériaux divers (pieux, branchages, roseaux, filets, etc.), cet engin est habituellement installé dans la zone de balancement des marées. Il est à distinguer des filets maillants fixes qui, à marée descendante, peuvent éventuellement laisser passer librement les poissons non maillés au-dessous de leur ralingue inférieure.

IT | barriera

NL | barrière

Afgesloten gedeelte van het water waar de vis gemakkelijk in kan zwemmen, maar niet terug kan, door afsluiting van het volume.

Vereist, net als de kom, permanente bewaking.

PT | barreira; barragem; estacada

Arte de pesca constituída por diversos materiais (canas, ramos de árvore, panos de rede, etc.), geralmente construída em zonas de maré. Difere dos tapa-esteiros porque, quando a maré baixa, permite que os peixes não emalhados ou enredados possam passar livremente por baixo dela.

3037

ES butrón; buitrón; botrino; butirón

Utilizado en ríos, estuarios o zonas de mucha corriente. Generalmente de forma cónica o piramidal, se fija mediante anclas o estacas, colocándolo de acuerdo con la dirección y fuerza de la corriente. La boca se suele mantener abierta mediante un bastidor, que a veces puede estar suspendido de una embarcación.

DA forankret hamme

Konisk net, der holdes åbent af bomme. Sættes i strømmende vand.

DE Ankerhamen; Steerthamen; Pfahlhamen

Hamen mit konischem Netzsack, der vom vor Anker liegenden Kutter im strömenden Wasser ausgesetzt wird; der Pfahlhamen, fest im Grund stehend, arbeitet nach dem gleichen Prinzip im Tidenbereich.

Das Hamengestell hat eine Öffnung von ca. 4 x 7 m und der konische Steert eine Länge von 8 bis 20 m.

GR δίχτυ στοιβάγματος· δίχτυ αναπλωρίσματος

Εργαλείο που μπορεί να χρησιμοποιηθεί μόνο σε ποταμούς, εκβολές ή περιοχές με ισχυρά ρεύματα. Συνήθως σχήματος κώνου ή πυραμίδας, τα δίχτυα αυτά είναι στερεωμένα σταθερά με άγκυρες ή πασσάλους τοποθετημένους σύμφωνα με τη διεύθυνση και φορά του ρεύματος. Τα στόμιά τους διατηρούνται ανοιχτά συνήθως με ένα πλαίσιο το οποίο μπορεί ή δεν μπορεί να μεταφερθεί από ένα σκάφος

EN stow net; swing net

Conical net held open by one or more horizontal beams below an anchored boat.

For sprats, whitebait.

FR filet à l'étalage; diable

Filet de forme conique ou pyramidale, maintenu ouvert par des perches et utilisé sur un bateau mouillé dans le courant, en estuaire ou dans les eaux côtières.

Les filets à l'étalage peuvent aussi être montés sur pieux, placés sur l'estran et découvrant à marée basse.

IT rete fissa a corrente

NL staande kuil; ankerkuil

Op stroom verankerd trechtervormig net, dat opengehouden wordt door scheerborden, drijvers of gewichten, een frame of palen; wordt zowel voor de binnenvisserij als de zeevisserij toegepast en kan vanaf het land of een schip worden bediend.

Gewoonlijk gebruikt voor de vangst van garnalen.

PT butirão

Arte de pesca tipo armadilha fixa que só pode ser utilizada em rios, estuários ou zonas com correntes fortes. Normalmente têm forma cónica ou piramidal, são fundeadas por meio de âncoras ou estacas e dispõem-se de acordo com a direcção e a força das correntes. A respectiva boca mantém-se aberta por meio de um quadro suspenso ou não de uma embarcação.

3037

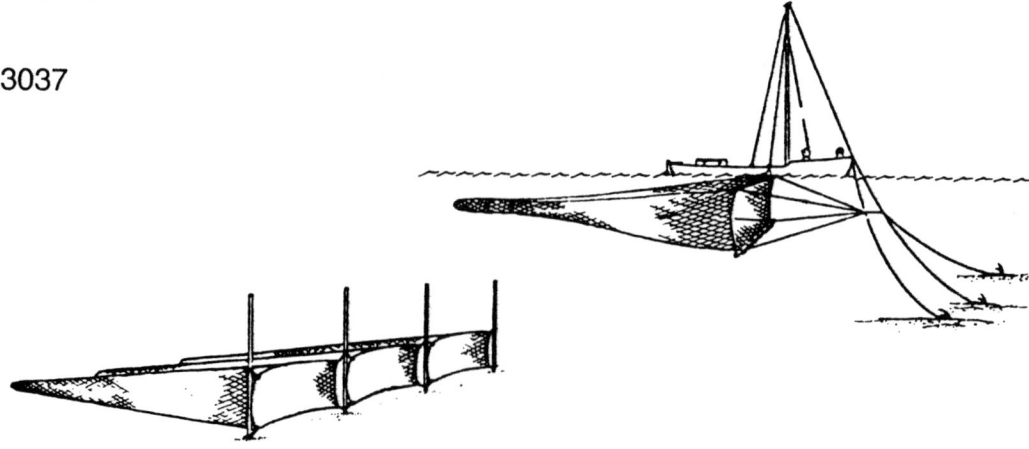

3038

ES | **garlito**

Trampa, normalmente utilizada en aguas poco profundas, formada por una manga de malla, de forma cilíndrica o cónica, montada sobre aros y otras estructuras rígidas, con alas o bandas para dirigir a los peces hacia la entrada.

Estas trampas se fijan al fondo con anclas, lastres o estacas y pueden utilizarse separadamente o en grupos.

DA | **ruse; garnruse**

Rusen er et tragtformet redskab, som består af selve rusen med en eller flere kalve og et par arme. Desuden anvendes som regel et radgarn, som fører fisken hen til rusens munding. Rusen sættes på bunden og fastgøres med pæle eller ankre.

DE | **Garnreuse; Spannreuse; Bügelsack; Korb ***

Fanggerät mit Netzflügeln, dessen Öffnung dem Ufer zugekehrt ist. Wird an Buhnen eingesetzt.

* Diese Bezeichnung wird an der Weser und Elbe benutzt.

GR | **βολκός**

Χρησιμοποιούμενες κανονικά σε ρηχά νερά, οι παγίδες αυτές αποτελούνται από κυλινδρικές ή κωνικές θήκες τοποθετημένες σε δακτύλιους ή άλλες άκαμπτες κατασκευές, καλυμμένες τελείως από δικτύωμα και συμπληρωμένες από φτερά ή οδηγούς που κατευθύνουν τα ψάρια προς το άνοιγμα των θηκών.Οι βολκοί, σταθεροποιημένοι στον πυθμένα με άγκυρες, έρμα ή ελάσματα, μπορούν να χρησιμοποιηθούν χωριστά ή σε ομάδες

EN | **fyke net**

Trap normally used in shallow water which consists of cylindrical or cone-shaped bags mounted on rings or other rigid structures, completely covered by netting and completed by wings or leaders which drive the fish towards the opening of the bags.

The fyke nets, fixed on the bottom by anchors, ballast or stakes, may be used separately or in groups.

FR | **verveux**

Piège utilisé normalement en eau peu profonde, constitué par des poches de capture, de forme cylindrique ou conique, montées sur des cercles ou autres structures rigides, entièrement recouvertes de filet, et complétées par des ailes ou guideaux qui rabattent les poissons vers l'ouverture des poches.

Les verveux, fixés sur le fond par des ancres, lests ou piquets, peuvent être employés isolément ou groupés.

IT | **cogollo; bertovello ***

Rete trappola con bocca rigida, un braccio di incanalamento ed alcuni ingressi consecutivi tronco conici per impedire la fuga del pesce catturato. Quest'ultimo viene prelevato salpando solo l'ultima parte dell'attrezzo, quella cioè dopo l'ultimo ingresso.

* In genere senza braccio d'incanalamento.

NL | **fuik**

Kegelvormig vistuig voor gebruik in binnenzeeën en riviermonden, bestaande uit netwerk dat om hoepels is gespannen, met een of meer enkels, en voorzien van vleugels en schutwant voor het binnenleiden van de vis.

In ons land veel gebruikt voor de vangst van aal of haring.

PT | **galricho**

Arte de pesca tipo armadilha fixa, normalmente utilizada em águas pouco profundas, constituída por uma manga de rede de pesca com forma cónica ou cilíndrica, montada sobre aros e/ou outras estruturas rígidas de madeira ou metal, dotadas ou não de asas destinadas a encaminhar os peixes para a respectiva boca.

São artes de pesca fundeadas por meio de âncoras, lastros ou estacas, utilizadas isoladamente ou em grupos.

3038

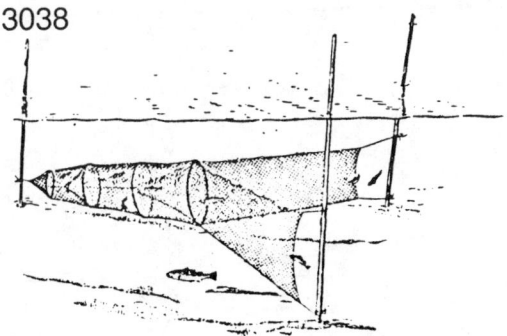

3039

ES orinque; orinque de boya

Cabo que sujeta una boya a un ancla fondeada o un arte calada en el fondo (p.e. nasa, red de enmalle, palangre).

DA bøjereb

Reb, der forbinder en bøje med et anker eller et fiskeredskab.

DE Bojenreep

Verbindungsleine zwischen einem auf Grund liegenden Objekt und der Markierungsboje.

GR σχοινί σημαδούρας

Σχοινί που συνδέει τη σημαδούρα με το τμήμα του εργαλείου που πρέπει να στηριχθεί ή επισημανθεί

EN buoy rope; buoy line

Rope connecting buoy to that part of the gear being supported or marked.

FR orin de bouée

Cordage reliant un engin mouillé ou calé au fond (casier, filet maillant, palangre) à une bouée signalant sa position à la surface.

IT grippia; cavo del galleggiante; caloma; orza

Pezzo di sagola o di cavetto che unisce il piombo di una rete o di una nassa alla boa o al gavitello che le segnalano.

NL boeireep

Touw waarmee de boei aan het kruis van het anker wordt verbonden.

PT arinque

Cabo geralmente de fibra que liga uma bóia de sinalização a uma arte de pesca ou âncora e que eventualmente pode ser utilizado para a alar.

3040

ES nasa

Trampa en forma de cajas o cestas hechas de diversos materiales (varillas de madera, mimbres, varillas de metal, red metálica, etc.), con una o más aberturas o entradas.

Estas trampas generalmente se colocan en el fondo, con o sin cebo, individualmente o en andanas, y están unidas mediante una sirga a una boya que indica su situación en la superficie.

DA tejne; kurv

Fiskeredskab formet som en beholder af træ, metal eller lignende, evt. med net. Har en eller flere ofte tragtformede indgange.

Anvendes normalt på bunden til fangst af krebsdyr og fisk. Forsynes i visse fiskerier med agn.

DE Ko̍rbreuse

Aus Holzstäben oder Holzgeflecht, vorn trichterförmig erweiterter beköderter oder unbeköderter Korb. Der Leitkorb mündet mit einer Einkehlung in den Fangkorb, der mit einem Pflock verschlossen ist.

GR κιούρτος· κοφινέλο

Οι παγίδες αυτές, σχεδιασμένες να παγιδεύουν ψάρια ή μαλακόστρακα, έχουν το σχήμα κλουβιού ή καλαθιού, είναι κατασκευασμένες από διάφορα υλικά (ξύλο, βέργα λυγαριάς, μεταλλικές ράβδους, συρματόπλεγμα, κλπ.) και έχουν ένα ή δύο ανοίγματα ή εισόδους. Συνήθως τοποθετούνται στον πυθμένα με ή χωρίς δόλωμα, μεμονωμένα ή σε σειρές συνδεδεμένες με σχοινιά σε σημαδούρα δείχνοντας έτσι στην επιφάνεια τη θέση τους

EN pot; fish pot

Trap, designed to catch fish or crustaceans, in the form of cages or baskets made from various materials (wood, wicker, metal rods, wire netting, etc.) and with one or more openings or entrances.

It is usually set on the bottom, with or without bait, singly or in rows, connected by ropes (buoy lines) to buoys showing its position on the surface.

3040

FR **casier; nasse**

Piège destiné à la capture des poissons ou crustacés, en forme de cages ou de paniers fabriqués avec des matériaux divers (bois, osier, plastique, tiges métalliques, grillage, etc.) et comportant une ou plusieurs ouvertures ou goulets d'entrée.

Muni ou non d'appâts, il est en général mouillé sur le fond, isolément ou en filière, relié par des filins («orins») à des bouées indiquant sa position à la surface.

IT **nassa**

Attrezzo da pesca a forma di trappola o di cesto fatto di giunchi intrecciati, o di rete metallica o anche di listerelle di legno, che viene calato in mare dopo essere stato convenientemente inescato. È munito di un'apertura congegnata in modo che i pesci o i crostacei, che entrano per nutrirsi dell'esca, non possano piú uscirne.

Viene usato soprattutto per la pesca delle aragoste e di altri crostacei.

NL **korf; kubbe**

Vistuig, grotendeels van hout of ijzer, met een of meer enkels.

PT **nassa; covo; alcatruz; murejona**

Arte de pesca tipo armadilha fixa que se utiliza para capturar peixes, moluscos ou crustáceos, com a forma de caixa, cesto ou pote; pode ser construída com diversos materiais (madeira, varas de metal, barro, rede de pesca, rede de metal, rede de plástico) e possui uma ou mais aberturas ou entradas (boca e endiche).

São colocadas no fundo com ou sem isco, isoladas ou em teias, e ligadas a um ou mais cabos de alagem referenciados à superfície por meio de bóias.

3040

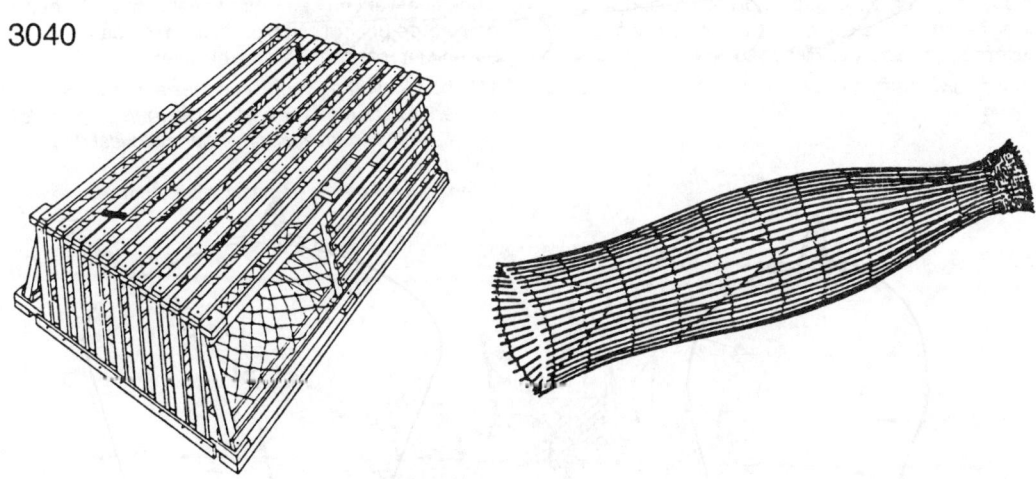

3041

ES almadraba fija descubierta; red calada *

Por regla general, una red grande anclada o sujeta a estacas, abierta en la superficie y provista de diversos sistemas para dirigir y retener a los peces. Suele estar dividida en compartimientos con el fondo cubierto de red.

* En Japón, estas artes se denominan redes caladas.

DA ikke overdækket bundgarn

Betegner en speciel form for bundgarn.

DE nicht bedeckte stationäre Reuse

Verankertes und an Pfählen gehaltenes, an der Oberfläche offenes Netz großer Abmessung mit diversen Vorrichtungen zur Zurückhaltung der Fische. Ein sogenanntes Leittuch lenkt die Fische in die Reuse, die am Boden mit Netztuch verschlossen ist.

Gebräuchlich für den Ostsee-Heringsfang.

GR στάσιμο και ακάλυπτο δίχτυ-ενέδρα

Δίχτυα, συνήθως μεγάλα, αγκυροβολημένα ή στερεωμένα σε πασσάλους, ανοιχτά στην επιφάνεια και εξαρτισμένα με διάφορους τύπους συσκευών για τη συγκέντρωση και συγκράτηση του αλιεύματος. Είναι, ως επί το πλείστον, διαιρεμένα σε θαλάμους κλειστούς στη βάση με δικτύωμα

Στην Ιαπωνία αυτή η ομάδα αναφέρεται σαν στάσιμα δίχτυα

EN stationary uncovered pound net; set net *

Usually large net, anchored or fixed on stakes, open at the surface and provided with various types of fish herding and retaining devices. It is mostly divided into chambers closed at the bottom by netting.

* In Japan this group is usually referred to as set-nets.

FR filet-piège fixe non couvert; filet fixe *; madrague **

Filet habituellement de grandes dimensions, ancré ou fixé sur des pieux, ouvert à la surface et muni de divers dispositifs de rabattement et de retenue du poisson. Généralement divisé en compartiments fermés à sa base par une nappe de filet.

*Au Japon, ce groupe est généralement désigné par le terme filets fixes.
**Utilisé couramment autrefois en Méditerranée.

IT rete trappola non coperta

NL onbedekte kom

Vistuig bestaande uit een samenstel van geleidenetwerk en een vanggedeelte, opgehouden door op de bodem verankerde palen en aan de bovenkant niet door netwerk gesloten.

- De bodem van de vangkamer bestaat vaak uit netwerk waarmee de vangst kan worden opgehaald.
- Een „weir" is gemaakt van ander materiaal dan textiel.
- Een „pound net" is gemaakt van netgarens.

3041

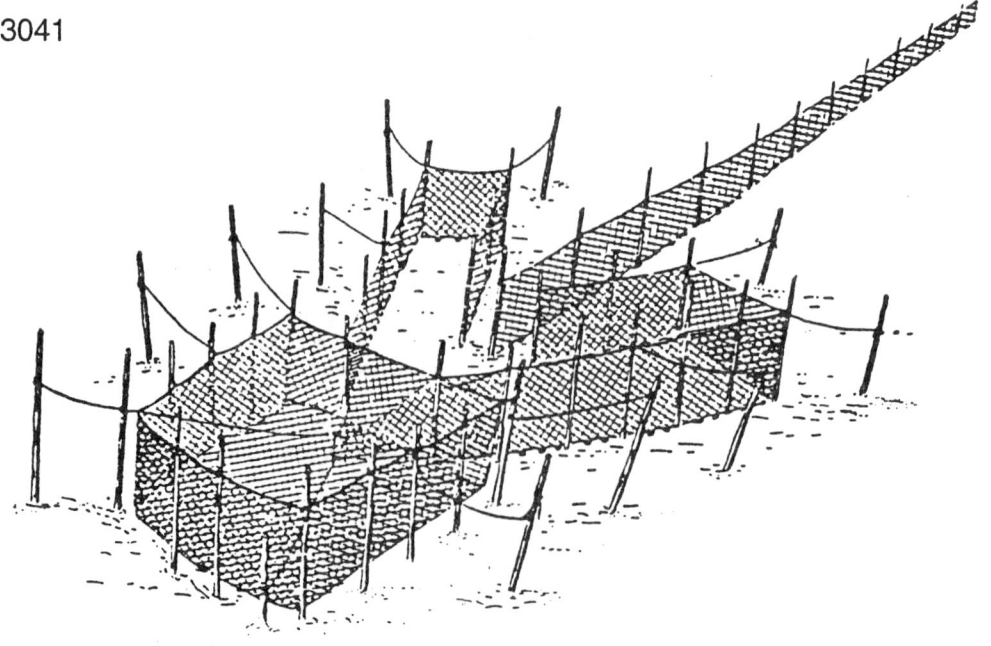

3041

PT armação; almadrava; almadraba

Arte de pesca, tipo armadilha fixa de grande extensão, constituída por redes verticais sustentadas por estacas ou cabos e âncoras, que partem de terra e entram pelo mar dentro, definindo uma série de canais, barreiras e câmaras através das quais os peixes são conduzidos, após terem penetrado na armadilha, até ao copo onde são apanhados.

Este processo de pesca destina-se à captura de atuns e/ou sardinha e consiste basicamente em colocar em permanência na água, em certas épocas do ano e junto à costa em local apropriado no itinerário normal de passagem dos cardumes, um complicado sistema de redes fixas.

3042

ES almadraba

Arte fijo complicado, tanto por su gran extensión —a veces de muchos kilómetros—, como por la serie de partes y compartimientos que comprende. En síntesis, está constituida por unas largas redes verticales, que parten de tierra y se adentran en el mar, las «raberas» —la más próxima a tierra es la «rabera de tierra» o de «dentro», y la más alejada la «rabera de fuera»— y por un cuadro, subdividido en diversos compartimientos, que se intercala entre ambas raberas, a cuyo cuadro son conducidos los atunes por aquéllas, y en el que son capturados.

DA bundgarn; ruse

Samlebetegnelse for et omfangsrigt redskab, der består af en bundgarnsrad, som placeres vinkelret ud fra kysten, og et eller flere bundgarnshoveder yderst. Raden består af net, der ophænges på pæle. Underkanten af raden holdes til bunden af en kæde eller af blystykker. Hovedet består ligeledes af net, der er lukket i bunden. Det har en tragtformet indgang, der fører fiskene ind til et indelukke, men udformningen kan variere.

DE Reuse

Fanggerät der Binnen- und Küstenfischerei zum Fang von Grundfischen und pelagisch in Flachwasserregionen lebenden Fangobjekten.

GR δίχτυ-ενέδρα

Γενικός όρος για κάθε αγκυροβολημένο και πασσαλωμένο δίχτυ, περιλαμβανομένου ενός οδηγού και ενός ή δύο περιφραγμάτων, π.χ. διχτυωτός σάκος, πασσαλωμένο δίχτυ, δίχτυ-κατσαρόλα

EN pound net

General term for any moored or staked net comprising a leader and one or more enclosures, e.g. bag net, stake net, kettle net.

3042

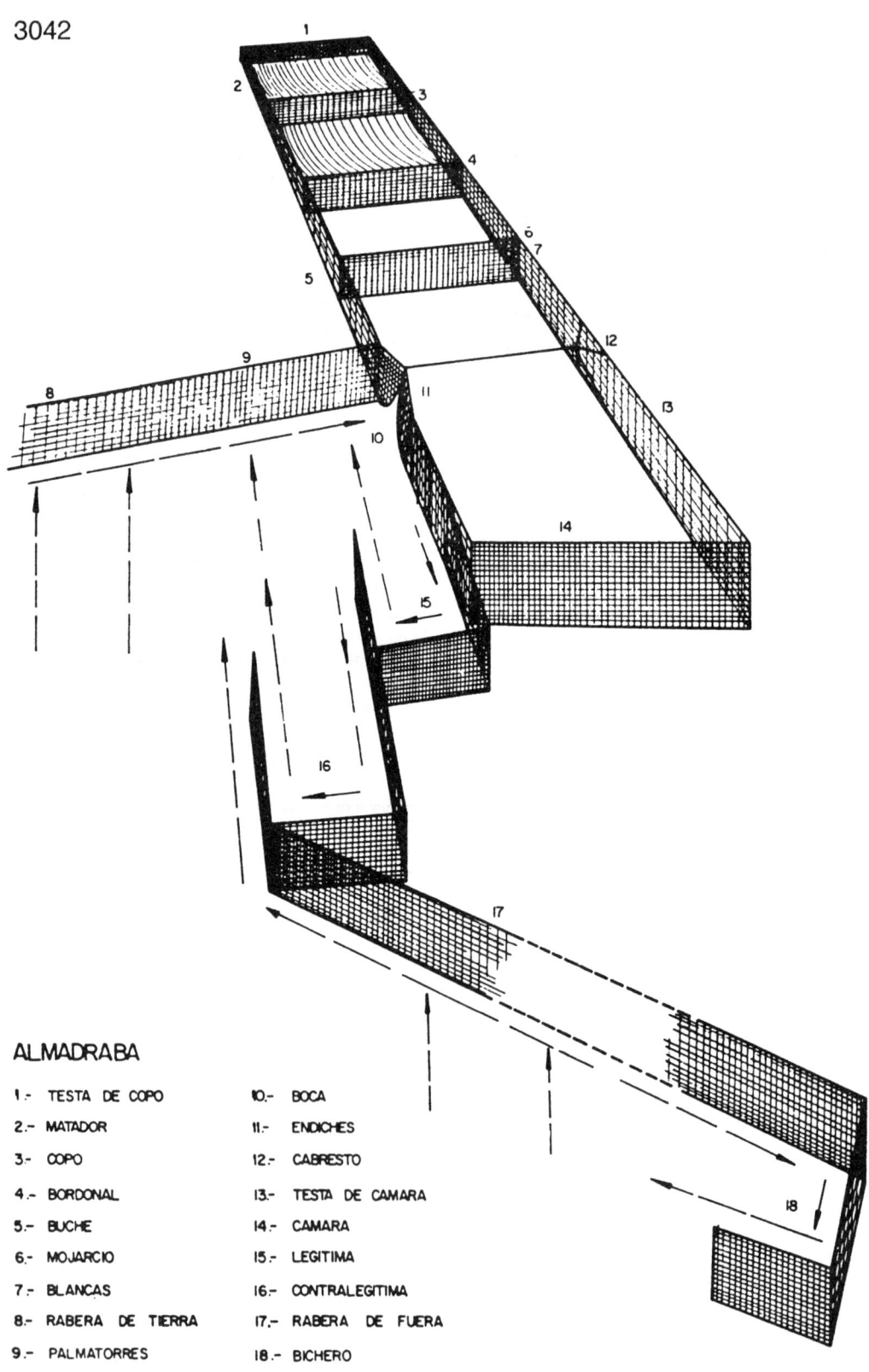

ALMADRABA

1.-	TESTA DE COPO	10.-	BOCA
2.-	MATADOR	11.-	ENDICHES
3.-	COPO	12.-	CABRESTO
4.-	BORDONAL	13.-	TESTA DE CAMARA
5.-	BUCHE	14.-	CAMARA
6.-	MOJARCIO	15.-	LEGITIMA
7.-	BLANCAS	16.-	CONTRALEGITIMA
8.-	RABERA DE TIERRA	17.-	RABERA DE FUERA
9.-	PALMATORRES	18.-	BICHERO

3042

FR | **filet-piège**

Engin généralement de grandes dimensions, mouillé dans les eaux côtières peu profondes. De structure complexe, ce filet comporte en général une ou des barrières guidant les poissons vers une ou plusieurs chambres de capture.

IT | **rete a postazione fissa**

Rete trappola generalmente di grandi dimensioni che viene calata in punti particolari ed ivi lasciata per tutta la stagione di pesca. Il recupero del pescato è effettuato ad intervalli regolari issando solo la parte terminale della rete, quella dove appunto viene raccolto il pesce.

NL | **kom**

Vistuig gebruikt voor het vangen van migrerende vis langs de kustlijn, bestaande uit een netwerk bevestigd aan palen ter geleiding van de vis (obstructie) naar de vangkamer, die afsluitbaar is.

Men moet het vistuig bewaken om te bepalen wanneer de vangkamer moet worden afgesloten.

PT | **armação; almadrava; almadraba**

Arte de pesca, tipo armadilha fixa de grande extensão, constituída por redes verticais sustentadas por estacas ou cabos e âncoras, que partem de terra e entram pelo mar dentro, definindo uma série de canais, barreiras e câmaras através das quais os peixes são conduzidos, após terem penetrado na armadilha, até ao copo onde são apanhados.

Este processo de pesca destina-se à captura de atuns e/ou sardinha e consiste basicamente em colocar em permanência na água, em certas épocas do ano e junto à costa em local apropriado no itinerário normal de passagem dos cardumes, um complicado sistema de redes fixas.

3043

ES | **arte de trampa**

Arte fija formada por una serie de mamparas de mallas distribuidas en forma de laberintos que conducen a los peces hacia una cámara de la que no pueden retroceder.

Las artes de trampa (almadrabas, corrales, etc.) se calan generalmente en el fondo y a pequeñas profundidades, permanecen caladas durante un largo período de tiempo e incluso de forma permanente.

DA | **fælde**

Samlebetegnelse for faststående redskaber, bestående af serier af net-tragte. Hver især åbner de ind i den næste, og fører fisken frem til en afsnøret pose. Tragtene holdes åbne af store ringe af jern eller andet materiale.

DE | **Fischfalle**

Oberbegriff für Fangbauten, Fischzäune, Labyrinthbauten, Fallen im eigentlichen Sinn, Schwerkraftfallen, Schlingen aus diversen Materialen und Reusen unterschiedlichster Art.

GR | **παγίδα**

— Κομμάτι σύρματος που περιελίσσεται γύρω από το δακτύλιο του ανοίγματος έτσι ώστε να σχηματίζει μονόδρομη παγίδα.
— Εργαλείο ψαρέματος που μπορεί να τοποθετηθεί σ' ένα καθορισμένο σημείο και εκεί αφήνεται και λειτουργεί παθητικά. Γενικά είναι κατασκευασμένο από ένα θάλαμο και ένα ή περισσότερα στόμια που επιτρέπουν την είσοδο αλλά όχι και την έξοδο των οργανισμών που έχουν συλληφθεί

EN | **fish trap; trap; trap net**

An inshore fishing apparatus consisting of a series of funnels, with their mouths kept open by hoops, opening into each other and finally closing into a sack forming a trap.

This type of net, of which there are many variations, is fastened to the bottom.

3043

FR | **piège**

Engin fixe le plus souvent calé sur le fond. Comporte en général une chambre munie d'une ou plusieurs ouvertures de forme spécialement étudiée pour permettre l'entrée mais non la sortie des animaux à capturer.

IT | **trappola**

Attrezzo da pesca che viene calato in un punto determinato ed ivi lasciato passivamente. Generalmente è costituito da una camera con una o più bocche che permettono l'ingresso, ma non l'uscita degli organismi da catturare.

NL | **val**

Vistuig waarin de vis wordt gelokt d.m.v. aas, en waaruit de vis zeer moeilijk kan ontsnappen.

PT | **armadilha**

Termo genérico dado às artes de pesca nas quais o peixe entra e de onde não pode sair facilmente pelos seus próprios meios.

3044

ES | **red combinada de enmalle-trasmallo**

Arte, que se cala en el fondo y que está formada por una red de enmalle cuya parte inferior se sustituye por una red atrasmallada. De esa manera, se pueden capturar peces de fondo en la parte inferior de la red (trasmallo) y especies semidemersales o pelágicas en la parte superior (enmalle).

DA | **kombineret garn og toggegarn**

En speciel form for bundsat garn, hvor den øverste del er almindeligt net, mens den nederste er udformet som et toggegarn.

DE | **kombiniertes Kiemennetz/Trammelnetz; kombiniertes Dreiwandnetz/Einwandnetz**

Kiemennetz. Aufstellung am Grund als sperrende Wand. Das dreiwandige Netz oder Spiegelnetz ist an den äußeren Wänden aus stärkerem Garn und größeren Maschen gestrickt. Die innere Wand ist aus feinerem Garn und engeren Maschen geknüpft. Das einwandige Netz, über dem Spiegelnetz angeordnet, ist eine einfache Netzwand mit Spannschnüren zwischen Ober- und Untersimm zur Verkürzung der Netzbreite. Sie bewirkt in den Maschen eine gewisse Lose.

Diese Netzkombination eignet sich zum Fang am Grunde (z. B. Plattfische) und im Pelagial lebender Fische.

GR | **συνδυασμός απλαδιών και μανωμένων διχτυών**

Εργαλείο βυθού κατασκευασμένο από ένα απλάδι, το κατώτερο τμήμα του οποίου έχει αντικατασταθεί από ένα μανωμένο δίχτυ. Μπορεί να πιάσει βενθικά είδη στο κατώτερο μανωμένο δίχτυ μαζί με βενθοπελαγικά ή πελαγικά ψάρια στο ανώτερο τμήμα του που είναι απλάδι

3044

EN combined gillnet-trammel net

Bottom-set gear made with a gillnet, the lower part of which is replaced by a trammel net. It may catch bottom fish in the lower trammel net part, together with semi-demersal or pelagic fish in the upper gillnet part.

FR trémail et filet maillant combiné

Engin calé sur le fond constitué d'un filet maillant dont la partie inférieure est remplacée par un trémail. Il peut capturer les poissons de fond dans sa partie inférieure en trémail, ainsi que les poissons semi-démersaux ou pélagiques dans sa partie supérieure en filet maillant.

IT incastellata

Combinazione di rete da imbrocco e rete a tremaglio.

NL gecombineerd kieuwnet en schakel

Net bestaande uit een kieuwnet waaronder een schakel is aangebracht.

PT rede mista de emalhar-tresmalho

Arte de pesca fundeada que é constituída por uma rede de emalhar cuja parte inferior é substituída por um tresmalho. Com este tipo de arte de pesca podem assim capturar-se peixes de fundo na sua parte inferior (tresmalho) e peixes semidemersais e pelágicos na sua parte superior.
Desconhece-se o seu uso em Portugal.

3044

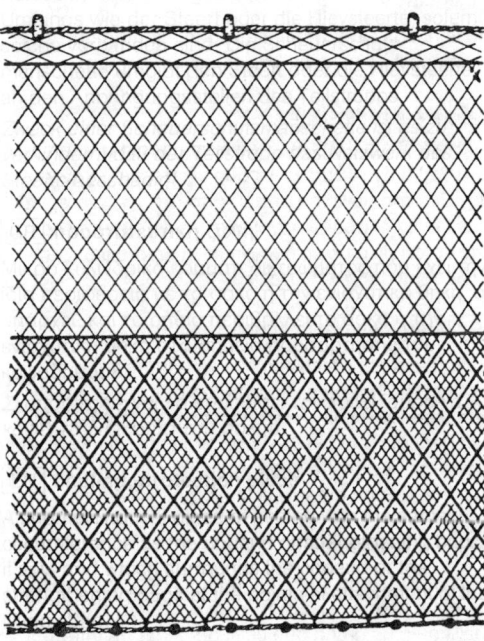

3045

ES red atrasmallada; trasmallo

Red, que se cala en el fondo y que está formada por tres redes superpuestas, dos exteriores de malla clara y una central montada más floja. Los peces se enredan en la red interior, de malla más tupida, después de atravesar las paredes exteriores.

DA toggegarn; grimegarn

Garntype, der består af tre lag. På ydersiderne net med store masker, i midten net med små masker. Der er så rigeligt med småmasket net, at fisken, når den går på nettet, kan trække en lomme af småmasket net med ud gennem det stormaskede net.

Effektivt redskab til fangst af fisk, der har svært ved at sidde fast i et normalt et-laget garn, f.eks. fladfisk.

DE Trammelnetz; Ledderungsnetz; Dreiwandnetz; Fesselnetz; dreifaches Wandnetz; Spiegelnetz

Besteht aus einer großmaschigen, straffen und zwei kleinmaschigen, losen Maschenwänden.

GR μανωμένο δίχτυ

Σύστημα διχτυών βυθού κατασκευασμένο από τρία επάλληλα δικτυώματα, τα δύο εξωτερικά έχοντας μεγαλύτερο μέγεθος ματιού από ό,τι το εσωτερικό που είναι χαλαρά αναρτημένο. Τα ψάρια μπλέκονται στο εσωτερικό με μικρό μάτι δικτύωμα αφού περάσουν από το εξωτερικό

EN trammel; trammel net

Bottom-set net which is made with three walls of netting, the two outer walls being of a larger mesh size than the loosely hung inner netting panel. The fish get entangled in the inner small meshed wall after passing through the outer wall and push themselves into the second outer wall, thus forming a bag.

FR trémail; tramail

Filet calé au fond constitué de trois nappes de filets, les deux nappes externes, aumées, étant d'un maillage plus grand que celui de la nappe interne, flue, montée avec beaucoup de flou. Les poissons s'emmêlent dans la nappe interne à petites mailles après avoir traversé une nappe externe.

IT tramaglio; tremaglio

Rete da posta consistente in un panno centrale a maglie relativamente strette, di altezza variabile tra uno e tre metri, fiancheggiato dai due lati, per tutta la sua lunghezza, da altri due panni di rete a maglie molto più grandi ma di altezza minore (dette pareti) in modo che quando sono fissati tutti e tre insieme sulle lime da piombo e da sugheri, quello centrale a maglie strette non risulta teso ma in bando.

NL schakel

Staand vistuig bestaande uit drie achter elkaar hangende netten, waarvan de twee buitenste (laddernetten) wijdmazig, en het binnennet fijnmazig en loshangend; de vis zwemt door het eerste laddernet, stuit tegen het binnennet en duwt dit door het tweede laddernet, waardoor een zak gevormd wordt door een deel van het binnennet.

PT tresmalho

Arte de pesca fundeada junto ao fundo, constituída por três panos de rede verticais sobrepostos, os dois exteriores com malhagem superior à do pano interior, o qual tem uma maior altura. Os peixes enredam-se no pano interior após terem atravessado os panos exteriores.

Os panos exteriores designam-se por albitanas e o interior por miúdo.

3045

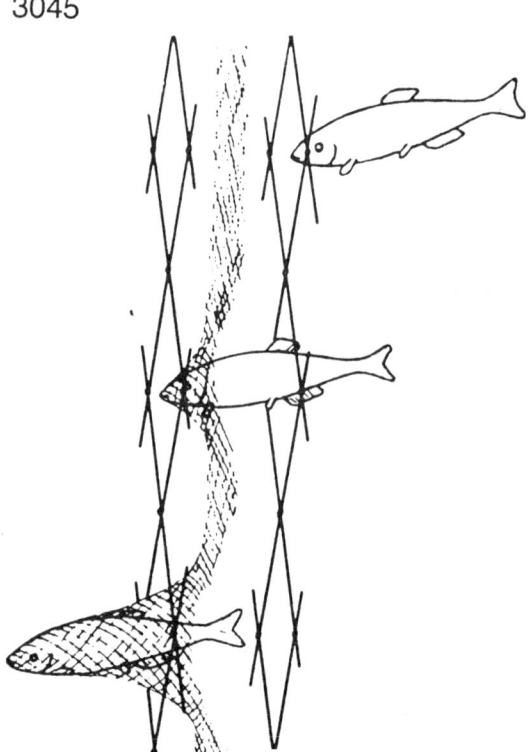

42

3046

ES red de enmalle fija; red de enmalle en estacas

Red calada sobre estacas clavadas en el fondo, utilizada esencialmente en aguas costeras.

Se recogen los peces cuando baja la marea.

DA garn fastgjort til pæle; hildingsgarn fastgjort til pæle

Anvendes i kystområder med tidevand, hvor garnet kan røgtes ved ebbe.

DE einwandiges Kiemennetz

An langen dünnen, in den Grund getriebenen Stangen befestigtes Netz; angesetzt entlang der Uferzone im Tidenbereich.

Fangentnahme bei Niedrigwasser.

GR σταθερό απλάδι· απλάδι σε πασσάλους

Αυτά τα δίχτυα είναι συνδεδεμένα σε πασσάλους στερεωμένους στον πυθμένα και χρησιμοποιούνται κυρίως σε παράκτια νερά

Το αλίευμα συλλέγεται κατά την αμπώτιδα

EN fixed gillnet; gillnet on stakes

Net mounted on stakes driven into the bottom, used essentially in coastal waters.

The fish are collected at low tide.

FR filet maillant fixe; filet maillant sur perches; filet maillant sur pieux

Ce filet monté sur des perches ou pieux plantés au fond est employé essentiellement dans les eaux côtières.

Les poissons sont démaillés à marée basse.

IT rete da posta a pali

Rete montata su pali e usata in acque costiere.

NL staand kieuwnet; vastgezet kieuwnet; vast kieuwnet

Kieuwnet dat op de bodem is verankerd door palen of stokken.

PT tapa-esteiros

Arte de pesca fixa utilizada essencialmente em águas costeiras, constituída por panos de rede fixados verticalmente da superfície ao fundo em estacas cravadas no fundo em zonas abrangidas pela acção das marés.

Os peixes recolhem-se na maré baixa.

3046

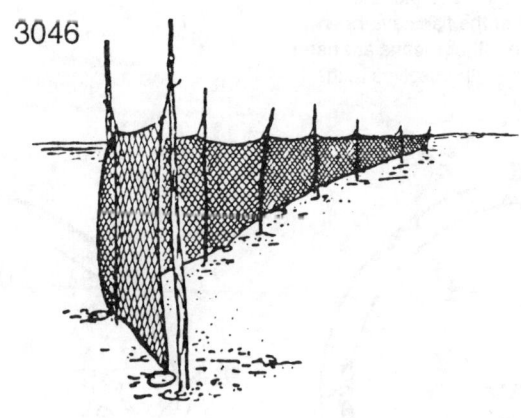

3047

ES red de enmalle de cerco; red de batir de cerco

Red utilizada, por regla general, en aguas poco profundas con la relinga superior en la superficie. Una vez que los peces han quedado cercados por la red, se hace ruido o, de otra manera, se les obliga a enmallarse o enredarse en los paños que los rodean.

DA omkredsende garn

Garn eller hildingsgarn til omkredsning af fisk.

Sættes rundt om en fiskeforekomst, hvorefter fiskene skræmmes med lyd eller andet til at flygte ud i nettet, hvor de sætter sig fast.

DE umschließendes Kiemennetz

Einwandiges Setznetz, das, zum Kreis geschlossen, im Flachwasserbereich eingesetzt wird. Der eingekreiste Fisch wird durch Scheuchwirkung in die Netzmaschen getrieben.

GR απλάδι περικύκλωσης

Εργαλείο που χρησιμοποιείται γενικά σε αβαθή νερά με το επάνω καζίλι να παραμένει στην επιφάνεια. Αφού τα ψάρια περικυκλωθούν από το δίχτυ, χρησιμοποιείται θόρυβος ή άλλο μέσο για να τα ωθήσουν να μπλεχτούν στο δίχτυ ή να παγιδευτούν σε μάτι του διχτυώματος που τα περιβάλλει

EN encircling gillnet

Gear generally used in shallow water with the floatline remaining at the surface. After the fish have been encircled by the net, noise or other means are used to force them to gill or entangle themselves in the netting surrounding them.

FR filet maillant encerclant

Engin généralement utilisé en eau peu profonde, dont la ralingue de flotteurs reste à la surface. Après avoir encerclé les poissons par le filet, on a recours au bruit ou autres procédés pour les forcer à se mailler dans la nappe qui les entoure.

IT rete da posta circuitante; rete da imbrocco circuitante

Rete da posta calata a cerchio o ad arco di cerchio.

NL omringend kieuwnet

Kieuwnet, waarvan de drijflijn aan het oppervlak drijft, dat een gedeelte van het wateroppervlak omringt en waar de te vangen vis ingejaagd wordt.

PT rede de emalhar envolvente

Arte de pesca geralmente utilizada em águas pouco profundas e com a tralha das cortiças à superfície. Após o cardume ser cercado, força-se o emalhar ou enredar dos peixes através de ruído ou outros processos.

3047

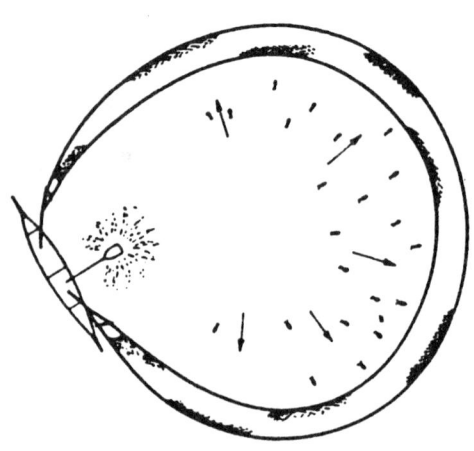

3048

ES red de enmalle de deriva

Red que se mantiene en la superficie o a cierta distancia bajo la superficie por numerosos flotadores, y que se deja a la deriva a merced de las corrientes, por sí sola o, más frecuentemente, junto con la embarcación a la que está ligada.

DA drivgarn

Et garn, der holdes ved overfladen eller i en vis afstand herfra af talrige flydere. Driver med strømmen eller oftere med fartøjet, hvortil det er fastgjort.
Anvendes til fangst af pelagiske fisk.

DE Treibnetz; Schwimmnetz

Im Wasser frei vom Grund hängendes Kiemennetz, das mit dem Boot verbunden in der Strömung treibt und mit Hilfe der Schwimmleine an der Oberfläche oder in gewünschter Tiefe gehalten wird.

GR παρασυρόμενα απλάδια· παρασυρόμενα δίχτυα

Παραμένοντας στην επιφάνεια ή σε συγκεκριμένη θέση κάτω από αυτή χάρη σε πολυάριθμους πλωτήρες, τα δίχτυα αυτά παρασύρονται ελεύθερα από το ρεύμα, ανεξάρτητα ή συνήθως μαζί με τη βάρκα στην οποία είναι προσδεδεμένα

EN drifting gillnet; driftnet

Net kept on the surface, or at a certain distance below it, by numerous floats. It drifts freely with the current, separately or, more often, with the boat to which the net is attached.

FR filet maillant dérivant; filet dérivant

Filet maintenu à la surface, ou à une certaine distance en dessous de celle-ci, grâce à de nombreux flotteurs, qui dérive librement avec le courant, isolément ou, le plus souvent, avec le bateau auquel il est amarré.

IT rete da posta derivante

Rete da posta lasciata all'azione dei venti e delle correnti.

NL drijfnet; drijfvleet

Kieuwnet dat als een gordijn in het water hangt, zwevend gehouden door aan de bovensim bevestigde drijvers, waarbij de reep onder de netten loopt, en dat vrij met de stroom kan meegaan.
In riviermonden gebruikt voor de vangst van zalm, elft en fint; op zee gebruikt voor haringvisserij.

PT rede de emalhar de deriva; rede de emalhar derivante; volanta

Arte de pesca mantida à superfície, ou a uma certa distância abaixo dela, por meio de numerosas bóias. Estas redes derivam livremente ao sabor das correntes, isoladamente ou, mais correntemente, em conjunto com a embarcação a que se encontram amarradas.

3048

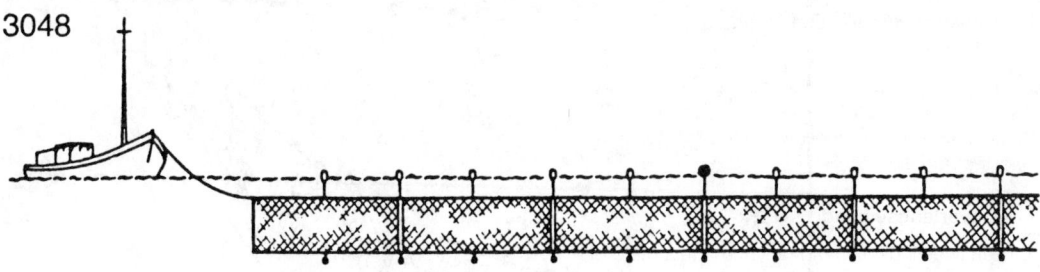

3049

ES red de enmalle de fondo; red de enmalle calada; volanta

Red que se fija en el fondo, o a cierta distancia de él, por medio de anclas o lastres suficientemente pesados para neutralizar los flotadores.

Tipo de red de enmalle de fondo.

DA forankret garn; bundsat garn

Betegner et garn, der er forankret til bunden, og hvor fiskene sætter sig fast i de enkelte masker (ofte ved gællerne).

Se 3056. På dansk skelnes ikke mellem 3049 og 3055.

DE Stellnetz; stationäres Kiemennetz; verankertes Kiemennetz

Am Boden eingesetztes Kiemennetz.

GR στάσιμο απλάδι βυθού· στάσιμο απλάδι· αγκυροβολημένο απλάδι

Δίχτυα στερεωμένα στον πυθμένα ή σε συγκεκριμένη απόσταση απ' αυτόν, με τη βοήθεια αγκυρών ή έρματος ικανού βάρους ώστε να εξουδετερώνεται η άνωση των πλωτήρων

3049

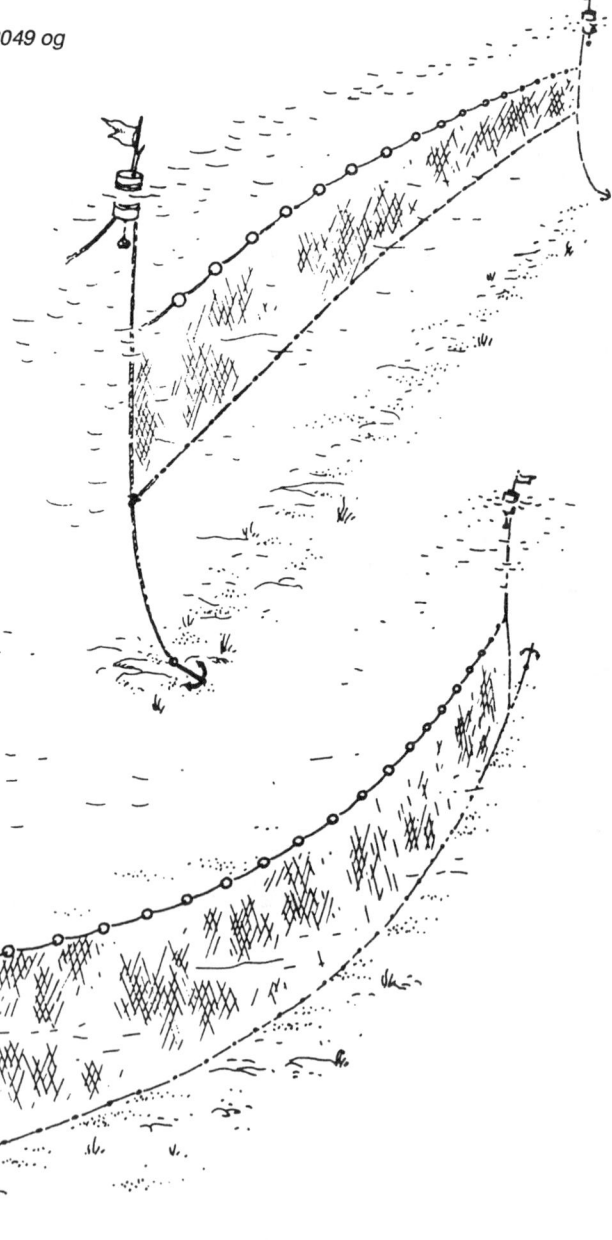

3049

EN bottom-set gillnet; set gillnet; anchored gillnet

Net fixed to the bottom, or at a certain distance above it, by means of anchors or ballast sufficiently heavy to neutralize the buoyancy of the floats.

FR filet maillant de fond; filet maillant calé; filet maillant ancré

Filet maillant posé sur le fond, ou à une certaine distance de celui-ci, au moyen d'ancres ou de lests d'un poids suffisant pour neutraliser la flottabilité des flotteurs.

IT rete da posta calata sul fondo; rete da posta ancorata; rete da posta fissa

Rete da posta calata o ancorata sul fondo o a una certa distanza dal fondo.

NL geankerd kieuwnet

Kieuwnet dat d.m.v. ankers of gewichten op een bepaalde plaats en diepte wordt gehouden.

PT rede de emalhar fundeada

Arte de pesca fixa ao fundo, junto a ele ou a uma certa distância dele, por meio de âncoras ou lastros com peso suficiente para neutralizar a força de flutuação das bóias.

3050

ES paño de red

Cada una de las distintas secciones que componen un arte y están formados por un conjunto de mallas semejantes tejidas del mismo hilo.

DA netmateriale; net

Materiale, som er udgangspunktet ved fremstilling af fiskeredskaber.

DE Netztuch

Ein aus Maschen bestehendes Flächengebilde von beliebiger Form und Größe,
— hergestellt aus einem Faden oder mehreren Fadensystemen, die miteinander verknotet oder auf andere Weise miteinander verbunden sind,
— hergestellt aus anderem Material und auf andere Weise, z. B. durch Stanzen oder Ausschneiden aus Folien.

GR δικτύωμα

Κατασκευή ματιών οποιουδήποτε σχήματος και μεγέθους:
— συγκροτούμενη από ένα νήμα ή από ένα ή περισσότερα συστήματα νημάτων συνυφασμένα ή ενωμένα
— λαμβανόμενη με άλλο τρόπο, για παράδειγμα με κόψιμο στην πρέσα ή με τεμαχισμό υλικών που είναι σε φύλλα ή με εξαγωγή

EN netting; webbing

A meshed structure of indefinite shape and size
- composed of one yarn or of one or more systems of yarns interlaced or joined, or
- obtained by other means, for example by stamping or cutting from sheet material or by extrusion.

3050

FR nappe de filet; alèze; alèse

Assemblage de mailles, de forme et de dimension quelconques,
- constitué d'un fil d'un ou deux systèmes de fils noués, entrecroisés ou liés, ou
- obtenu par d'autres moyens, par exemple: par estampage ou découpage de matière en feuille ou par extrusion.

IT pezza di rete

Insieme di maglie di forma e dimensioni qualsiasi ottenuto da un filo, monofilo, filato e da uno o due sistemi di fili, monofili, filati intrecciati o annodati oppure con altri metodi (per esempio per stampaggio o taglio di materiali in film o per estrusione).

NL netwerk; want

Samenstel van netgarens die aan elkaar geknoopt of anderszins verbonden worden en zo mazen vormen.

PT peça de rede

Conjunto de malhas com forma e dimensões variáveis.

Trata-se do material fornecido em bruto pelas fábricas e cujas dimensões são normalmente estandartizadas de acordo com o tipo de artes de pesca a que se destinam.

3051

ES andana; flota

De redes.

DA lænke; garnlænke

Et antal garn, som bindes sammen og sættes som en enhed under fiskeriet.

DE Fleet

Fanggerät der Treibnetzfischerei in Gestalt einer schwimmenden Netzwand von 3 000 bis 5 000 m Länge, die aus einzelnen Netztüchern zusammengesetzt ist.

GR στόλος διχτυών

Οποιοσδήποτε αριθμός διχτυών που είναι ενωμένα από άκρη σε άκρη και λειτουργούν σαν ένα πλήρες σύστημα
Απλάδια· στάσιμα δίχτυα

EN fleet

Any number of nets joined end to end and operated as a complete outfit.

FR tésure; tessure

Dispositif de pêche formé par un nombre variable de filets dérivants, mis bout à bout.

IT numero di reti da posta calate

Corredo di reti che viene calato in una singola operazione di pesca.
Un peschereccio può calare più reti da posta in funzione della loro lunghezza e della rapidità di operazione.

NL vleet

Bij de drijfnetvisserij de gezamenlijk aaneengeknoopte, tot één geheel verbonden netten, die per vaartuig worden uitgezet.

PT caçada; andana; andaina

Dispositivo de pesca constituído por um número variável de redes de emalhar ou de tresmalhos reunidos entre si pelas respectivas extremidades.

3052

ES lastre

Para dar peso o estabilidad a la red.

DA synk; vægt

De vægte eller lodder af bly, kæde, sten, som påsættes en fiskeline.

DE Ballast; Beschwerung

GR έρμα· σαβούρα· βαρίδι· βάρος· μολύβι

Το κάτω καζίλι (γραντί), κατά μήκος της κατώτερης πλευράς ενός αλιευτικού διχτυού, φέρει βάρος από μολύβι, πέτρα ή σιδερένια αλυσίδα

EN ballast;* sinker; weight **

*Weight used to sink a fishing line.
**One of the weights spaced along the footrope of a fishing net.

FR lestage; lest

Poids de nature et forme variables, fixés à la partie inférieure de l'engin, pour assurer sa tenue au fond ou son déploiement vertical.

IT piombo; peso

Rete.

NL zinker

Stuk steen of lood dat het net doet zinken.

PT lastragem; lastro

Pesos de natureza e formas variáveis, fixados na parte inferior da arte de pesca, assegurando o seu contacto com o fundo ou a sua abertura vertical.

3052

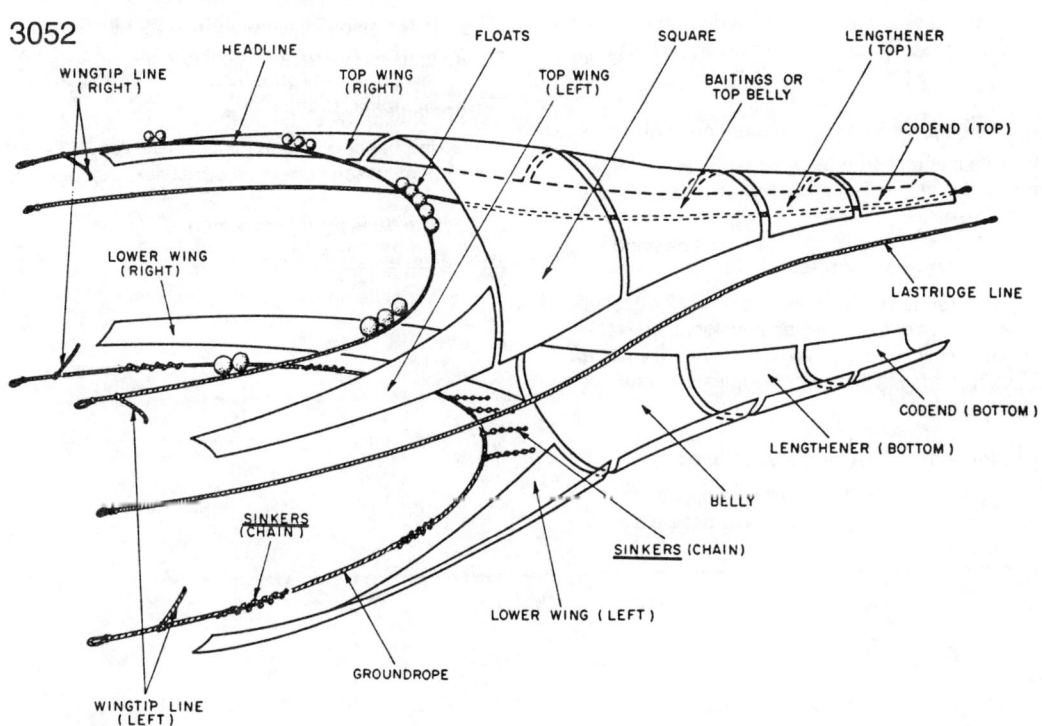

3053

ES flotabilidad

Propiedad que tienen ciertos cuerpos sumergidos en un líquido, en determinadas condiciones, de aflorar a la superficie.

DA opdrift; flydeevne

Den opadrettede kraft et legeme i en væske udsættes for, som er resultaten af vægten af det vand, legemet fortrænger, og legemets egenvægt.

DE Schwimmfähigkeit

GR πλευστότητα

EN buoyancy; floatability

The resultant of upward forces, exerted by a liquid upon a floating body equal to the weight of the liquid displaced by this body.

FR flottabilité

Force dirigée vers le haut résultant de la combinaison des deux forces qui agissent sur un corps flottant, son poids et la poussée d'Archimède.

IT galleggiabilità

Capacità che un corpo ha di stare parzialmente emerso dall'acqua essendo libero di spostarsi verticalmente.

NL drijfvermogen; drijfbaarheid

De resulterende kracht op een geheel of gedeeltelijk in het water gedompeld lichaam (volgens de wet van Archimedes worden op zo'n lichaam het gewicht en de opdrijvende kracht van het verplaatste water uitgeoefend).

PT força de flutuação; flutuabilidade

Resultante da acção de duas forças opostas que actuam sobre um corpo flutuante, o peso e a impulsão.

3054

ES a profundidad intermedia; a profundidad media; entre dos aguas; pelágico

DA pelagisk

Betegnelse for noget, der befinder sig midt i vandsøjlen uden kontakt med hverken overfladen eller havbunden.

DE im Pelagial; pelagisch

GR μεσόνερα

EN in midwater; pelagical

Substantially below the surface and substantially above the bottom of the sea.

FR entre deux eaux; pélagique

Immergé totalement, en pleine eau, sans contact avec le fond ou la surface.

IT a mezz'acqua; tra due acque; pelagico

Essere immerso senza toccare il fondo.

NL pelagisch

Betrekking hebbend op de middenpartij van de waterkolom tussen bodem en oppervlak.

PT entre duas águas; pelágico

Zona entre a superfície e o fundo.

3055

ES | red de enredo

Red montada con mucho embando que captura los peces más envolviéndolos que por enmalle.

DA | garn

Betegner et garn, der er ført så løst, at fiskene er i stand til at vikle sig ind i garnet.
Se 3056. På dansk skelnes ikke mellem 3049 og 3055.

DE | Verwickelnetz; verwickelndes Netz

Setznetz mit lose hängendem Netzwerk, z. B. Dreiwandnetz.

GR | δίχτυ μπλεξίματος

Χαλαρά αναρτημένο κάθετο δίχτυ που συλλαμβάνει τα ψάρια κυρίως επειδή μπλέκονται σ' αυτό παρά επειδή παγιδεύονται μέσα στο μάτι του π.χ. ειδικό απλάδι, δίχτυ για σαλάχια

EN | entangling net

Loosely hung vertical net that catches fish by entangling rather than enmeshing, e.g. tangle net, ray net.

FR | filet emmêlant; folle

Filet maillant monté avec beaucoup de flou, capturant les animaux plutôt par emmêlement que par maillage proprement dit.

IT | rete da posta impigliante

Rete da posta a basso rapporto d'armamento che cattura gli organismi marini che vi restano impigliati più che ammagliati.

NL | warnet; warrelnet

Staand net bestaande uit loshangend netwerk van licht en soepel garen waarin vissen en schaaldieren met vinnen en stekels verward raken.

PT | rede de enredar; rasca

Arte de pesca constituída por numerosos panos de rede colocados verticalmente tipo rede de emalhar, mas nos quais os peixes se enredam mais do que se emalham.

3056

ES | red de enmalle

Con este tipo de arte los peces quedan enmallados en los paños de red, que pueden ser uno solo (redes de enmalle) o tres (redes atrasmalladas). A veces en un mismo arte se combinan varios tipos de red (por ejemplo, red atrasmallada y red de enmalle). Estas pueden utilizarse solas o, más frecuentemente, en andanas («flotas» de redes).
Según su diseño, lastre y flotabilidad, pueden servir para pescar en la superficie, a profundidad intermedia o en el fondo.

DA | garn; hildingsgarn

Rektangulær netvæg, enkeltlaget, fremstillet af tyndt materiale. Garn bringes til at stå lodret i vandet ved at sætte opdrift på overtællen og synk på undertællen. Omfatter drivgarn 3048 og bundsatte garn 3049 og 3055.
Hildingsgarn er en gammel betegnelse for et redskab af tyndt materiale, hvor fiskene sætter sig fast i maskerne, se 3055.

DE | Setznetz; Kiemennetz

Einwandiges Netz mit beschwertem Untersimm, senkrecht im Wasser hängend.

GR | απλάδι

Είδος διχτυού με ορθογώνιο σχήμα, συνήθως κατασκευασμένο από λεπτό σπάγγο, το οποίο συλλαμβάνει το ψάρι συγκρατώντας το στο μάτι, π.χ. παρασυρόμενο δίχτυ, στάσιμο απλάδι

EN | gillnet; gill net

Usually rectangular in shape, made of thin twine, which catches fish by holding them in the meshes, e.g. drift net, set gill net.
Held vertically in the water by floats and weights.

FR | filet maillant; filet droit

De forme rectangulaire et fabriqué en fils fins, ce type de filet se tient verticalement dans l'eau, tendu entre les flotteurs de la ralingue supérieure et les plombs de la ralingue inférieure. La largeur des mailles est calculée pour retenir le poisson par la tête ou l'avant du corps.

IT | rete da imbrocco; rete a imbrocco

Rete da posta formata da una sola pezza di rete in cui il pesce resta ammagliato; la dimensione della maglia varia con la specie del pesce che si intende catturare; il pesce penetra con la testa nella maglia e vi resta prigionero, impossibilitato ad andare avanti e a tornare indietro.

3056

NL **gillnet; kieuwnet**

Staand net, gebruikt voor de vangst van vis die enkel met zijn kop door de mazen kan, waarbij door de druk op de kieuwdeksels achterwaarts ontsnappen verhinderd wordt.

PT **rede de emalhar**

Arte de pesca constituída por numerosos panos de rede rectangulares que se colocam verticalmente na água estendidos entre um cabo superior a que são entralhados (tralha das cortiças) e dotado de numerosas bóias e um cabo inferior a que também são entralhados (tralha dos chumbos) e dotado de numerosos lastros. A respectiva malhagem é calculada de acordo com a forma e dimensões dos peixes que se pretende capturar.

3056

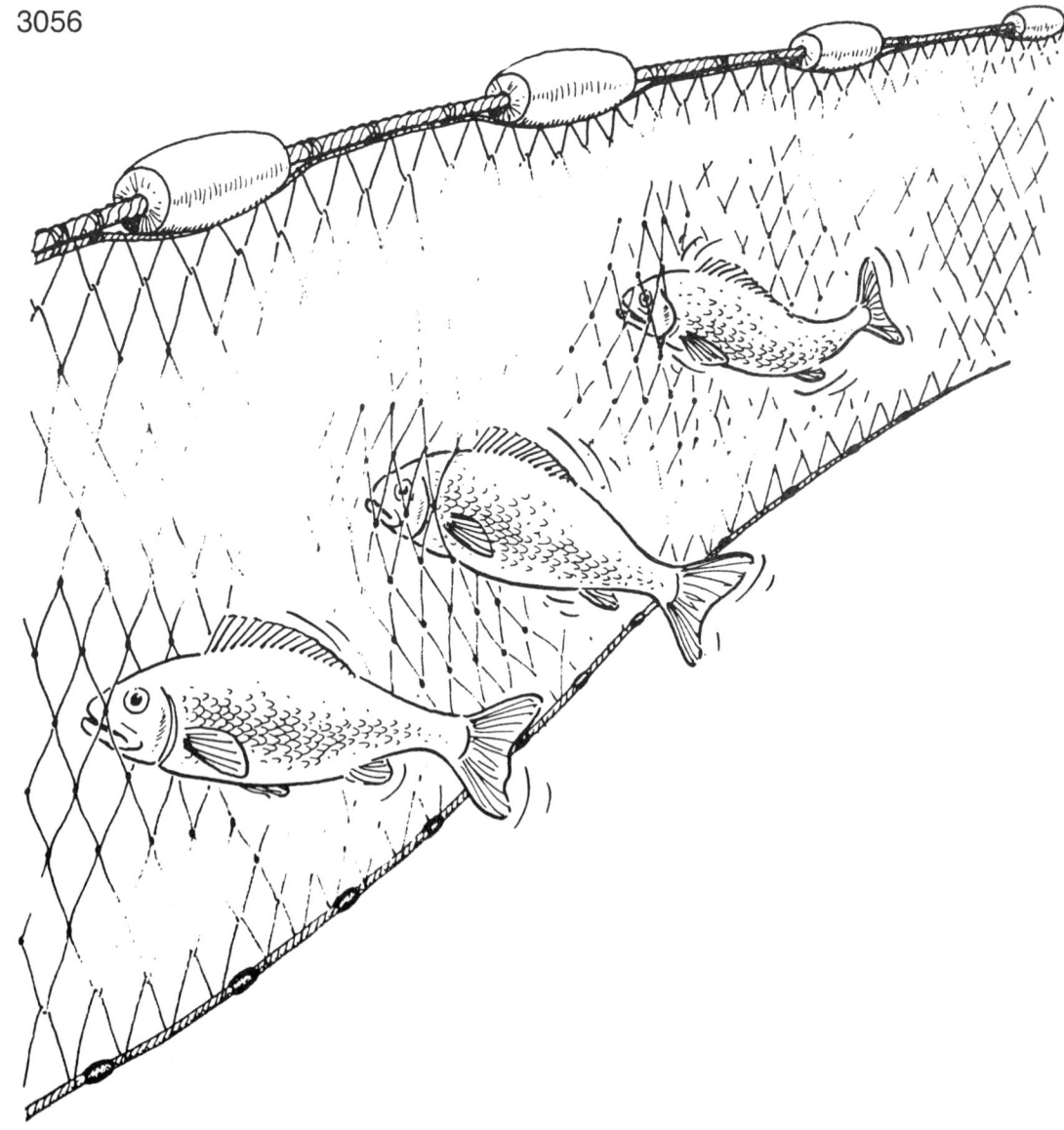

3057

ES cabrero

DA faldnet med fast ramme

Anvendes ikke i Danmark.

DE Stülpgerät

Greifnetz, das im Ufergewässer oder vom Boot aus ausgeworfen wird. Die kreisrunde Netzfläche mit dem bleibeschwerten Simm stülpt sich beim Absinken über den Fang. In Abständen sind Schnüre an den Simm geknüpft, die innerhalb des Netzmantels nach oben zusammenlaufen und in der Schnürleine enden. Durch Zug der Schnürleine schließt sich das Netz in Form eines Beutels.

GR δίχτυ-φανάρι

Δίχτυ ανοιχτής παγίδας κωνικού σχήματος, όπου το κατώτερο άνοιγμα είναι μεγαλύτερο. Ρίχνεται με το χέρι σε αβαθή νερά καλύπτοντας τον πυθμένα για να πιάσει το ψάρι. Αφού πιαστούν τα ψάρια, απομακρύνονται από το ανώτερο άνοιγμα

EN lantern net

Type of open trap of a conical form, the lower opening being the larger. Cast by hand in shallow water, it covers the bottom to catch the fish. Once captured the fish are removed by way of the upper opening.

FR filet lanterne

Filet en forme de nasse ouverte aux deux extrémités, de forme grossièrement tronconique, dont l'ouverture inférieure est la plus large. Utilisé à la main en eaux peu profondes, il est plaqué sur le fond rapidement pour surprendre le poisson. Une fois capturé, celui-ci est retiré à la main par l'ouverture supérieure.

IT rete lanterna

Non usata in Italia.

NL lantarennet

Kegelvormig of doosvormig vistuig, bestaande uit netwerk gespannen over een frame met openingen aan de onder- en bovenzijde; door het vistuig op de bodem te drukken ontstaat een afgesloten ruimte, waaruit de vis niet meer kan ontsnappen. De vis kan met de hand worden gegrepen, ingesloten worden door het dichttrekken van een netje aan de onderzijde, ingespierd worden met een speer of gevangen worden in het netwerk zelf (kieuwnet).

PT arte de arremeço tipo campânula

Arte de pesca de arremeço tipo armadilha com forma tronco-cónica, aberta nas duas extremidades e cuja abertura inferior é de maior dimensão. É manuseada por um só homem, em águas pouco profundas e é colocada repentinamente sobre o fundo para surpreender os peixes, os quais, uma vez capturados, são retirados à mão pela abertura superior.

3057

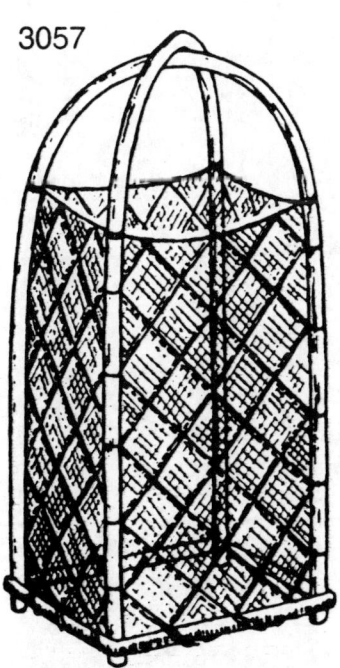
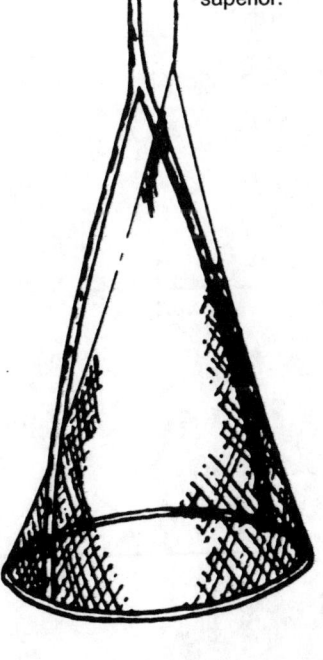
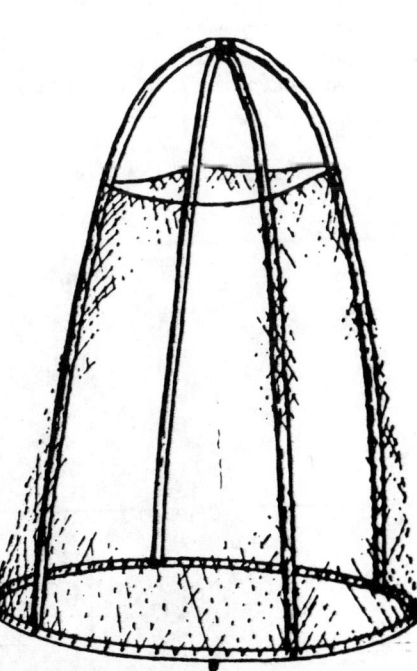

3058

ES cabrero de cañas

DA faldnet med fast ramme

Bikubeformet fiskeredskab af vidje eller lignende, der kastes på lavt vand med den største åbning nedad.
Anvendes ikke i Danmark.

DE Stülpgerät

GR δοχείο κάλυψης

Χρησιμοποιείται όπως και το δίχτυ-φανάρι, αλλά είναι κατασκευασμένο από καλάμι

EN cover pot

Falling gear of wicker construction, like a beehive with an opening at the top; the gear is thrown over the animal by the fisherman wading in shallow water and the prey inside the pot is taken out through the opening.

FR panier coiffant

Engin retombant, de fonctionnement similaire à celui du filet lanterne, mais fabriqué en vannerie.

IT cesto a caduta

NL stolpmand; stolpnet; stulpmand

Vistuig, gemaakt van twijgen, in de vorm van een afgeknotte kegel met openingen aan de onder- en bovenzijde; de vis wordt door het plaatsen van de mand op de bodem ingesloten, waarna hij op verschillende manieren (met de hand, een elger of spies) uit de mand kan worden verwijderd.
Wordt momenteel in Nederland niet meer gebruikt.

PT

Arte lançada de estrutura rígida.

3058

3059

ES esparavel

Red que se arroja desde la ribera o desde una embarcación, y que atrapa a los peces al caer, encerrándolos.
Generalmente se emplea sólo en aguas poco profundas.

DA kastenet

Cirkulært net, der kastes ud over en fiskeforekomst på lavt vand.
Anvendes ikke i Danmark.

DE Wurfnetz; Wurfgarn

Greifnetz, das als Stülpgerät bzw. Hand- oder Standwurfnetz verwendet wird.

GR πεζόβολος· δίχτυ πόντισης

Δίχτυ το οποίο, ποντιζόμενο από την ακτή ή από σκάφος, συλλαμβάνει τα αλιεύματα καθώς πέφτει τυλίγοντάς τα μέσα σ'αυτό
Η χρήση του συνήθως περιορίζεται σε ρηχά νερά

EN cast net

Circular net, thrown in such a way as to fall flat upon the water and, dropping rapidly to the bottom, encloses any fish that might happen to be beneath it.

3059

FR | **épervier**

Filet lancé du rivage ou d'une embarcation, capturant les poissons en retombant et en se refermant sur eux.
Son emploi est généralement limité aux eaux peu profondes.

IT | **rete da lancio; giacchio; rezzaglio**

Rete da pesca di forma circolare con serie di piombi sul bordo della circonferenza e una fascia di maglie più larghe; al centro del cerchio è fissata una sagola che è trattenuta dal pescatore dopo il lancio.

NL | **werpnet**

Kegelvormig net met een d.m.v. een loodjeslijn verzwaarde wijde onderrand en bovenaan dichtgehouden; bij het uitwerpen spreidt de loodjeslijn zich in een grote cirkel op het water uit en zinkt naar de bodem waarbij de vis die zich binnen de cirkel bevindt, wordt ingesloten; ophalen geschiedt door de loodjeslijn naar het centrum te trekken, waarbij zich een gesloten zak vormt.
Naast dichttrekken komt het ook voor, dat het net in een ring wordt opgetrokken naar de top van de kegel.

PT | **tarrafa de mão; chumbeira**

Arte de pesca de arremeço manuseada por um só homem, em águas pouco profundas, a pé ou de uma pequena embarcação. É uma arte muito original, com forma cónica ou de funil e que se abre em círculo quando é arremeçada. A respectiva malhagem vai aumentando desde o vértice até à base, a qual é guarnecida com chumbos.

3060

ES | **red de caída**

DA | **faldnet**

Samlebetegnelse.

DE | **fallendes Netz**

Oberbegriff für Stülpgeräte, diverse Wurfnetze und Schleifgarne.

GR | **εκτοξευόμενο εργαλείο**

EN | **falling gear**

Gear which is thrown down on the fish or other animals to be caught — which are thus taken from above.

FR | **engin retombant**

Les engins de cette catégorie capturent les poissons en retombant et en se refermant sur eux. Ils peuvent être lancés du rivage ou d'une embarcation, le plus souvent à la main, en général dans des eaux peu profondes.

IT | **rete da lancio**

NL | **vallend net**

Kegelvormig net, opgehangen aan de top boven het wateroppervlak, dat men over eronder zwemmende vis kan laten vallen waarna de onderkant snel boven water kan worden getrokken om de vis te vangen.

PT | **arte de pesca de arremeço; arte lançada**

3059

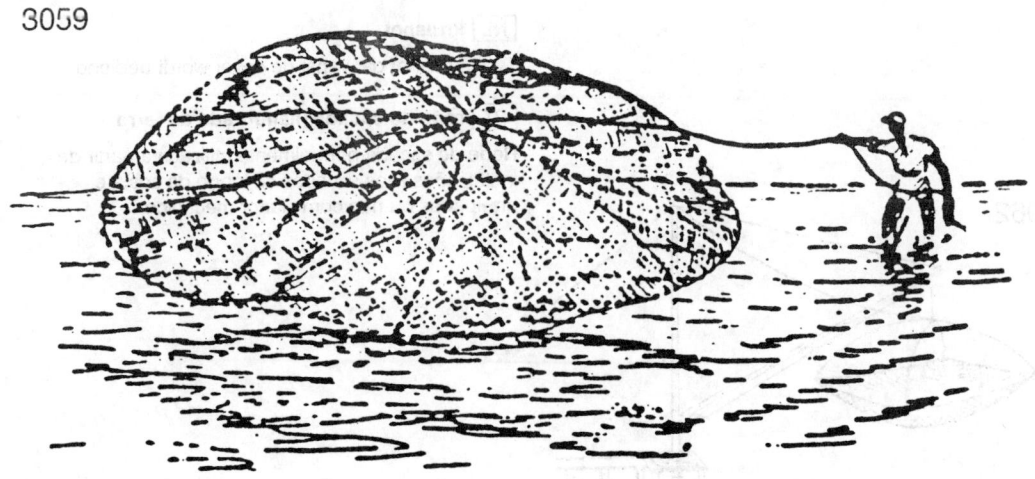

3061

ES halador mecánico

Red.

DA garnhaler

Maskine, der bruges til at hale garnene om bord med.

DE Netzeinholer; Netzeinholmaschine

Gerät zum Holen und Aufwickeln diverser Netze; Netztrommel zum Auftrommeln der Stander und der pelagischen Netze.

GR βαρούλκο ανύψωσης δικτυού

EN net hauler; net lifter

Hydraulically powered reel specially designed for hauling gillnets.

FR vire-filet; treuil de relevage

Treuil servant au halage des filets maillants.

IT verricello salpareti

Verricello usato per ricuperare le reti.

NL ophaalsysteem

Lier, kabel en uithouder.

PT alador

Qualquer aparelho de força auxiliar, que não um guincho de pesca, utilizado para alar redes de pesca, linhas de pesca e cabos.

3062

ES red izada maniobrada desde la costa; balanza

En general, instalación fija situada en la costa, a veces con un sistema mecánico de halado.

DA faststående løftenet; vippegarn

Løftenet anvendt fra stationære installationer ved bredden eller kysten, til tider mekaniseret.

DE stationäres Hebenetz; stationäres Senknetz

Handsenknetz, Standsenknetze, Senktuch und Wasserräder. Letztere im strömenden Wasser eingesetzt.

GR μπέντουλας ακτής

Ο χειρισμός αυτών των διχτυών γίνεται συνήθως από σταθερές εγκαταστάσεις τοποθετημένες κατά μήκος της ακτής, όπου το ανυψωτικό σύστημα είναι μερικές φορές μηχανικό

Δίχτυ που ποντίζεται στο νερό κρεμασμένο από την ακτή

EN shore-operated lift net

Lift net usually operated from stationary installations situated along the shore, the lifting system sometimes being mechanized.

FR filet soulevé manœuvré du rivage; carrelet

Filet le plus souvent utilisé à partir d'une installation fixe, située le long du rivage et comportant ou non une mécanisation du système de relevage.

IT quadra

Rete da raccolta che viene calata e salpata agendo sulle pulegge applicate ai 4 pali piantati sul fondo che ne assicurano l'apertura.

NL kruisnet

Type kruisnet dat vanaf de oever wordt bediend.

PT rede de sacada manobrada de terra

Rede de sacada normalmente utilizada a partir de instalações fixas situadas ao longo da costa e equipadas ou não com alador mecânico.

3062

3063

ES red izada maniobrada desde embarcación

Arte formado por paños cuadrados, que se puede bajar o subir desde embarcaciones que la sujetan por los cuatro vértices.

DA »blanket net«

Type af løftenet til brug på fartøjer.
Anvendes ikke i Danmark

DE Blankettnetz; Rahmennetz

Hängt am Ausleger des Fangbootes.
In Deutschland unbekannt.

GR δίχτυ-κουβέρτα

Δίχτυ αναρτημένο από το ένα άκρο του σε προώστη σκάφους και το οποίο μαζεύεται από τον πυθμένα με ένα σχοινί που έλκεται από το κατάστρωμα

EN blanket net

An impounding net suspended by one end from an outrigger in a boat and pulled in from the bottom by a line drawn from the deck.
Type of lift net.

FR filet couverture; globe

Filet soulevé de grandes dimensions, sans cadre de soutien, nécessitant pour sa manœuvre l'emploi de plusieurs embarcations ou d'installations fixes adaptées.
Le globe, employé en Méditerranée, est représentatif de cette catégorie.

IT rete da raccolta a più natanti

NL

Vistuig bestaande uit een groot rechthoekig netwerk, dat aan een aantal hoeken boven het water wordt uitgetild, zodat vis die erboven zwemt wordt gevangen.
- Soort liftnet met zodanige afmetingen dat een frameconstructie niet meer gebruikt kan worden.
- Wordt bediend vanaf de oever(s), een schip of vanaf een op de bodem verankerde constructie.

PT rede de sacada manobrada de embarcações

Rede de sacada com forma de bolsa de grandes dimensões que é manobrada a partir de uma ou mais embarcações por intermédio de cabos e retrancas.

3064

ES estero; tapa-esteros; entallada; huelga; estacada

Red sujeta por estacas clavada en el fondo que se dispone en forma semicircular cortando la salida de un brazo de mar, y en cuya parte central hay un copo para la captura de peces.

DA ruse fæstnet til pæle

Anvendes f.eks. i flodmundinger, hvor netrader på tværs af strømmen ender ved ruseåbningen.

DE Pfahlreuse

In variabler Höhe an Pfählen zu befestigende Netzreuse, bei der eine quer zur Strömung gestellte Netzwand in der Reusenöffnung endet.

GR πασσαλωμένο δίχτυ

Δίχτυ παρεμφερές με το διχτυωτό σάκο, υποστηριζόμενο από πασσάλους

EN stake net; stake gillnet

Net, similar to a bag net, supported by stakes.

FR filet fixe sur pieux; haut-parc

Filet maillant soutenu par une série de pieux, découvrant à marée basse.

IT rete da posta a pali

NL paalfuik; staalbomer

Vistuig op stroom verankerd aan stalen palen.
Werd vroeger in Nederland gebruikt in de riviervisserij.

PT estacada

Arte de pesca tipo armadilha fixa constituída por uma estrutura de varas, estacas e outros materiais que se dispõem em semicírculo cortando a saída de um braço de mar e em cuja parte central existe uma espécie de saco para a captura dos peixes.

3065

ES red de saco; arte de copo

DA poseformet net

Samlebetegnelse for en række fiskeredskaber, hvor fangsten opsamles i en pose.

DE Sacknetz

Allgemeine Bezeichnung für Netze, wie sie z. B. als Zugnetz in der Krabbenfischerei oder als Standnetz in der Hamenfischerei verwendet werden. Aufnahme des Fangs im Netzsack.

GR διχτυωτός σάκος

Ονομασία για όλα τα δίχτυα, σταθερά ή κινητά, όπου τα ψάρια συλλαμβάνονται σε μία τσέπη: τράτες, γαριδόδιχτυα, δίχτυα στοιβάγματος, δίχτυα επιλογής, δίχτυα στεφάνια, γρίποι

EN bag net; bag shaped net

A name for all the nets, fixed or moving, in which the fish are received into a pocket: trawl nets, shrimp nets, stow nets, filter nets, hoop nets, seines.

FR filet à poche

Désigne tous les filets, fixés ou traînés, dans lesquels les poissons sont récoltés dans une poche: chaluts, filets à l'étalage, sennes, etc.

IT rete a sacco

Rete fissa o mobile in cui il pesce è trattenuto in un sacco; p.e. rete da traino, rete fissa a corrente, sciabica.

NL zaknet

Net dat door zijn bouw nagenoeg een zak vormt zoals b.v. een trawl, haam, lampara, zegen, botnet, kornet, sleepnet, spieringtoger.

PT rede com saco

Qualquer tipo de arte de pesca fixa ou não e na qual os peixes capturados são acumulados num saco.

3066

ES red izada maniobrada desde embarcación; mediomundo

Aro metálico, de hasta seis metros de diámetro, provisto de su correspondiente red en forma de bolsa, que pende de cuatro cables o vientos (bolinas), que se reúnen en un cabo único terminal.

DA synkenot; synkenet

Synkenoten er firkantet; den sænkes på bunden eller passende langt ned og hales op med et tov i hvert hjørne fra op til fire både, når fiskestimerne kommer over noten.

DE Senktuch

In horizontaler Ebene, von bis zu vier kleineren Booten ausgesetztes Netztuch. Gehört zur Kategorie der Senk- und Hebenetze.

GR μπέντουλας σκάφους

Αυτή η ομάδα περιλαμβάνει τους διχτυωτούς σάκκους και τα δίχτυα-κουβέρτες που τα χειρίζονται από διάφορα σκάφη

Δίχτυ που ποντίζεται στο νερό κρεμασμένο από σκάφος

EN boat-operated lift net

This group comprises the bag nets, usually operated from one boat, and the blanket nets, operated from several boats.

3066

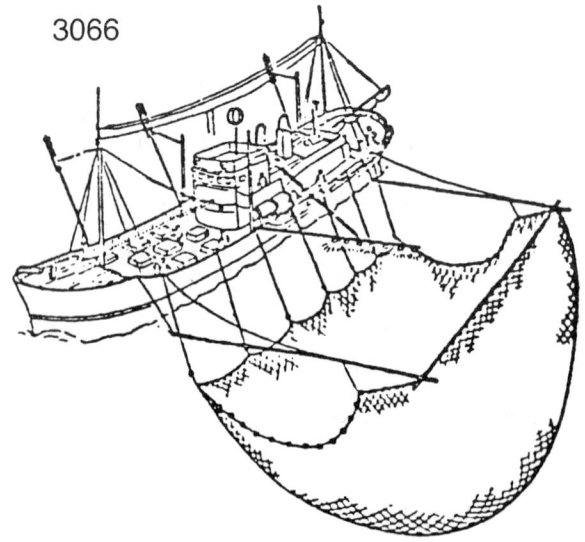

3066

FR filet soulevé manœuvré en bateau

Ce groupe comprend notamment les filets sacs («basnig») manœuvrés habituellement à partir d'un bateau, et les filets couvertures manœuvrés à partir de plusieurs embarcations.

IT rete da raccolta manovrata da natanti

Rete da raccolta che viene calata e salpata da bordo di natanti.

NL vanaf een schip bedienbaar kruisnet

Niet in Nederland gebruikt.

PT rede de sacada manobrada de embarcações

Rede de sacada com forma de bolsa de grandes dimensões que é manobrada a partir de uma ou mais embarcações por intermédio de cabos e retrancas.

3067

ES red izada portátil; balanza

Pequeña red de mano que no necesita instalación fija.

DA håndløftenet

DE Handsenknetz

Gehört zur Kategorie der Senk- und Hebenetze.

GR φορητός μπέντουλας

Μικρός μπέντουλας που τον χειρίζονται με το χέρι, χωρίς σταθερή εγκατάσταση

EN portable lift net; hand lift net

Small lift net operated by hand, with no fixed installation.

FR filet soulevé portatif; balance

Filet soulevé de petites dimensions manœuvré à la main, sans installation fixe.

IT bilancia

Rete da raccolta che viene calata e salpata da un punto solo costituito dall'incrocio tra 2 pertiche le cui estremità, collegate ai 4 angoli della rete, ne assicurano l'apertura.

NL draagbaar kruisnet

Klein kruisnet dat door één man bediend wordt.

PT rede de sacada portátil

Pequena rede de sacada manobrada à mão, sem qualquer tipo de instalação fixa.

3068

ES | cebo

Sustancia sápida que se da a los peces para atraerlos y capturarlos.

DA | madding; agn

Enhver substans, der ved lugt eller udseende lokker byttet til at bide på en krog eller til at gå ind i en fælde.

DE | Köder

Zum Fang von Fischen verwendete Lockspeise.

GR | δόλωμα

Οποιαδήποτε ουσία που με την εμφάνισή της ή τη μυρωδιά της δελεάζει το ψάρι ώστε να πιαστεί σ' ένα αγκίστρι ή να μπει σε ένα δοχείο ή παγίδα

EN | bait

Any substance that by its appearance or smell is used to lure fish to take a hook or enter a pot or trap.

FR | appât; esche; boëtte; amorce; strouille; bromige

Toute substance dont l'apparence ou l'odeur est utilisée pour attirer le poisson, afin de l'inciter à mordre à l'hameçon ou à entrer dans une nasse ou un piège.

On peut distinguer l'appât, fixé directement à l'engin, de l'amorce, répandue dans l'eau pour attirer les poissons à proximité de l'engin.

IT | esca

Qualsiasi tipo di cibo, o di oggetto simile a un cibo, capace di attirare un pesce o un altro animale marino allo scopo di poterlo catturare.

NL | aas; lokaas

Aantrekkingsmiddel bestaande uit echt of kunstmatig voedsel bevestigd aan een haak van een vistuig of in een val aangebracht.

PT | isco; isca; engodo; engano

Qualquer substância que pela respectiva aparência ou cheiro é utilizada para atrair o peixe a ferrar um anzol ou a entrar numa armadilha.

3069

ES | red izada; mediomundo

Consiste en un paño de red horizontal o una bolsa en forma de paralelepípedo, pirámide o cono con la boca abierta hacia arriba. Utilizando luz o cebo para atraer a los peces, se sumerge a la profundidad deseada y luego se saca a mano o bien se hala mecánicamente, desde la costa o desde una embarcación. Los peces que se hallan sobre la red quedan retenidos en ella cuando el agua se escurre.

DA | løftenet

Firkantet net på en ramme, ophængt i fire stropper. Efter at have været nedsænket på en bestemt dybde hales nettet igen, enten ved hånden eller mekanisk, fra land eller fra en båd.

DE | Hebenetz; Senknetz

An rechteckigem Rahmen befestigtes Netztuch, aufgehangen an diagonal angeordneten Bügeln.

Handhabung des beköderten Hebenetzes über ein Hangergeschirr.

GR | μπέντουλας· σταφνοκάρι· αθερινολόγος

Τα ψάρια που μπορεί να έχουν προσελκυθεί από φως ή δόλωμα, συλλαμβάνονται μέσα σε δίχτυα αποτελούμενα από ένα οριζόντιο διχτυωτό φύλλο ή από ένα δίχτυ-σάκο, σε σχήμα παραλληλεπιπέδου, πυραμίδας ή κώνου, με το άνοιγμα προς τα πάνω. Αφού έχουν βυθιστεί στο ζητούμενο βάθος, τα δίχτυα ανυψώνονται ή έλκονται έξω από το νερό με το χέρι ή μηχανικά, από την ακτή ή από το σκάφος. Τα ψάρια που είναι πάνω από το δίχτυ συγκρατούνται μέσα σ' αυτό καθώς το νερό φεύγει

EN | lift net

Net consisting of a horizontal netting panel or a bag shaped like a parallelepiped, pyramid or cone with the opening facing upwards. After being submerged at the required depth, the net is lifted or hauled out of the water, by hand or mechanically, from the shore or from a boat. The fish which are above the net are retained in it when the water runs away.

3069

FR filet soulevé

Filet formé d'une nappe horizontale ou d'une poche de forme parallélépipédique, pyramidale ou conique à ouverture tournée vers le haut. Après avoir été immergé à la profondeur voulue, le filet est relevé ou viré hors de l'eau, à la main ou mécaniquement, à partir du rivage ou à bord. Les poissons qui se trouvent au-dessus du filet y sont retenus lorsque l'eau s'en écoule.

Voir également: carrelet, globe, balance, basnig.

IT rete da raccolta

Rete da pesca destinata, con moto dal fondo alla superficie, a catturare organismi marini.

NL kruisnet; totebel; batnet; hefnet; lift-net; zinknet

Een meestal op kalm water gebruikt vierkant visnet, geknoopt aan een raam of aan kruiselings over elkaar gebonden roeden, dat aan een loper hangt en d.m.v. een boom en een lier wordt bediend.

PT rede de sacada; sacada; rede de feva

Rede formada por um pano de rede horizontal ou por um saco com forma paralelipipédica, piramidal ou cónica e com abertura virada para cima. Depois de submergidas à profundidade desejada, as redes são aladas à mão ou mecanicamente a partir da costa ou de embarcações. Os peixes que se encontram por cima da rede são capturados quando as redes saem da água.

3070

ES rastra de mano; rastrillo; angazo

Pequeña y ligera red que se maneja en agua poco profunda, desde la costa o desde una embarcación.

DA håndbetjent skraber

DE Handdredge; Handschiebehamen

Handgerät zum Garnelenfang im seichten Wasser. Gerades Holz mit darüber im Halbkreis gespanntem Bügel und daran anschließendem Netzsack. Am geraden Holz schließt im rechten Winkel der Stiel an, der mit dem Netzbügel verbunden ist.

GR δράγα χειρός

Μικρή ελαφριά δράγα που σύρεται με το χέρι σε ρηχά νερά, από την ακτή ή πίσω από ένα σκάφος

EN hand dredge

Small, light dredge, pulled behind by hand in shallow waters, from the shore or from a boat.

FR drague à main

Drague légère et de faibles dimensions, manœuvrée manuellement en eaux peu profondes, à partir du rivage, à pied ou d'une embarcation.

IT draga a mano

Attrezzo da pesca tirato a mano che, penetrando nel fondo marino, nel suo progressivo avanzamento separa gli organismi marini dall'acqua, dalla sabbia e dal fango.

NL handkor; handdreg

Een met de hand bediende kor of dreg.

Meestal een soort schepnet waarmee de bodem kan worden afgekrabd.

PT draga de mão; arrasto de cintura; ancinho de mão

Pequena e leve draga com ou sem dentes e dotada de um saco de rede, manobrada à mão em águas pouco profundas a partir de uma embarcação ou a pé.

3071

| ES | depresor

Superficie plana colocada en la parte anterior de las dragas para aumentar su adherencia al fondo.
Rápido, draga.

| DA | depressor

Anordning til at holde et slæbt redskab nede i vandet, under anvendelse af vandets hydrodynamiske pres.
a. flad plade forrest på visse skrabertyper, der hjælper med til at presse skraberen mod bunden.
b. flyformet metallegeme, ophængt under eller lige foran et net, der skal slæbes i en vis dybde selv under høje slæbehastigheder.

| DE | Tauchbrett

Je nach Netzgröße mehr oder weniger schwere Stahl- oder Stahl/Holzkonstruktion mit Stahlsohle und Bügel. Einerseits mit der Kurrleine, andererseits mit dem Jager verbunden. Aus Scherwinkel und Zugkraft resultiert die horizontale Öffnung des Netzes.
Dredge.

| GR | σανίδα κατάδυσης

| EN | depressor; diving board

Board designed to counteract the ascent of the net.
Midwater trawl.

| FR | volet plongeur

Surface plane ou légèrement incurvée, placée à l'avant de certains types de dragues à coquillages. Le courant d'eau causé par le déplacement de la drague y crée une poussée vers le bas qui appuie la drague sur le fond.

| IT | depressore

Tavola in legno o metallo montata sulla bocca del rapido che mantiene l'attrezzo sul fondo.
Rapido, draga.

| NL | duikplaat

Scheerbord voor het naar beneden drukken van de kor.

| PT | depressor

Painel metálico ou de madeira situado no bordo inferior da boca das dragas e que pode ser ou não dotado de dentes; destina-se a revolver o fundo.
Draga.

3072

| ES | rastra para embarcación; draga

Consiste en un marco metálico, de forma rectangular, semicircular, etc., provisto de su correspondiente copo de red, que va rastreado desde una embarcación.
Está destinada a la captura de especies que viven pegadas al fondo o enterradas en él, principalmente de invertebrados, como los moluscos bivalvos, las esponjas, los corales, etc.

| DA | skraber

Skraber, som den anvendes på et fiskefartøj.
Se 3074.

| DE | Dredge; Schleppgerät

Vom Boot gezogene Dredge.
Eingesetzt als Muschelnetz mit Fassungsvermögen bis zu 500 kg.

| GR | δράγα σκάφους

Δράγα με ποικίλο βάρος και μέγεθος, αλλά συνήθως αρκετά βαριά, εφοδιασμένη με ή χωρίς σανίδες κατάδυσης

| EN | boat dredge

Dredge of varying weight and size, but usually fairly heavy, equipped with or without diving boards.

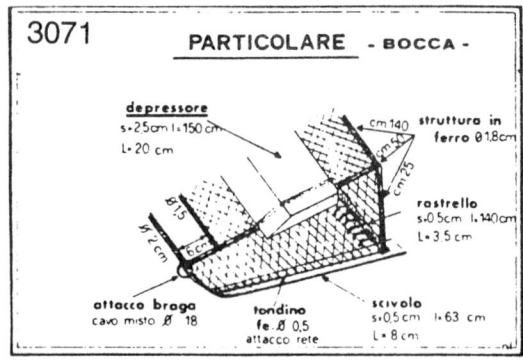

Particolare della bocca del « Rapido ».

3072

FR **drague remorquée par bateau**

Drague de poids et dimensions variables, mais en
général assez lourde, munie ou non de volets
plongeurs.

IT **draga tirata da natanti**

NL **vanaf een schip bediende kor**

PT **draga rebocada por embarcação; draga de
arrasto; ganchorra; berbigoeira**

Draga, manobrada a partir de uma embarcação e
possuindo dimensões e pesos variáveis, mas
geralmente bastante pesada, e equipada ou não com
patilha. Destina-se à captura de espécies que vivem
pegadas ao fundo ou nele enterradas e é constituída
por uma armação metálica rectangular, semicircular
ou triangular com ou sem dentes metálicos.

3072

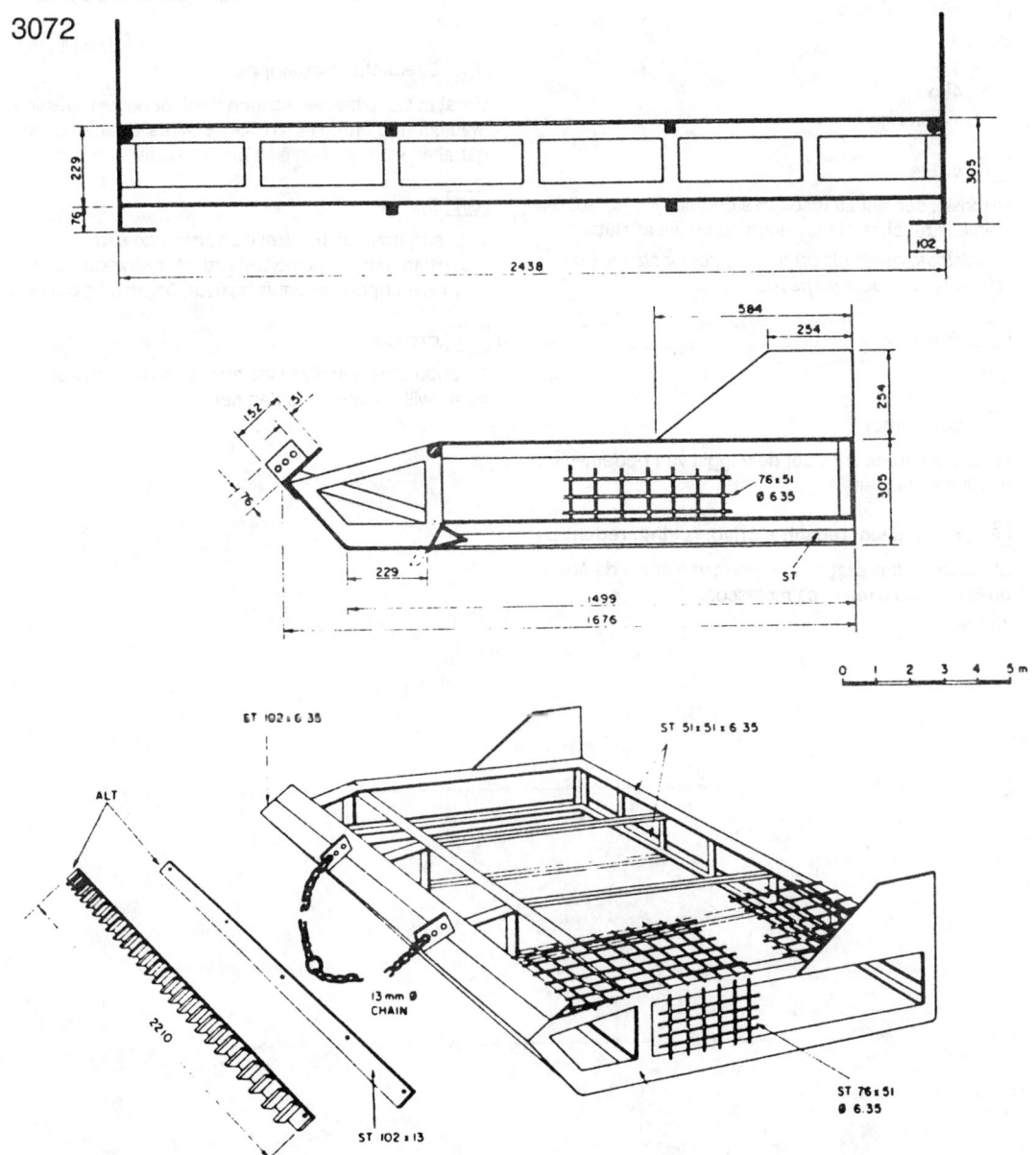

3073

ES criba; tamiz

Rastra; rastro

DA pose; kurv

Den del af en skraber, der opsamler fangsten.

DE Sieb

Teil einer Dredsche zur Abseiung und Aussortierung des Beifangs, z. B. in der Krabbenfischerei.

GR κόσκινο· σίτα

Δράγα

EN sieve

Part of a dredge.

FR crible

Dispositif servant au triage des coquillages ou autres animaux récoltés par les dragues ou les chaluts.

Les cribles, oscillants ou rotatifs, sont installés en général sur le pont de pêche.

IT setaccio

Draga.

NL zeef; zifter

Gedeelte van de kor waar de vangst van bodemvuil wordt gescheiden.

PT coifa; saco; redanho; crivo; grelha; redenho

Mecanismo das dragas que permite a saída da água, areia ou lodo e retem os moluscos.

Draga.

3074

ES rastra; rastro

Aparejo que se emplea para rastrear sobre el fondo, usualmente para recoger moluscos (mejillones, ostras, vieiras, almejas, etc.). Los moluscos quedan retenidos en una especie de saco o tamiz que deja salir el agua, el barro o la arena.

DA skraber

Slæbt fiskeredskab til fangst af bløddyr: muslinger og snegle. Består af en kort finmasket pose af kunststof eller jernringe spændt til en solid jernramme eller -kasse.

DE Dredsche; Schleppgerät

Gerät in Gestalt eines kleinen Schleppnetzes, dessen Maulöffnung durch einen starren Metallrahmen offen gehalten wird, an den sich ein Netzsack anschließt.

GR δράγα

Συσκευή που αποτελείται συνήθως από ένα σιδερένιο πλαίσιο με μορφή παραλληλόγραμμου, όπου είναι προσαρμοσμένος ένας διχτυωτός σάκος

EN dredge

An apparatus usually in the form of an oblong iron frame with an attached bag net.

3074

FR **drague**

Sac en filet ou panier en métal remorqué au moyen
d'une armature présentant une ouverture de forme et
de largeur variables, dont la partie inférieure est
munie d'une lame formant racloir. Cet engin sert
généralement à la récolte des coquillages.

IT **draga**

Attrezzo da pesca tirato a mano o da natanti che,
penetrando nel fondo marino, nel suo progressivo
avanzamento, separa gli organismi marini dall' acqua,
dalla sabbia e dal fango.

NL **kor; mosselkor; dreg**

Netwerk voorzien van een metalen of houten frame
en kettingmatten, voor het vangen van schelpdieren.

*Het begrip kor is meeromvattend en van origine bijna
synoniem aan trawl. Ervan afgeleid is b.v. ook de
boomkor, hetgeen beslist geen „dredge" is. Meestal
gebruikt men in de bedoelde betekenis het woord
mosselkor.*

PT **draga**

Termo genérico utilizado para referir
indiscriminadamente as dragas de arrasto,
ganchorras, berbigoeiras, arrastos de cintura,
ancinhos de mão.

*Trata-se de um saco de rede ou de um copo de
metal ou plástico que é rebocado por meio de uma
armadura metálica com abertura de forma e
dimensões variáveis, dotada na parte inferior de uma
lâmina com ou sem dentes. Trata-se de uma arte de
pesca destinada à captura de moluscos bivalves.*

3074

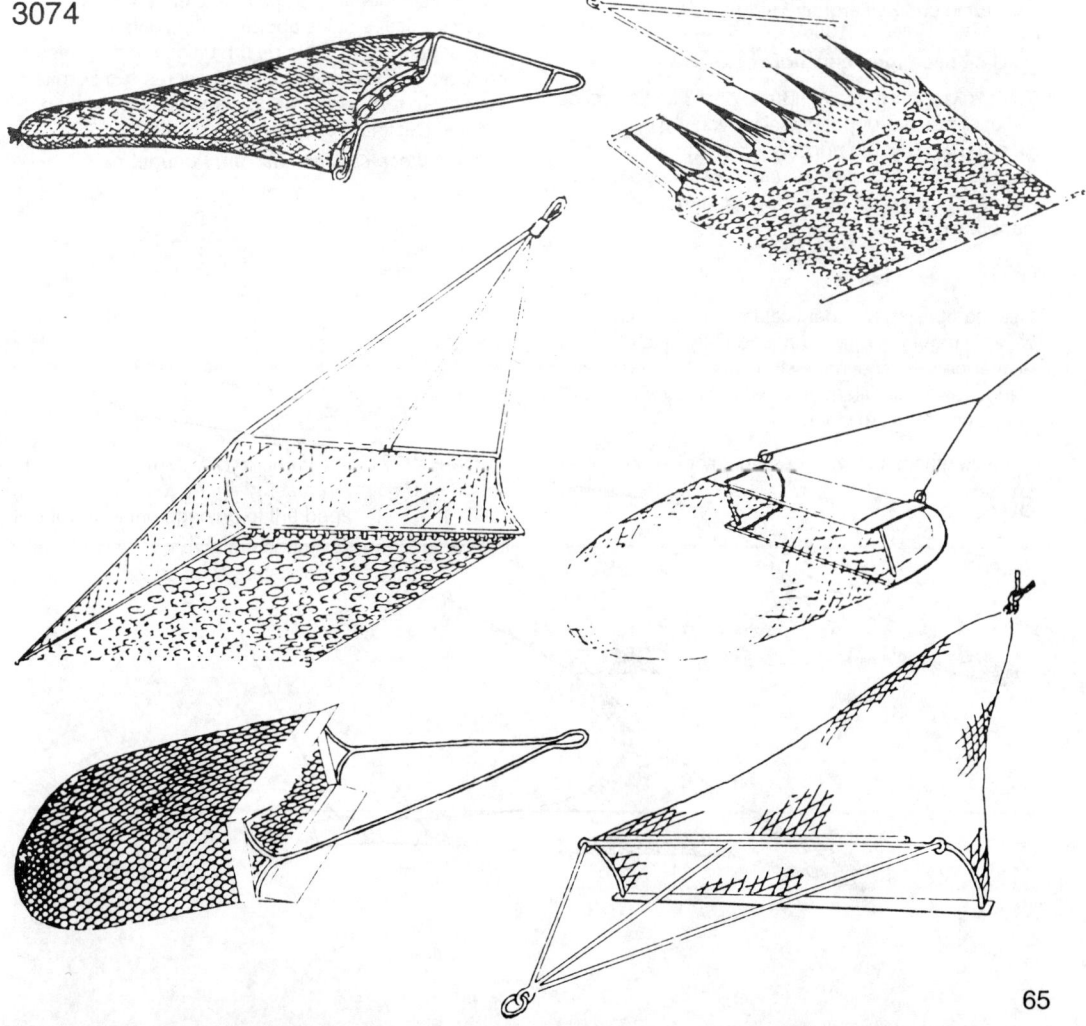

3075

ES redes gemelas de arrastre, con puertas

Aparejo que comprende dos redes de arrastre idénticas (gemelas) que trabajan juntas, y que se mantienen abiertas en sentido horizontal con un par de puertas. Las bandas interiores van sujetas a un calón que se arrastra simultáneamente con las puertas desde un solo pie de gallo.

DA dobbelttrawl

Trawlsystem, hvor der slæbes med to trawl ved siden af hinanden efter ét fartøj. Trawlene slæbes i tre wirer (eller to, når der anvendes en hanefod foran skovlene). De midterste vinger holdes til bunden af en tung slæde eller kædevægt. Spiles horisontalt med to skovle.

DE Scherbrett-Hosennetz; Doppelnetz; Dantrawl

Netz mit breiter horizontaler, aber geringer vertikaler Öffnung. Die Scherbretter sind dicht am Netz angeordnet. Dieses Netz ist mit drei Schleppleinen versehen und wird zum Fang in geringer Tiefe auf Seezunge und Kaisergranat eingesetzt.

GR δίδυμες τράτες με πόρτες

Το εργαλείο αυτό περιλαμβάνει δύο ολόιδιες τράτες (δίδυμη) που λειτουργούν μαζί, ανοιχτές οριζόντια με ένα μόνο ζευγάρι πόρτες. Τα εσωτερικά φτερά είναι προσδεδεμένα σε ένα είδος έλκηθρου που έλκεται ταυτόχρονα με τις πόρτες από ένα κοινό παραδέτη

EN otter twin trawls

Gear comprising two identical trawl nets ('twin') working together, opened horizontally by a single pair of otter boards. The inner wings are attached to a sledge towed simultaneously with the otter boards from a common crowfoot.

FR chaluts jumeaux à panneaux

Engin comprenant deux chaluts identiques («jumeaux») fonctionnant ensemble, ouverts horizontalement par une seule paire de panneaux. Les ailes intérieures sont fixées à une sorte de traîneau remorqué simultanément avec les panneaux au moyen d'une patte d'oie commune.

IT reti gemelle a divergenti

Trainate due per volta.

NL dubbele bordentrawl

Een samenstel van twee naast elkaar gesleepte trawlnetten die door middel van één stel visborden worden opengespreid, terwijl het ontmoetingspunt van de twee netten en de beide borden aan een gemeenschappelijk punt zijn verbonden.

PT redes de arrasto geminadas com portas

Esta particular arte de pesca comprende duas redes de arrasto idênticas trabalhando em conjunto (geminadas) e cujas aberturas horizontais são conseguidas por meio de um único par de portas de arrasto. As respectivas asas interiores são fixadas a um calão rebocado por um cabo comum às duas redes. Destinam-se à pesca de camarão.

Desconhece-se o seu uso em Portugal.

3075

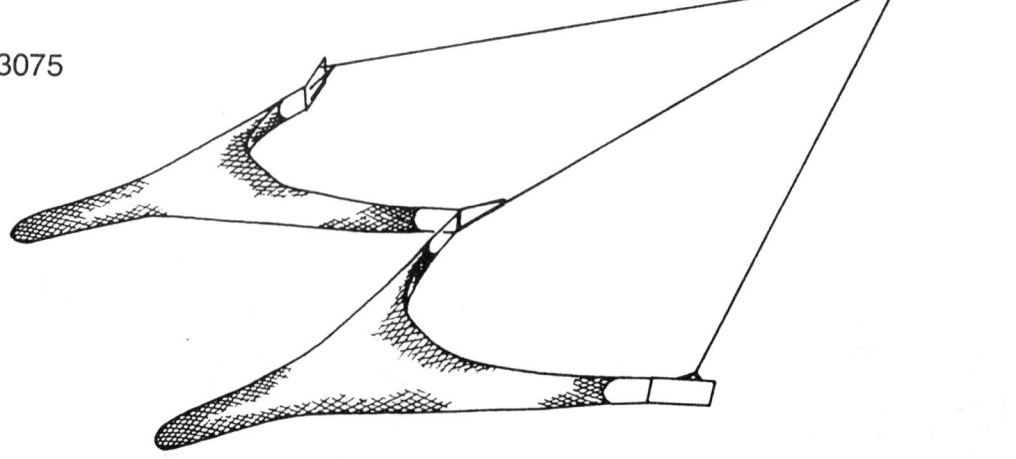

3076

ES red de arrastre de superficie

DA flydetrawl; pelagisk trawl

Her til brug i overfladen. Se 3083.

DE Oberflächentrawl

Pelagisches Netz, welches über Schwimmkörper und eine Verbindungsleine zur Fischleine in einer bestimmten Distanz zur Oberfläche gehalten wird. *Bei der Zweibootmethode wird in ähnlicher Weise verfahren. Die Verbindungsleine zu dem Schwimmkörper ist an der Position des Vorgewichts angesteckt.*

GR τράτα επιφάνειας

EN surface trawl

A floating trawl pulled by two boats, for use in shallow water only.

FR chalut de surface

Chalut pélagique remorqué par deux bateaux, conçu pour fonctionner uniquement dans les eaux superficielles.

IT agugliara; rete da traino superficiale

Rete da traino di superficie che viene trainata con la lima da sugheri fuori dell'acqua per evitare che le aguglie, con i loro caratteristici balzi fuori dall'acqua, possano evitare la cattura. *Rete usata praticamente solo in Adriatico.*

NL oppervlaktetrawl

Trawl waarmee met de bovenpees aan het wateroppervlak wordt gevist.

PT rede de arrasto pelágico funcionando à superfície

3077

ES red de arrastre pelágico a la pareja

Red que se arrastra con dos barcos asegurando así la abertura horizontal de la misma. Está diseñada y aparejada para trabajar a profundidad media. *En esta categoría están incluidas también las redes de arrastre de superficie.*

DA flydetrawl til partrawling; pelagisk partrawl

Trawl til brug i midtvand. Den slæbes af to fartøjer, der holder nettet åbent horisontalt.

DE pelagisches Zweischiffschleppnetz

GR μεσοπελαγική τράτα με ζευγαρωτά σκάφη· τράτα με ζευγαρωτά σκάφη για μεσόνερα

Ρυμουλκούμενα από δύο σκάφη, έτσι ώστε να εξασφαλίζεται το οριζόντιο άνοιγμα του διχτυού, τα δίχτυα αυτά είναι σχεδιασμένα και εξαρτισμένα ώστε να χρησιμοποιούνται σε μεσόνερα *Οι τράτες επιφάνειας περιλαμβάνονται σ'αυτή την κατηγορία*

EN midwater pair trawl

Towed by two boats, thus ensuring the horizontal opening of the net, this net is designed and rigged to work in midwater. *Surface trawls are also included in this category.*

FR chalut-bœuf pélagique

Remorqué par deux bateaux, ce qui assure l'ouverture horizontale du filet, cet engin est conçu et gréé pour travailler entre deux eaux. *C'est dans cette catégorie que l'on groupe également les chaluts de surface.*

IT rete da traino pelagica a coppia; volante

Rete da traino che opera senza contatto col fondo e la cui apertura orizzontale è assicurata da traino con due natanti.

NL pelagisch spannet; spantrawl

Vistuig bediend en opengehouden door twee schepen, waarmee pelagisch wordt gevist. *Kenmerkend is het ontbreken van de visborden om het net te openen.*

PT rede de arrasto pelágico de parelha

Rebocada por dois barcos cujo afastamento assegura a respectiva abertura horizontal, esta arte de pesca de arrasto é planeada e armada para trabalhar entre duas águas. *As redes de arrasto para trabalhar à superfície estão incluídas neste tipo de artes de pesca.*

3078

ES forma hidrodinámica

DA hydrodynamisk form

En udformning af et legeme, som tager hensyn til vandets hydrodynamiske kræfters påvirkning af legemet, når det bevæges gennem vandet.

DE hydrodynamische Form

GR υδροδυναμικό σχήμα

EN hydrodynamic shape

FR forme hydrodynamique

IT forma idrodinamica

NL hydrodynamische vorm

De vorm van een lichaam die bij het verplaatsen van dit lichaam in een vloeistof een minimale weerstand of een maximale lift oplevert.

PT forma hidrodinâmica

3079

ES red de arrastre pelágica de puertas; red de arrastre pelágica con un barco

Red que se arrastra desde una sola embarcación. La abertura horizontal de la red se controla mediante las puertas, que usualmente son de forma hidrodinámica y no tocan el fondo.

DA énbåds flydetrawl; flydetrawl med skovle

DE pelagisches Scherbrettnetz; pelagisches Einschiffschleppnetz

Netz, das von einem Fischereifahrzeug aus ausgesetzt und gehievt wird. Seine horizontale Öffnung wird durch die Scherkraft der hydrodynamisch geformten Bretter gewährleistet, die vertikale Öffnung durch Auftriebskörper (Kugeln) und das Höhenbrett. Es kann in Grund- und Oberflächennähe eingesetzt werden.

GR μεσοπελαγική τράτα με πόρτες· τράτα με πόρτες για μεσόνερα

Τράτα που σύρεται από ένα μόνο σκάφος. Το οριζόντιο άνοιγμα του διχτυού ελέγχεται από πόρτες, συνήθως υδροδυναμικού σχήματος, οι οποίες κανονικά δεν ακουμπούν στον πυθμένα.

EN midwater otter trawl; pelagic one-boat trawl

Trawl towed by a single boat. The horizontal opening of the net is controlled by otter boards, usually of a hydrodynamic shape, and which normally do not touch the ground.

3079

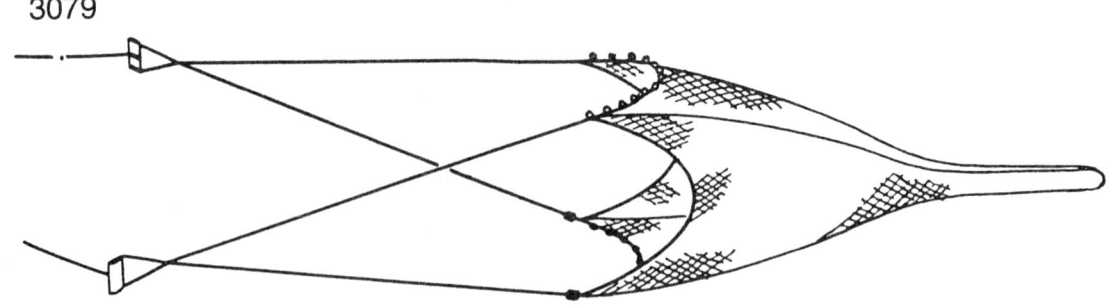

3079

FR chalut pélagique à panneaux; chalut pélagique à un seul bateau

Chalut remorqué par un seul bateau. L'ouverture horizontale du filet est assurée par des panneaux divergents, en général de forme hydrodynamique et qui normalement ne touchent pas le fond.

IT rete da traino pelagica a divergenti; rete da traino pelagica a un solo natante

Rete da traino che opera senza contatto col fondo e la cui apertura orizzontale è assicurata dai divergenti o porte.

NL pelagische trawl met visborden; pelagische enkelscheepstrawl

Pelagische trawl bediend door één schip met visborden aan het voortuig.

PT rede de arrasto pelágico com portas; rede de arrasto pelágico manobrada por uma embarcação

Rede de arrasto rebocada por uma só embarcação. A abertura horizontal da rede é assegurada por duas portas de arrasto, geralmente com forma hidrodinâmica, que normalmente não tocam no fundo.

3080

ES sonda de red; netsonde; batitelémetro

Ecosonda cuyo proyector va montado sobre la relinga superior de la red. Su función es la de dar a conocer la profundidad a que trabaja la red, su abertura vertical y el proceder del pescado ante la boca de la misma.

DA netsonde

En slags ekkolod, hvor svingeren er anbragt på trawlens overtælle, mens skriveren eller monitoren er på fartøjet. Signalerne overføres ved kabel eller akustik. Viser trawlens højde i forparten og position i vandsøjlen samt eventuelle forekomster af fisk.
Er et vigtigt hjælpemiddel i flydetrawlsfiskeriet.

DE Netzsonde

Elektroakustisches Ortungssystem mit Oben- und Untenschwinger, deren Signale zur Oberfläche und zum Grund die Lage des Netzes im Pelagial auf einem schreibenden Gerät sichtbar machen. Moderne Geräte sind zusätzlich mit einer Temperatursonde ausgestattet. Die Übertragung der Echos erfolgt über das Netzsondenkabel. Auf dem schreibenden Gerät werden neben der Oben- und Untenlotung die Netzöffnung, der in das Netz einschwimmende Fang und der Füllstand zur Vermeidung von Netzschäden durch überfüllte Netze angezeigt. Die Wassertemperatur wird auf der Temperaturlinie geschrieben und ist direkt in Grad Celsius ablesbar.

GR ενδείκτης βάθους διχτυού· ηχοβολιστικό διχτυού· βυθόμετρο διχτυού
Για τον έλεγχο του βάθους αλιείας των μεσοπελαγικών τρατών.

EN net sounder; net sonde

Transducer fixed to the trawl headline and linked to a recorder on the vessel by cable or acoustic link. Provides indication of depth of a pelagic trawl, also shows fish above and below the trawl, and in some cases, water temperature.

3080

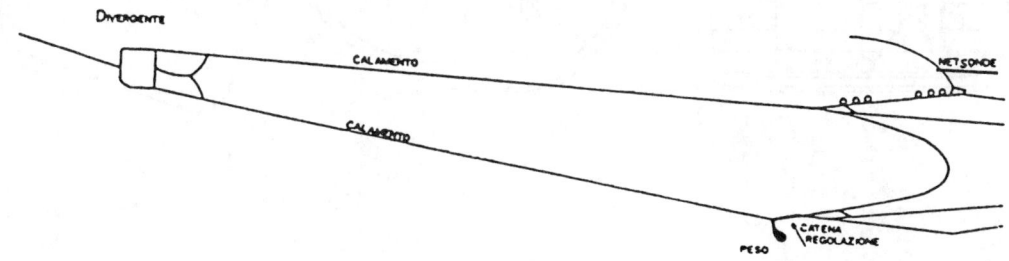

3080

FR sondeur de filet; netsonde

Sondeur fixé au filet, relié (par câble ou voie acoustique) à un enregistreur placé à bord du bateau. Il fournit des informations sur l'ouverture verticale et la géométrie du filet, sa distance par rapport au fond et/ou à la surface, ainsi que la présence de bancs de poissons entrant dans le filet. *La liaison sondeur de filet-bateau peut aussi être utilisée pour transmettre des indications complémentaires, telles que la température de l'eau de mer à l'entrée du filet et l'importance de la capture.*

IT netsonde; sonda rete; net sounder

Strumento munito di uno o più trasduttori elettro-acustici, montato sulla lima dei sugheri di una rete volante e collegato con gli strumenti di bordo mediante cavo oppure onde acustiche. Indica al pescatore lo stato di apertura verticale della rete e la sua posizione rispetto al fondo ed alla superficie; può anche rivelare se il pesce è entrato nella bocca della rete o se è passato sotto o sopra.

NL netsonde

Een op de bovenpees van een trawl geplaatst echolood ter bepaling van de verticale netopening en/of de positie van het net in de waterkolom, alsmede de aanwezigheid van vis in de netmond. *Is door middel van de netsondekabel verbonden met het schip, waar het signaal op een schrijver of een beeldbuis is af te lezen.*

PT sonda de rede

Ecossonda cujo transdutor é colocado no centro do cabo de pana das redes de arrasto pelágico. É utilizada para verificar a profundidade de actuação da rede, a respectiva abertura vertical e o comportamento do peixe na boca da rede.

3081

ES puerta

Elemento de variada forma; plano de contorno rectangular, plano de contorno oval, plano en forma angular, superficie convexa, etc., cuya misión esencial es la de abrir la boca del arte en sentido horizontal.

DA trawlskovl; skovl

Svære plader af jern eller træ. Sættes i en vis afstand foran trawlen og stilles på en vinkel, der sikrer, at vandets hydrodynamiske kræfter spiler trawlen åben horisontalt.

DE Scherbrett; Scheerbrett

Bestandteil des Einschiffschleppnetzes, Teil vom Vorgeschirr, dient als Seitenscherbrett zur horizontalen Spreizung des Schleppnetzes und unterstützt als Höhenscherbrett in Verbindung mit den Auftriebskörpern die Bildung der vertikalen Netzöffnung, hat bei pelagischen Schleppnetzen außerdem Stabilitätsaufgaben.

GR πόρτα τράτας

Δύο διατμητικά εξαρτήματα που κρατούν ανοιχτά οριζοντίως τα φτερά και το στόμιο μιας τράτας.

EN otter board; trawl board; trawl door; board; door

Shearing device, two of which hold open horizontally the wings and mouth of a trawl.

3081

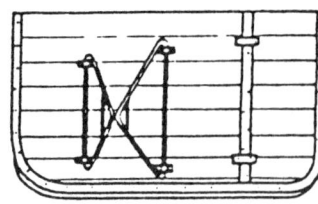 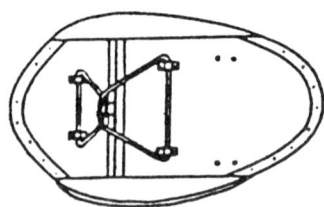 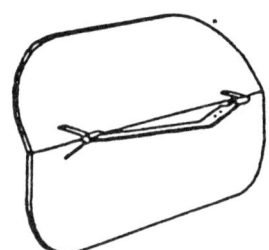

3081

FR **panneau de chalut; panneau; planche; panneau divergent**

De construction et de forme variables selon le type de pêche, les panneaux ont une surface et un poids adaptés à la puissance du chalutier. Le gréement d'un chalut comporte une paire de panneaux qui assurent par leur divergence l'ouverture horizontale du filet.

IT **divergente; porta**

Pannelli in legno, ferro od altro materiale che, opportunamente «armati», quando vengono trainati in acqua assicurano l'apertura orizzontale alle reti da traino.

NL **visbord**

Een aan het einde van de vislijn bevestigd onderdeel van een trawl, in de vorm van een rechthoekig of ovaal vlak of iets gebogen stalen of houten bord; deze kunnen zowel pelagisch als op de bodem worden gebruikt; in het laatste geval zijn ze doorgaans voorzien van slijtstrippen.

Wanneer ze op de bodem worden gebruikt, vergroten ze de visnamigheid doordat ze bodemmateriaal loswervelen.

PT **porta de arrasto**

Dispositivo com forma variada (rectangular, oval, em V, plano ou convexo) cuja função é abrir a boca da rede de arrasto no sentido horizontal.

3082

ES **barra; vara; bao**

Arte de arrastre de vara.

DA **bom**

Den tværbom af jern, der holder nettet udspændt i en bomtrawl. Er i enderne forsynet med meder, som trawlen kører på bunden med.

DE **Kurrbaum**

Querbaum, überwiegend aus Stahl, an dessen beiden Enden der Kurrschuh (Bügel, Klaue, Kufe oder Schlitten) angebracht ist und an dessen Obersimm das Netz befestigt wird. Bewirkt die horizontale Spreizung des Netzes.

GR **δοκός· ζυγός· καμάρι**

Ξύλινο ή ατσάλινο δοκάρι που κρατά οριζόντια ανοιχτό το δίχτυ μιας τράτας.

EN **beam**

Wood or steel spar which holds the net of a beam trawl open horizontally.

FR **perche**

Barre en bois ou tube en métal assurant l'ouverture horizontale du chalut à perche.

IT **asta**

Asta in legno o in ferro con due slitte all'estremità che assicura l'apertura orizzontale della sfogliara.

NL **korboom**

Rondhout of spar, dan wel stalen buis om de boomkor open te houden.

PT **vara**

Vara de madeira ou metal nas extremidades da qual são montados dois patins e cuja finalidade é assegurar a abertura horizontal das redes de arrasto de vara.

3083

ES red de arrastre pelágico

Red de arrastre, por lo general mucho mayor que las redes de fondo; está diseñada y aparejada para trabajar a profundidad media e, incluso, en aguas de superficie. La sección delantera de la red a menudo está hecha de unas mallas muy grandes o de cabos, que conducen los peces hacia el fondo de la red. La profundidad de pesca se controla generalmente por medio de una ecosonda de red. Puede ser remolcada por una o dos embarcaciones.

DA flydetrawl; pelagisk trawl

En trawl, der er specielt designet og rigget til at blive brugt i midtvand eller ved overfladen. Den er ofte meget stor og har i forparten store masker eller stræbere.

DE pelagisches Schleppnetz

Wesentlich größeres Netz als das Grundschleppnetz. Durch seine spezielle Konstruktion kann der Bereich Grund- und Oberflächennähe befischt werden. Das Vornetz besteht aus sehr großen Maschen oder auch aus Leinen (Spaghettinetz), weil es hier nur auf die scheuchende Wirkung ankommt.

GR μεσοπελαγική τράτα· πελαγική τράτα

Οι τράτες αυτές, συνήθως αρκετά μεγαλύτερες από τις τράτες βυθού, είναι σχεδιασμένες και εξαρτισμένες ώστε να χρησιμοποιούνται στα μεσόνερα, περιλαμβανομένων και των νερών επιφάνειας. Τα μπροστινά τους διχτυωτά τμήματα πολύ συχνά έχουν πολύ μεγάλα μάτια ή σχοινιά, τα οποία συναθροίζουν τα κοπάδια των ψαριών προς τις πίσω περιοχές του διχτυού. Το βάθος αλιείας συνήθως ελέγχεται με ενδείκτη βάθους. Είναι δυνατό να σύρονται από ένα ή δύο σκάφη.

EN midwater trawl; pelagic trawl; floating trawl

Trawl, usually much larger than the bottom trawl, designed and rigged to work in midwater, including surface water. The front net sections are very often made with very large meshes or ropes, which herd the fish schools towards the net aft sections. The fishing depth is usually monitored by means of a net sounder (net sonde). They may be towed by one or two boats.

FR chalut pélagique

Filet, en général beaucoup plus grand que les chaluts de fond, conçu et gréé pour fonctionner entre deux eaux, y compris dans les eaux proches de la surface. Les pièces de la partie antérieure du filet sont le plus souvent réalisées en très grandes mailles, ou en cordages, qui rabattent les bancs de poissons vers la partie postérieure du filet. Le contrôle de la profondeur de pêche se fait le plus souvent au moyen d'un sondeur de filet (netsonde). Peut être traîné par un ou deux bateaux.

IT rete da traino pelagica

NL pelagisch net; pelagische trawl; pelagisch trawlnet

Net voor de vangst van pelagische vis.

PT rede de arrasto pelágico

Rede de arrasto, geralmente muito maior que as de arrasto pelo fundo; é planeada e armada para actuar entre duas águas, nomeadamente perto da superfície. As suas secções anteriores possuem normalmente grande malhagem ou são formadas unicamente por cabos e conduzem os cardumes para a parte posterior da rede. A verificação da profundidade a que actuam é conseguida através da sonda de rede. Pode ser rebocada por uma ou duas embarcações.

3084

ES red de arrastre de fondo a la pareja; red de pareja de fondo

Red que se remolca con dos barcos a la vez, por lo que la distancia entre los mismos asegura la abertura horizontal de la red.

DA bundtrawl til partrawling

DE Zweischiffgrundschleppnetz

Fanggeschirr, das in der Gespannfischerei mit Grundschleppnetz ohne Scherbretter verwendet wird. Die Knüppel sind über vorgeschaltete Jager mit dem Vornetz verbunden. Die horizontale Netzöffnung wird durch den Abstand der schleppenden Fahrzeuge bewirkt.

GR τράτα βυθού με ζευγαρωτά σκάφη

Τράτα που σύρεται από δύο σκάφη συγχρόνως. Η απόσταση μεταξύ των σκαφών εξασφαλίζει το οριζόντιο άνοιγμα του διχτυού.

EN bottom pair trawl

Trawl towed by two boats at the same time, the distance between the boats ensuring the horizontal opening of the net.
Pair trawl used on the seabed.

FR chalut-bœuf de fond

Chalut remorqué simultanément par deux bateaux dont l'écartement assure en même temps l'ouverture horizontale du filet.

IT rete a strascico a coppia; rete a strascico in coppia

Reta da traino che opera sul fondo marino e la cui apertura orizzontale è assicurata dal traino con due natanti.

NL bodemspannet

Trawlnet, bediend vanuit twee schepen, dat op de bodem wordt gebruikt.
Zie pair trawling.

PT rede de arrasto pelo fundo de parelha

Rede de arrasto pelo fundo rebocada por duas embarcações cujo afastamento assegura a abertura horizontal respectiva.

3085

ES red de arrastre de fondo de puertas

Red que se arrastra con una sola embarcación: se mantiene abierta en sentido horizontal mediante puertas relativamente pesadas y armadas con una zapata de acero, destinada a resistir un contacto marcado con el suelo.

DA enbåds bundtrawl

Bundtrawl med skovle.

DE Grundscherbrettnetz

GR τράτα βυθού με πόρτες

Τράτα που σύρεται από ένα μόνο σκάφος. Το οριζόντιο άνοιγμα επιτυγχάνεται με πόρτες σχετικά βαριές και εφοδιασμένες με ένα ατσάλινο πέλμα, σχεδιασμένο για έντονη επαφή με τον πυθμένα.

EN bottom otter trawl

Trawl towed by a single boat; its horizontal opening is obtained by the use of other boards which are relatively heavy and equipped with a steel sole designed to withstand rough contact with the bottom.

FR chalut de fond à panneaux

Chalut remorqué par un seul bateau dont l'ouverture horizontale est assurée par des panneaux divergents, relativement lourds et munis d'une semelle d'acier prévue pour un contact accentué avec le fond.

IT rete a strascico a divergenti

Rete da traino che opera sul fondo marino e la cui apertura orizzontale è assicurata dai divergenti o porte.

NL bodemtrawl

Trawl waarvan de onderpees en de visborden tijdens het vissen de bodem niet verlaten.

PT rede de arrasto pelo fundo com portas

Rede de arrasto pelo fundo rebocada por uma só embarcação e cuja abertura horizontal é assegurada pelas portas de arrasto relativamente pesadas e munidas de uma sapata de aço destinada a suportar um contacto acentuado com o fundo.

3086

[ES] **armamento simple**

[DA] **enkelt rig**

Betegner, at et fartøj kun anvender en trawl.
Til forskel fra en dobbelttrawl rig.

[DE] **einfaches Geschirr**

[GR] **απλός εξαρτισμός· απλή αρματωσιά**

[EN] **single rig**

Gear consisting of a single trawl net.

[FR] **gréement simple**

Gréement ne comportant qu'un seul chalut, par
opposition au gréement double dans lequel deux
chaluts ou davantage sont remorqués simultanément.

[IT] **attrezzatura singola**

[NL] **enkelvoudige optuiging; enkelvoudig
voortuig**

Optuiging bedoeld voor het slepen van slechts één
net per boot.

[PT] **armamento simples**

Cada embarcação apenas reboca uma só rede.

3085

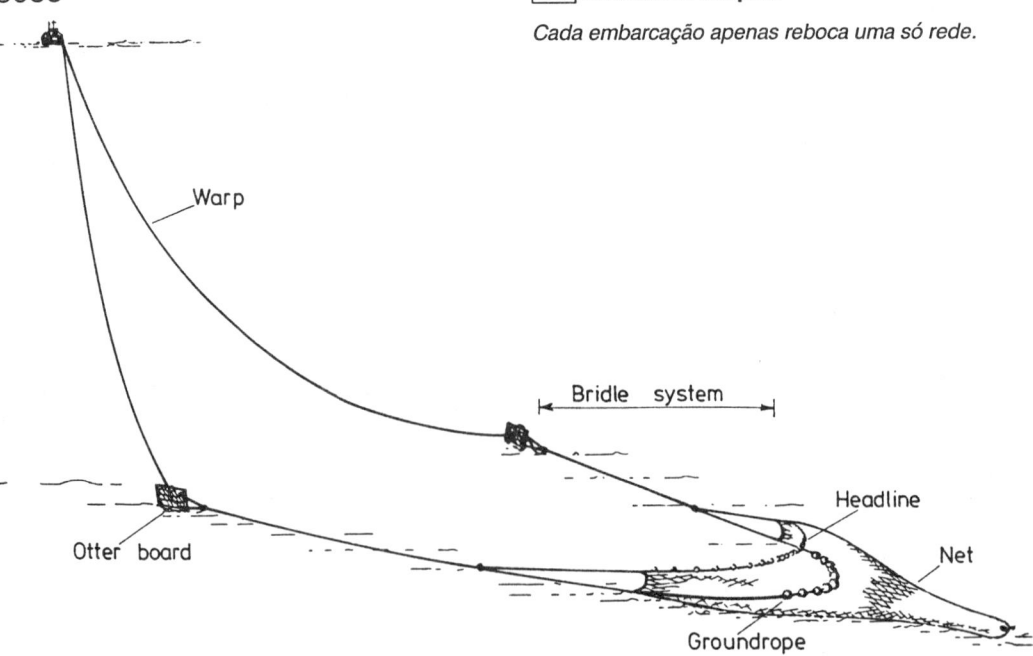

The principal features of a bottom otter trawl

74

3087

ES red de arrastre de vara

Red de arrastre, cuya abertura horizontal se mantiene con una vara, de madera o metal, que puede tener una longitud de 10 m o más. Las redes de arrastre de vara se utilizan principalmente para la pesca del lenguado y del camarón.
Tipo de red de arrastre de fondo.

DA bomtrawl

Trawlredskab, som spiles horisontalt ved hjælp af en tværbom. Anvendes til fiskeri af rejer og fladfisk. Kan være forsynet med svære kæder foran nettets undertælle, som tjener til at skræmme fiskene op fra bunden og ind i nettet.

DE Baumkurre

In der Plattfisch- und Krabbenfischerei eingesetztes trichterförmiges Netz, dessen horizontale Maulöffnung durch den Kurrbaum aufrechterhalten wird. Der Netzsack besteht aus dem Vornetz, das am Obersimm befestigt ist, dem Grundtau, Hinternetz und Steert.
Das mit schweren Ketten bestückte Grundtau dringt in den Grund ein und fördert vordergründig das Fangergebnis; es schädigt allerdings die Nahrungsbasis der Nutzfische.

GR δοκότρατα

Το οριζόντιο άνοιγμα αυτής της τράτας επιτυγχάνεται με μία δοκό, κατασκευασμένη από ξύλο ή μέταλλο, η οποία μπορεί να έχει μήκος 10 μ. ή μεγαλύτερο. Οι δοκότρατες χρησιμοποιούνται κυρίως για την αλιεία πλατύψαρων και γαρίδων.

EN beam trawl

The horizontal opening of this trawl is provided by a beam, made of wood or metal, which may be 10 m long or more. Beam trawls are used mainly for flatfish and shrimp fishing.
Type of bottom trawl.

FR chalut à perche

L'ouverture horizontale de ce chalut est assurée par une perche, en bois ou en métal, dont la longueur peut atteindre et dépasser une dizaine de mètres. Les chaluts à perche sont surtout utilisés pour la pêche des poissons plats et des crevettes.
Type de chalut de fond.

IT sfogliara; carpasfoglie

Rete da traino a bocca fissa, con una corona di piombi strascicanti sul fondo, penzoloni dall'asta che mantiene l'apertura orizzontale. Tale rete è usata principalmente per la pesca delle sogliole.

NL boomkor

Trechtervormig net voor de vangst van platvissen en garnalen, opengehouden door de zgn. korboom; de vissen komen in het net terecht door ze op te jagen met kettingen of met een elektrisch visstimuleringssysteem.
Wordt ook wel voor rondvis gebruikt.

PT rede de arrasto de vara

Nesta rede de arrasto a respectiva abertura horizontal é assegurada por uma vara de madeira ou metal cujo comprimento pode atingir ou mesmo ultrapassar a dezena de metros e em cujas extremidades existem dois patins de ferro cuja base contacta com o fundo.
Tipo de arrasto pelo fundo.

3087

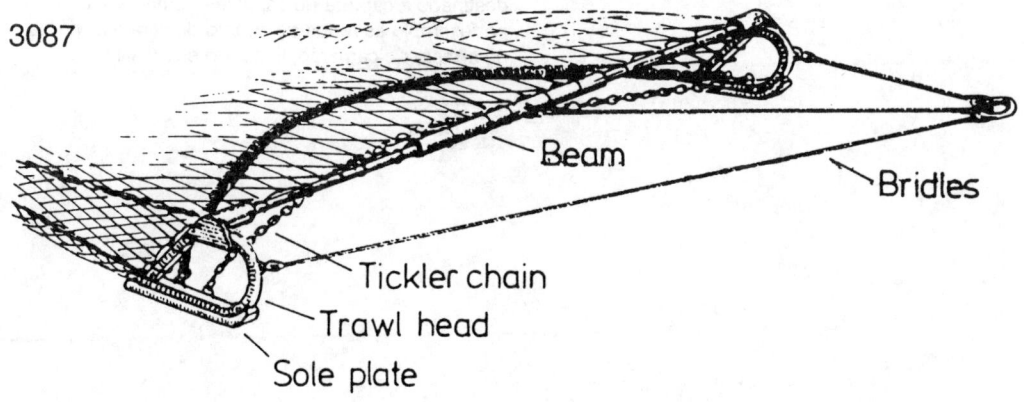

Beam

Bridles

Tickler chain

Trawl head

Sole plate

3088

ES red de arrastre de boca alta; red de arrastre de mucha abertura vertical; red de arrastre de gran abertura; red de arrastre semipelágica

Tipo de red de arrastre de fondo apropiado principalmente para la captura de especies semidemersales o pelágicas.

DA trawl med høj åbning; bundtrawl med høj åbning

DE hochstauendes Grundschleppnetz

GR τράτα υψηλού ανοίγματος

Τύπος τράτας βυθού η οποία είναι κατάλληλη βασικά για τη σύλληψη βενθοπελαγικών ή πελαγικών ειδών.

EN high-opening trawl

A type of bottom trawl which is mainly suitable for the capture of semi-demersal or pelagic species.

FR chalut à grande ouverture verticale; chalut de fond à grande ouverture verticale; GOV

Type de chalut de fond convenant principalement à la capture des espèces semi-démersales ou pélagiques.

IT rete a strascico a grande apertura verticale; rete francese *

Rete a strascico la cui apertura verticale in pesca è due o tre volte quella di una rete normale.
* Termine dialettale.

NL rondvisnet; rondvistrawlnet; hoog-openend net; rondvistrawl

Bodemtrawl met vergrote verticale opening, verkregen door lift en/of door opdrijfwerk, en voorzien van aanvullend netwerk in de zijden.

3089

ES red de arrastre de fondo de poca abertura vertical

Tipo de red de arrastre de fondo, por ejemplo red de arrastre de vara y red de arrastre de camarón, lenguado y cigala.

DA trawl med lav åbning; bundtrawl med lav åbning

DE Schleppnetz mit geringer vertikaler Öffnung

GR τράτα χαμηλού ανοίγματος

Τύπος τράτας βυθού η οποία είναι ειδικά προσαρμοσμένη για τη σύλληψη βενθικών ειδών.
Όπως η δοκότρατα και οι τράτες για γαρίδες, γλώσσες και καραβίδες.

EN low-opening trawl

A type of bottom trawl specially designed for the capture of demersal species.

FR chalut à faible ouverture verticale

Type de chalut de fond spécialement adapté à la capture des espèces démersales.
Les chaluts à perche, ainsi que la plupart des chaluts à crevette, à sole ou à langoustine appartiennent à ce type.

IT rete a strascico a debole apertura verticale

NL platvistrawlnet

Bodemtrawl met kleine verticale en grote horizontale opening, verkregen d.m.v. lange vlerken.
Een veel gebruikt type wordt „loggernet" genoemd.

PT rede de arrasto de pequena abertura vertical

Tipo de rede de arrasto pelo fundo especialmente destinado à captura de espécies demersais, como por exemplo as redes de arrasto de vara e as redes de arrasto de camarão, linguado e lagostim.

3090

ES red de arrastre de fondo; red de arrastre bentónica

Red diseñada y aparejada para pescar cerca del fondo.

Según el tipo utilizado, se puede distinguir entre redes de boca baja — por ejemplo, redes de arrastre de vara y redes de arrastre de camarón, lenguado y cigala— y redes de arrastre de boca alta, apropiadas principalmente para la captura de especies semidemersales o pelágicas. En las redes de arrastre de fondo, el borde inferior de la boca está protegido normalmente por una relinga gruesa lastrada con cadenas y a menudo cubierta de rodillos de goma o diábolos.

DA bundtrawl

En trawl, der er specielt designet og rigget til at blive slæbt på bunden og holde en god kontakt med denne. Langs undertællen er ofte fæstnet en eller anden form for vægt, der samtidig tjener som beskyttelse for nettet, afhængig af bundtypen.

Forekommer i utallige varianter, hver især tilpasset lokale bundforhold, fiskeforekomster og traditioner.

DE Grundschleppnetz

Schleppnetz für den Fang von Krabben, Plattfischen und anderer am Boden lebender Arten. Die Scherbretter sind unmittelbar an den Flügeln angesetzt. Die Beschwerung der Grundleine erfolgt mit Ketten. Es wird auf flachem, weichem Grund eingesetzt.

GR τράτα βυθού· βενθική τράτα

Τράτα σχεδιασμένη και εξαρτισμένη έτσι ώστε να σύρεται κοντά στο βυθό.

Ανάλογα με τον χρησιμοποιούμενο τύπο, μπορεί να διακρίνει κανείς: τράτες χαμηλού ανοίγματος, ειδικά προσαρμοσμένες για τη σύλληψη βενθικών ειδών, όπως οι δοκότρατες και οι τράτες για γαρίδες, γλώσσες και καραβίδες, και τράτες υψηλού ανοίγματος κατάλληλες κυρίως για τη σύλληψη βενθοπελαγικών ή πελαγικών ειδών. Στις τράτες βυθού, το κάτω χείλος του στόμιου του διχτυού προστατεύεται από ένα χοντρό σχοινί ερματισμένο με αλυσίδα μολυβιών και καλυμμένο συχνά με ελαστικούς δίσκους, καρούλια, κλπ.

EN bottom trawl; demersal trawl

Trawl designed and rigged to work near the bottom.

According to the type used, one may distinguish: low-opening trawls, specially designed for the capture of demersal species, such as beam trawls and shrimp, sole or nephrops trawls; and high-opening trawls, suitable mainly for the capture of semi-demersal or pelagic species. In bottom trawls, the lower edge of the net opening is normally protected by a thick groundrope ballasted with chain sinkers and often covered with rubber discs or bobbins.

FR chalut de fond

Chalut conçu et gréé pour travailler près du fond.

Selon le type utilisé, on distingue les chaluts à faible ouverture verticale, spécialement adaptés à la capture des espèces démersales, comme les chaluts à perche et les chaluts à crevette, à sole ou à langoustine, et les chaluts à grande ouverture verticale, convenant principalement à la capture des espèces semi-démersales ou pélagiques. Dans les chaluts de fond, le bord inférieur de l'ouverture du filet est normalement protégé par un épais bourrelet, lesté de morceaux de chaîne et souvent garni de rondelles de caoutchouc ou de diabolos.

IT rete a strascico

Rete da traino che opera sul fondo marino.

NL bodemtrawl

Trawl waarvan de onderpees tijdens het vissen de bodem niet verlaat, gebruikt voor het vangen van demersale vissoorten.

Demersale vissoorten zijn platvissen, zoals tong, schar, tarbot en griet, en de rondvissen, zoals kabeljauw, schelvis en wijting.

PT rede de arrasto pelo fundo

Rede de arrasto planeada e armada para trabalhar junto ao fundo.

De acordo com o respectivo tipo, existem as redes de pequena abertura vertical especialmente destinadas à captura de espécies demersais como por exemplo as redes de arrasto de vara e as redes de arrasto para linguados e para lagostins, e as redes de grande abertura vertical principalmente destinadas à captura de espécies semidemersais e pelágicas. Nas redes de arrasto pelo fundo o bordo inferior da boca da rede é normalmente constituído por um cabo de aço forte, forrado ou não, denominado arraçal e lastrado com correntes de ferro e muitas vezes munido com rodelas de borracha (bolachas), roletes, diábolos.

3091

aparejo doble

En algunos casos, como en la pesca de arrastre de camarones o peces planos, puede armarse el barco con tangones para arrastrar dos (o hasta cuatro) redes al mismo tiempo.

DA **dobbeltrig**

Betegnelse for den form for fiskeri, hvor et fartøj med udriggerbomme fisker med en (nogle gange to) trawl på hver side.
Anvendes bl.a. til fiskeri efter fladfisk og i troperne til fiskeri efter rejer.

DE **doppeltes Geschirr**

GR **διπλός εξαρτισμός· διπλή αρματωσιά**

Σε ορισμένες περιπτώσεις, όπως το καλάρισμα τράτας για γαρίδες ή πλατύψαρα, η μηχανότρατα μπορεί να εξαρτισθεί ειδικά με προώστες προκειμένου να σύρει δύο (ή ακόμη και τέσσερις) τράτες συγχρόνως.

EN **double rigging; double rig**

In certain cases, as in trawling for shrimp or flatfish, the trawler can be specially rigged with outriggers to tow two (or even four) trawls at the same time.

FR **gréement double**

Dans certains cas, comme pour le chalutage de la crevette ou des poissons plats, le chalutier peut être gréé spécialement de tangons pour remorquer deux (ou même quatre) chaluts simultanément.

IT **attrezzatura doppia**

NL **dubbele optuiging; dubbel voortuig**

Optuiging voor het vissen met meer dan één vistuig per boot.

PT **armamento duplo**

Em certos casos, como por exemplo para a arrasto de camarões ou de peixes chatos, a embarcação (arrastão) pode ser especialmente armada com retrancas para rebocar simultaneamente duas (ou mesmo quatro) redes de arrasto.

3092

ES **red de pareja**

DA **partrawl**

Trawl slæbt af to fartøjer. Afstanden mellem disse er bestemmende for den horisontale åbning af nettet. Der anvendes ikke skovle.

DE **Gespannschleppnetz; Zweischiffschleppnetz; Gespann-Netz**

Schleppnetz, das von zwei Booten gleichzeitig gezogen wird. Es hat keine Scherbretter, und die Scherwirkung wird durch das Auseinanderfahren der Boote erzielt.

GR **τράτα με ζευγαρωτά σκάφη**

EN **pair trawl; bull trawl**

Trawl towed by two boats whose separation controls the horizontal opening of the net. Otter boards are not used.

FR **chalut-bœuf; chalut à deux bateaux; gangui ***

Chalut remorqué simultanément par deux bateaux dont l'écartement assure l'ouverture horizontale du filet. Dans le gréement de cet engin, les panneaux divergents ne sont pas utilisés.
* Terme méditerranéen.

IT **rete da traino a coppia**

Rete da traino la cui apertura orizzontale è assicurata dal traino con due natanti.

NL **spannet**

Trawl bediend en opengehouden door twee schepen.

PT **rede de arrasto de parelha; parelha**

Rede de arrasto rebocada por duas embarcações e cuja abertura horizontal é assegurada pelo afastamento das embarcações.

3093

ES pesca de arrastre

Procedimiento de pesca que consiste en remolcar una red con una o dos embarcaciones, buscando atravesar los bancos de peces, para que éstos penetren en su interior y queden apresados en el copo.

DA trawlfiskeri; trawling

Mest udbredte fiskeriform. Redskabet slæbes efter et eller to fartøjer på bunden eller oppe i vandet.

DE Schleppnetzfischerei; Fang mit Schleppnetz

Wichtigste Fischfangmethode der Hochseefischerei. Dabei wird das Schleppnetz von einem oder zwei Fangschiffen durch Kurrleinen mit einer zum Fangen notwendigen Schleppgeschwindigkeit geschleppt.

GR καλάρισμα

Η σύρση της τράτας στον πυθμένα της θάλασσας για τη σύλληψη αλιεύματος. Αυτό μπορεί να γίνει μόνο σε περιοχές όπου ο πυθμένας είναι αρκετά ομαλός και χωρίς εμπόδια.

EN trawling

Most commonly used method of catching fish, which involves dragging a trawl net through the water.

FR chalutage; pêche au chalut; pêche chalutière

Méthode de pêche au moyen d'un chalut remorqué sur le fond ou entre deux eaux. Selon le cas, le chalutage peut être effectué par un seul bateau (chalutage à panneaux ou chalutage à perche) ou par deux bateaux opérant simultanément (chalutage en bœuf).

IT pesca al traino

Pesca con reti da traino.

NL trawlvisserij; treilvisserij

Vismethode waarbij met trawlnetten wordt gevangen.

PT pesca de arrasto

Acto de rebocar uma rede de arrasto pelo fundo ou entre duas águas com a finalidade de capturar peixes, moluscos, crustáceos, etc.

3092

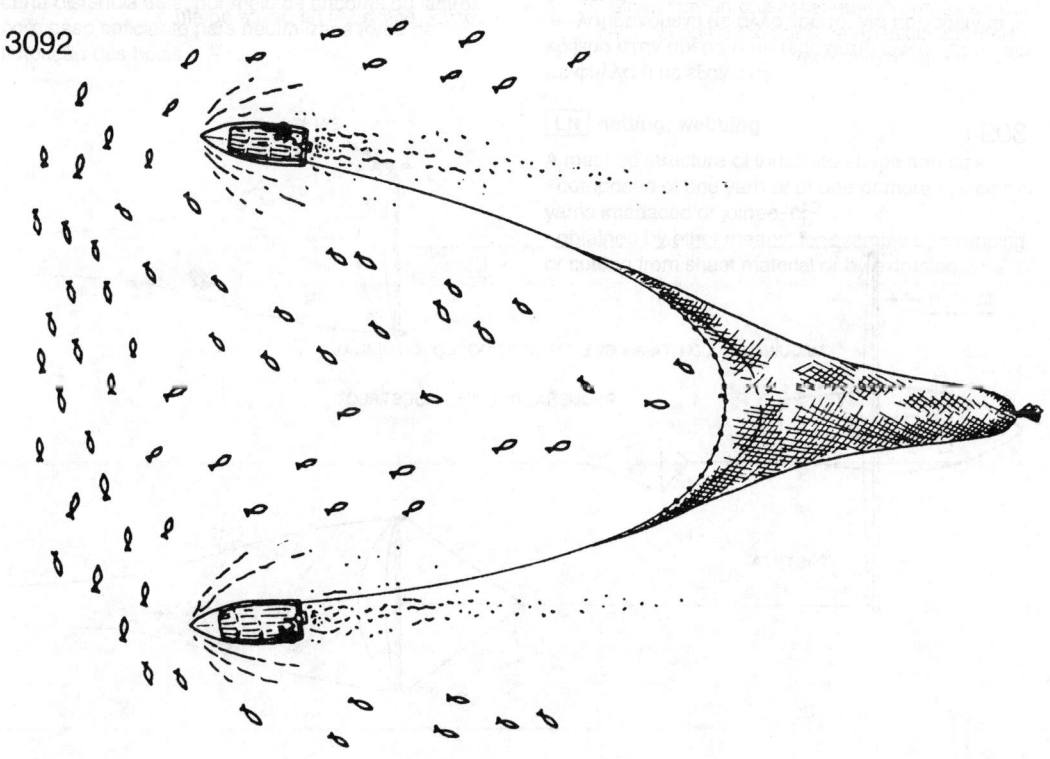

3094

ES tangón; puntal

Palo de un barco que puede sacarse hacia fuera cuando es necesario para algo.
Palo horizontal apoyado en el mástil.

DA udriggerbom

Kraftige bomme, der svinges ud til siden på fartøjer, der fisker med en dobbeltrig (et trawlsystem på hver side). Slæbewirerne går gennem blokke ved udriggerbommens spids.

DE Auslegerbaum; Ausleger

GR προώστης· βαρδαφόγος· μπούμα· κέρκος· δοράτιο

EN outrigger; outrigger boom

Strong boom to spread the fishing gear. The outrigger is usually fastened to the mast and extends out from the sides of the vessel towing two or more trawls by means of passing through blocks at the ends of the outrigger.

FR tangon

Espar mobile établi horizontalement à l'extérieur du navire, à la hauteur du pont supérieur et perpendiculairement à la coque.
Employé soit pour le chalutage au gréement double, soit pour la pêche aux lignes de traîne.

IT buttafuori

Asta di legno o di metallo che si protende fuori bordo.

NL giek; boom; spriet

Stalen of houten balk waarmee een last buiten boord of aan boord kan worden gezet, of waarmee een last vanaf de wal aan boord kan worden verplaatst.
Voor boomkorvaartuigen worden de twee vistuigen door twee gieken behandeld.

PT retranca; pluma; tangão

Pau de carga destinado a manobrar e conduzir os cabos que rebocam as redes de arrasto no caso de armamento duplo. Situados normalmente a meio navio e ligados à parte inferior de um mastro.
Em Portugal também se usa este termo para designar os suportes de onde partem as linhas de corrico utilizadas na pesca do atum.

3094

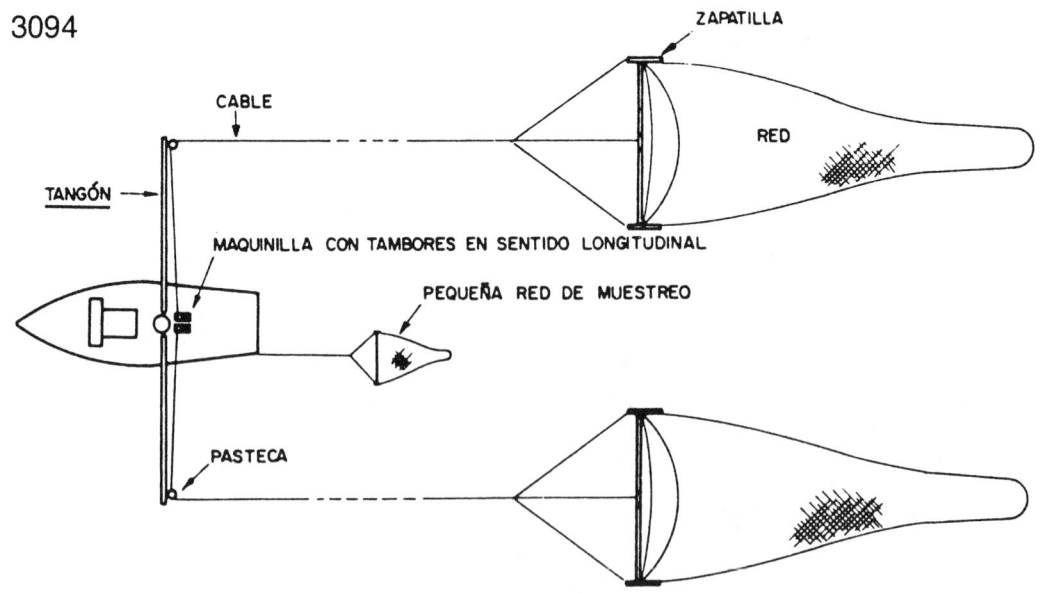

ZAPATILLA

CABLE

RED

TANGÓN

MAQUINILLA CON TAMBORES EN SENTIDO LONGITUDINAL

PEQUEÑA RED DE MUESTREO

PASTECA

3095

ES arrastrero de pesca por el costado; arrastrero de costado

DA sidetrawler

Et fartøj, der sætter og bjærger trawlen samt tager fangsten ind over siden. Styrehuset er da placeret agten for midten.

DE Seitentrawler; Seitenfänger

Bei Seitentrawlern befindet sich das Fangdeck im Bereich des Mittelschiffs. Das Schleppnetz wird über die Seitenreling ausgesetzt und eingeholt.

GR Μηχανότρατα πλάγιας σύρσης

Μηχανότρατα με πόρτες, όπου τα δίχτυα ρίχνονται και σηκώνονται από τα πλάγια του σκάφους.

EN side trawler; side-set trawler

Trawler in which the nets are set and hauled in from the sides of the vessel.

FR chalutier latéral; chalutier à pêche latérale

Chalutier à bord duquel le filage et le virage du chalut sont effectués par le côté.

IT peschereccio per traino laterale; peschereccio alla francese; peschereccio per strascico laterale

Peschereccio che traina la rete di lato.

NL zijtrawler; zijtreiler

Schip waarop de trawl vanuit de zijde wordt uitgezet en gehaald.

Dit scheepstype wordt in Nederland bijna niet meer gebruikt. Men gaat steeds meer over tot het vissen over het hek (hektrawler).

PT arrastão lateral; navio de arrasto lateral

Tipo de embarcação que utiliza redes de arrasto em que a respectiva calagem e alagem é efectuada lateralmente.

Em Portugal a manobra é realizada por estibordo.

3095

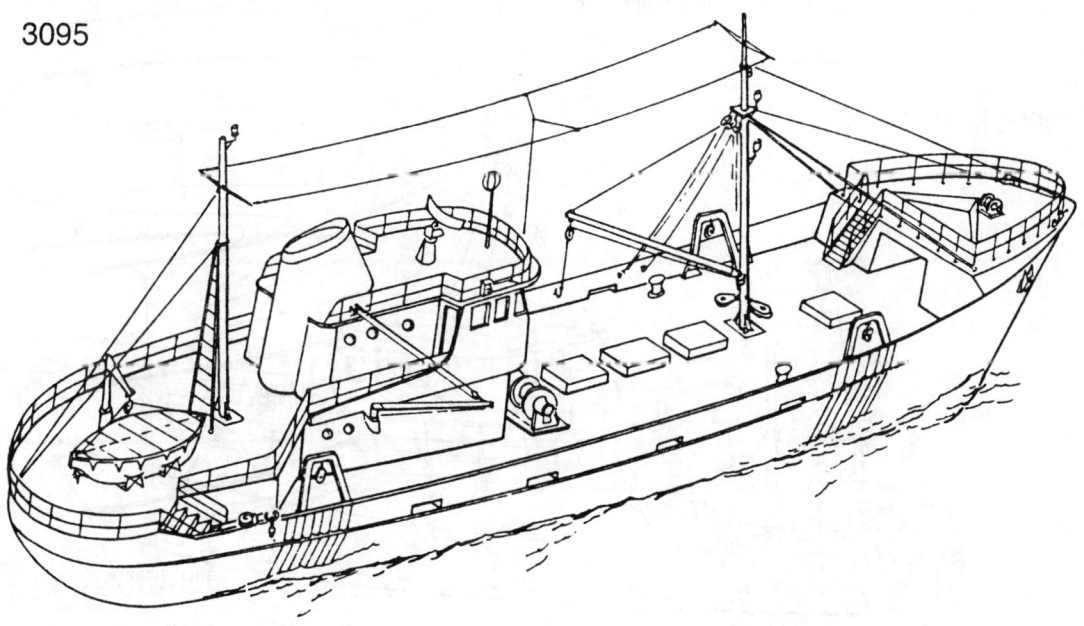

3096

ES buque de rampa a popa; barco de pesca por popa; arrastrero de popa; arrastrero con rampa a popa; rampero *

Arrastrero de popa abierta.

DA hæktrawler

Et fartøj, hvor trawlen kan sættes og hales over hækken (agterenden). På de fleste større hæktrawlere findes agter en egentlig rampe, som trawlen med fangsten kan glide op gennem.

DE Hecktrawler

Bei Hecktrawlern befindet sich das Fangdeck im Bereich des Achterschiffs. Das Schleppnetz wird über die Heckaufschleppe an Bord geholt bzw. ausgesetzt.

GR μηχανότρατα πρυμναίας σύρσης· μηχανότρατα οπίσθιας σύρσης· μηχανότρατα οπίσθιας έλξης

Μηχανότρατα εξοπλισμένη έτσι ώστε το ρίξιμο και σήκωμα των διχτυών να γίνεται από την πρύμνη.

EN stern trawler

Trawler rigged for setting and hauling in the net over the stern.

FR chalutier à pêche arrière

Chalutier à bord duquel le filage et le virage du chalut sont effectués par l'arrière.

IT peschereccio per traino poppiero

Peschereccio che traina la rete di poppa.

NL hektrawler; hektreiler

Schip waarop de trawl over het hek (spiegel) wordt uitgezet en binnengehaald.

Kenmerkend is de aanwezigheid van een portaalmast op het hek, de aanwezigheid van een grote lier voor vislijnen en voorlopers, alsmede een of meer nettentrommels.

PT navio de arrasto pela popa; arrastão pela popa; arrastão com rampa

Tipo de embarcação que utiliza redes de arrasto em que a respectiva calagem e alagem é efectuada pela popa, quer por meio de uma rampa quer por meio de um rolete.

3096

3097

ES | arrastrero

Buque o embarcación dedicada a la pesca de arrastre.

DA | trawler

Fartøj, der er udrustet til fiskeri med trawl.

DE | Trawler; Schleppnetzfischer

Fischereifahrzeug der Schleppnetzfischerei in der Hochsee- und Küstenfischerei.

GR | μηχανότρατα

Σκάφος κατάλληλο για τη σύλληψη ψαριών με καλάρισμα.

EN | trawler

Vessel fitted for catching fish by trawling.

FR | chalutier

Navire pêchant au chalut.

IT | peschereccio per traino

Peschereccio che opera con reti da traino.

NL | trawler; treiler

Vissersvaartuig gebouwd en ingericht voor de visserij met het trawlnet.

PT | navio de arrasto; arrastão

Tipo de embarcação destinada à pesca de arrasto.

3098

ES | armar; aparejar; habilitar

Realizar el montaje de un arte de pesca.

DA | rigge til

Udtryk for den proces at forsyne og montere et redskab eller fartøj med det fornødne tilbehør og udstyr til et bestemt fiskeri.

DE | ausrüsten

Ein Fanggerät mit dem notwendigen Zubehör versehen.

GR | αρματώνω· εξαρτίζω

Η διαδικασία συναρμολόγησης των εξαρτημάτων των αλιευτικών εργαλείων.

EN | rig (verb)

The process of fitting the necessary ropes and accessories so as to make a net ready for fishing.

FR | gréer

Garnir un engin de pêche de ses accessoires et de ses filins de manœuvre.
S'emploie également pour les voiles ou la mâture.

IT | armare

Preparare i vari componenti di un attrezzo da pesca.

NL | optuigen; takelen

Het geheel van handelingen in de ontwerp- en uitvoeringsfase, verband houdende met de optuiging van een visnet.

PT | armar; aparelhar

Modo como as diferentes partes que constituem uma arte de pesca, incluindo os respectivos cabos, são montados

3099

ES cabo de una red danesa

DA vodtov

De lange tove, der anvendes i snurrevodsfiskeriet. De lægges ud på bunden, så de omkredser et større areal og skræmmer ved indhalingen fisk ind med selve voddets bane. Er som regel forsynet med bly for at holde sig på bunden så længe som muligt under indhalingen.

DE Wadenleine

Leine zum Ausfahren und Einholen der Waden. *Nicht für Ringwaden.*

GR σχοινί δανέζικης τράτας

Ένα από τα δύο πολύ μακριά σχοινιά που χρησιμοποιούνται για την έλξη του διχτυού, αποτελούμενο από έναν αριθμό τμημάτων σχοινιών (κορκωμάτων), 120 οργυιές το καθένα, που είναι ενωμένα μεταξύ τους από άκρη σ' άκρη. *Δανέζικη τράτα.*

EN seine rope

One of the two very long ropes to haul the net, comprising a number of 120 fathom coils joined end to end. *Danish seine.*

FR cordage de senne danoise

Un des deux cordages utilisés pour le halage d'une senne danoise (ou senne de fond); habituellement très longs, ces cordages sont utilisés sur le fond de manière à assurer le plus grand rabattement possible du poisson vers l'ouverture du filet.

IT cavo della sciabica; calamento della sciabica

NL vislijn van de zegen

Lijn die over de bodem wordt gesleept met het doel vis die zich binnen het door de lijn omsloten oppervlak bevindt, bijeen te houden en naar het vistuig te geleiden. *Meestal is de lijn verzwaard om een continu bodemcontact te waarborgen.*

PT cala; cabo de calamento

Um dos dois longos cabos utilizados para alar a rede e que são constituídos por um certo número de cabos ligados topo a topo e que também têm por finalidade assegurar que o maior número possível de peixes seja reunido e conduzido para a boca da rede. *Rede de cerco dinamarquesa; xávega; chinchorro; chincha; redinha; mugiganga.*

3099

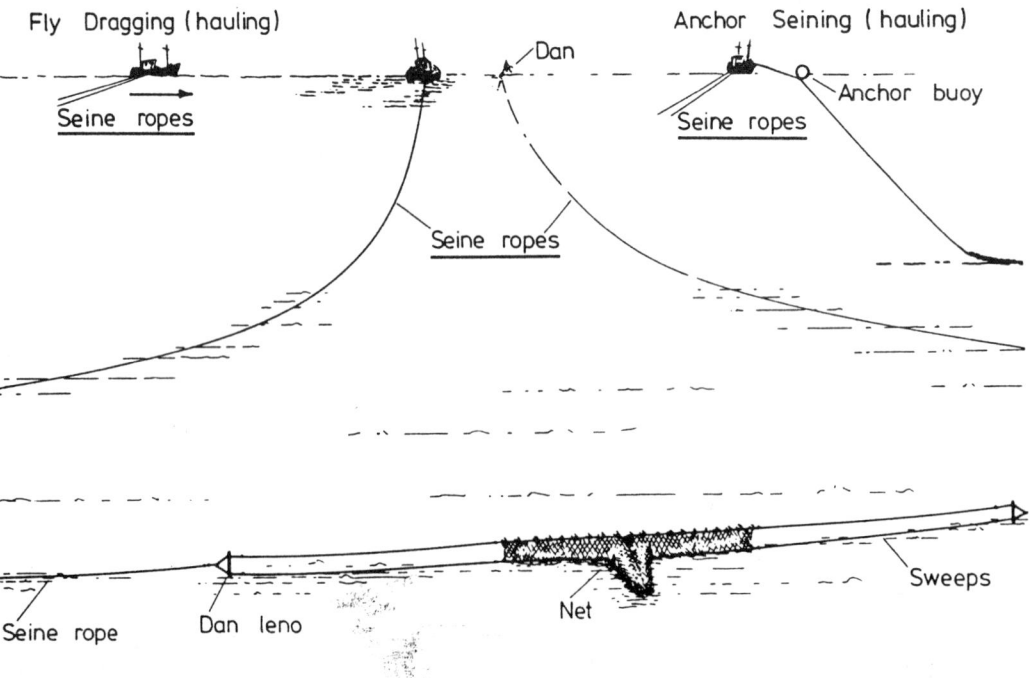

Fly Dragging (hauling)
Seine ropes
Dan
Anchor Seining (hauling)
Anchor buoy
Seine ropes
Seine ropes
Seine rope
Dan leno
Net
Sweeps

3100

ES halar; virar

Red.

DA hale; hive; hale ind; hive ind; hale indenbords; hive om bord

DE einholen; einhieven; hieven

Hieven der Kurrleine, des Netzes, einer Last.

GR έλκω· σύρω· ρυμουλκώ

Γενικά, τραβώ ένα σχοινί. Πιο συγκεκριμένα, χρησιμοποιείται όταν τραβάμε προς τα επάνω ένα σχοινί χωρίς τη βοήθεια τροχαλιών ή άλλων μηχανικών μέσων.

EN haul in (verb); board the net (verb)

To pull in fishing nets, lines, etc.

FR haler; virer

Tirer à bord un filet ou un engin de pêche à la fin d'une opération de pêche.

IT salpare

Tirare a bordo le reti.

NL halen; thuishalen; ophalen; binnenhalen

Het binnenhalen van het net op vissersvaartuigen.

PT alar; virar

Acto de recolher uma arte de pesca da água para bordo de uma embarcação ou para terra.

3101

ES cuerpo de la red

Red danesa.

DA krop

Den midterste del af en trawl eller et snurrevod mellem pose og forstykke.

DE

Umfaßt Bauchstück, Dachstück und Tunnel.

GR κύριο τμήμα· κύριο σώμα

Δανέζικη τράτα.

EN body

The centre which is usually the main part of a net or sections of a trawl, for instance, the bag portion of a seine net between the wings and the codend, consisting of belly and batings.

Danish seine, trawl.

FR corps

Partie centrale du filet située entre les ailes et la poche.

Chalut, senne danoise.

IT corpo

Sciabica danese.

NL samenstel van buik en rug

Gedeelte van het net gelegen tussen de vlerken en de kuil.

PT boca; corpo

Parte de uma rede envolvente-arrastante situada entre as asas e o saco.

Rede de cerco dinamarquesa, xávega, chinchorro, chincha. O termo boca é também empregue em Portugal para designar a abertura de qualquer arte de pesca, bem como o(s) malheiro(s) imediatamente anterior(es) ao saco das redes de arrasto.

3102

ES red de tiro desde embarcación

Red que consiste en dos alas, un cuerpo y un copo, es similar en muchos aspectos a el de las redes de arrastre. Se maniobra desde embarcaciones y generalmente trabaja sobre el fondo, en donde se arrastra por dos cabos, usualmente muy largos, colocados en el agua de manera que hagan converger el mayor número posible de peces hacia la boca de la red.

El tipo más representativo de esta categoría es la red danesa.

DA vod

Betegner her et vod brugt fra et fartøj.

DE Bootswade

Zugnetz mit zwei Flügeln, an die sich der kleine Netzsack anschließt. Die langen Wadenleinen sind mit dem Knüppel am Netzflügel befestigt. Sie haben einerseits ihren Festpunkt an der Ankerboje, andererseits am Kutter.

GR γρίπος συρόμενος από σκάφος

Ο σχεδιασμός αυτών των διχτυών που αποτελούνται από δύο φτερά, ένα κύριο τμήμα και ένα σάκο, είναι σε πολλά σημεία παρόμοιος με των τρατών. Χειριζόμενα από ένα σκάφος, γενικά σύρονται στο βυθό με δύο σχοινιά, συνήθως μεγάλου μήκους, τοποθετημένα στο νερό με τέτοιο τρόπο ώστε να εξασφαλίζουν την οδήγηση ή συνάθροιση προς το στόμιο του διχτυού όσο το δυνατό μεγαλύτερου αριθμού ψαριών.

Ο πιο αντιπροσωπευτικός τύπος αυτής της κατηγορίας είναι η δανέζικη τράτα.

EN boat seine

Net, consisting of two wings, a body and a bag, operated from a boat, generally used on the bottom, where it is hauled by two ropes, usually very long, set in the water so as to ensure that as many fish as possible are driven or herded towards the opening of the net.

The type most representative of this category is the Danish seine.

FR senne de bateau

Filet formé de deux ailes, d'un corps et d'une poche, manœuvré à partir d'un bateau, généralement utilisé sur le fond où il est halé par deux cordages, habituellement très longs, mis à l'eau de manière à assurer le plus grand rabattement possible du poisson vers l'ouverture du filet.

Le type le plus représentatif de cette catégorie est la senne danoise.

IT sciabica da natante

Sciabica che viene tirata da bordo di natanti.

NL bootzegen

Zegen bediend vanaf een vissersvaartuig.

PT rede envolvente-arrastante de alar para bordo; mugiganga

A concepção desta rede, formada por duas asas, boca e saco, lembra a das redes de arrasto. É manobrada a partir de uma embarcação, actua sobre o fundo e é rebocada e alada por dois longos cabos (cordas) que também asseguram a concentração e a condução do peixe para a boca da rede.

A rede mais representativa deste tipo de artes de pesca é a rede de cerco dinamarquesa. Outras são a xávega o chinchorro, e a chicha.

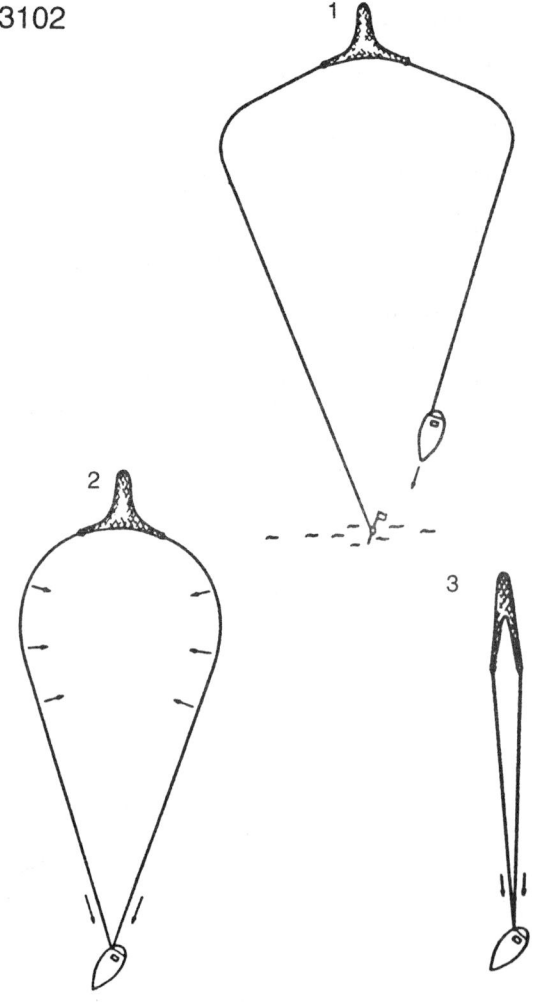

3102

3103

ES arte de playa; jábega; boliche; pubega

Tipo de red que normalmente se cala desde una embarcación y que se puede maniobrar desde la costa

Se hace una distinción entre los artes de playa con copo y los sin copo; estas últimas tienen una parte central de malla más pequeña y floja, que retiene los peces capturados.

DA landdragningsvod; landvod; strandvod

Vod, der anvendes fra land på den måde, at det med et fartøj sejles ud i en bue fra kysten og hales ind til denne ved linerne, enten ved hjælp af et på stranden opstillet spil eller ved håndkraft.

DE Strandwade

Ein vom Strand aus mit einem Boot im Bogen ausgefahrenes Zugnetz, als einfache Netzwand oder mit dazwischen gesetztem Netzsack. Einholen der Wade von Hand, wenn der zweite Flügel an Land gebracht ist.

GR γρίπος ακτής· πεζότρατα

Τύπος γρίπου που συνήθως ρίχνεται από ένα σκάφος και ο χειρισμός του γίνεται από την ακτή.

Υπάρχει διάκριση μεταξύ των γρίπων ακτής με σάκο και γρίπων ακτής χωρίς σάκο. Οι τελευταίοι, πάντως, έχουν ένα κεντρικό τμήμα με μικρότερα μάτια και πιο χαλαρό, το οποίο συγκρατεί το αλίευμα που συλλαμβάνεται.

EN beach seine; shore seine

A type of seine net, which is usually set from a boat and operated from the shore.

A distinction is made between beach seines with a bag and beach seines without a bag; the latter do have, however, a central part with smaller meshes and more slack, which retains the fish caught.

FR senne de plage; senne halée à terre; senne côtière

Type de senne qui est normalement mise à l'eau à partir d'une embarcation et est manœuvrée du rivage.

On distingue les sennes de plage avec poche et les sennes de plage sans poche; ces dernières comprennent cependant une partie centrale à plus petites mailles et montée avec davantage de flou, où est retenu le poisson capturé.

IT sciabica da spiaggia

Sciabica che viene tirata direttamente dalla spiaggia.

NL landzegen; strandzegen

Zegen bediend vanaf het strand.

PT rede envolvente-arrastante de alar para a praia; xávega; chinchorro; chincha; redinha

Rede que é calada a partir de uma embarcação e manobrada e alada a partir de terra. Pode ter ou não saco; no caso de não terem saco, a sua parte central, com malha de menor dimensão, é montada com grande folga de molde a originar uma bolsa onde se concentra o peixe capturado.

3103

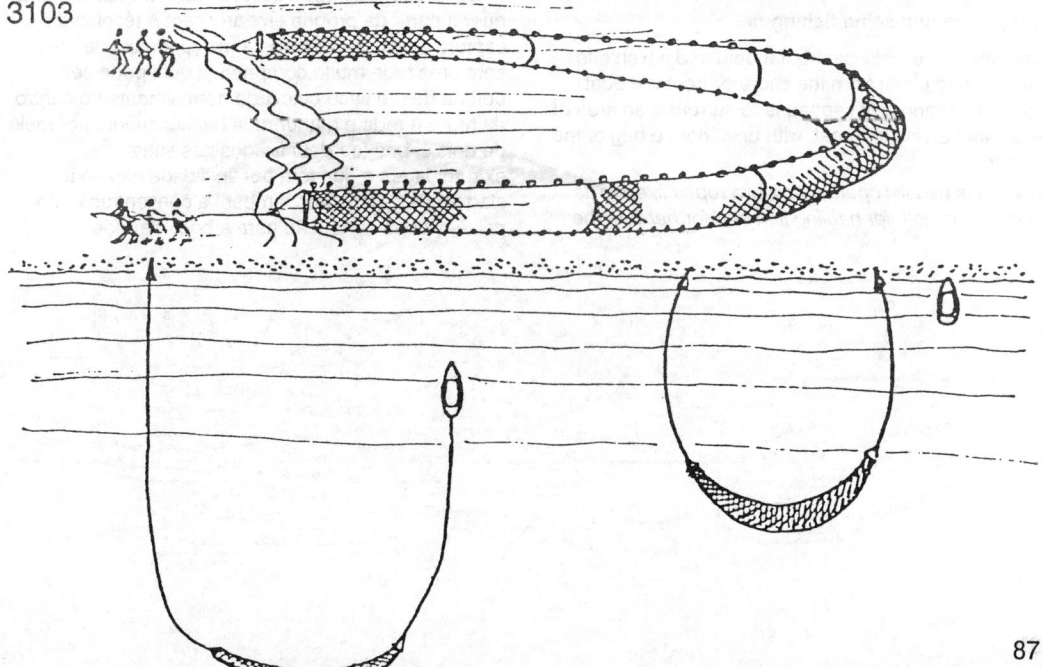

3104

ES red de jábega; jábega; jabéca

DA vod; not

Samlebetegnelse for et net, der fanger ved at blive sat rundt om en fiskeforekomst. Det sættes normalt fra et fartøj og kan så hales på land (strandnot; -vod) eller om bord på fartøjet selv (snurrevod; snurpenot). Nettet er langstrakt med eller uden en pose i midten og med lange vinger, hvortil der er fastgjort tove, som nettet hales i.

På dansk er vod tillige en ældre betegnelse for en trawl.

DE Wade

Fischfanggerät der Binnen-, Küsten- und Hochseefischerei. Wade kann eine Bezeichnung für Umschließungsnetze sein, wobei die Fangobjekte durch Umschließen mit einer Netzwand am Entweichen gehindert werden, oder für Zugnetze für relativ flache Gewässer der Binnen- oder Küstenfischerei.

GR γρίπος· γριπόδιχτο

Ο χειρισμός αυτών των διχτυών, που συνήθως ρίχνονται από ένα σκάφος, μπορεί να γίνεται είτε από την ακτή είτε από το ίδιο το σκάφος. Ο τρόπος σύλληψης είναι η περικύκλωση μιας υδάτινης περιοχής με ένα πολύ μακρύ δίχτυ, με ή χωρίς σάκο στο κέντρο.

Ο χειρισμός του διχτυού συνήθως γίνεται με δύο σχοινιά στερεωμένα στα άκρα του, που χρησιμοποιούνται και τα δύο για το βιράρισμα του διχτυού και τη συνάθροιση του αλιεύματος.

EN seine net; seine fishing net

Net, which is usually set from a boat and which can be operated either from the shore or from the boat itself. The manner of capture is to surround an area of water with a very long net, with or without a bag at the centre.

The net is usually operated by two ropes fixed to its ends, used both for hauling it in and for herding the fish.

FR senne; seine

Filet, qui est normalement mis à l'eau à partir d'une embarcation et qui peut être manœuvré soit du rivage, soit à partir du bateau lui-même. Le principe de la capture consiste à entourer une surface d'eau par un filet de grande dimension horizontale, muni ou non d'une poche placée le plus souvent au centre de l'engin.

Le filet est manœuvré habituellement par deux filins fixés à ses extrémités, servant à la fois au halage et au rabattement du poisson.

IT sciabica

Rete da pesca con grandi braccia e piccolo sacco che viene calata a semicerchio e che cattura il pesce nel suo progressivo avanzamento.

NL zegen

Vistuig waarbij gebruik gemaakt wordt van de opschrikwerking van over de zeebodem gesleepte kabels om de vis voor de netmond te concentreren, waarbij de vleugels het eerst uitgezet worden, en dat vanaf een gefixeerd punt (wal of schip) wordt binnengehaald (Deense zegen).
Langgerekt visnet, dat bij het vangen een gedeelte van het watervolume omringt, al dan niet voorzien van een zak.

Van de zegen in de tweede betekenis werd vooral vroeger gebruik gemaakt in de riviervisserij. Momenteel niet meer in gebruik.

PT rede envolvente-arrastante

Rede normalmente calada a partir de uma embarcação, que pode ser manobrada quer de terra quer a partir da própria embarcação. A técnica de captura consiste em cercar uma superfície de água com uma rede muito comprida, a qual pode ser dotada de um saco colocado normalmente no centro da rede. A rede é manobrada habitualmente por meio de dois cabos (cordas) fixados nas suas extremidades e que têm por finalidade não só a alagem da rede, como também a concentração do peixe e a sua condução para a boca da rede.

3105

ES cardumen de peces; banco de pescado

DA fiskestime

DE Fischschule; Fischschwarm

Ansammlung von Fischen gleicher Art, walzenförmig dicht beieinander in gleicher Richtung schwimmend. *Typisches, instinktives Verhalten, z. B. des Kabeljaus, zum Schutz gegen natürliche Feinde.*

GR κοπάδι ψαριών· μπάγκος ψαριών· πάγκος ψαριών

EN fish school; shoal

Significant aggregation of fish, usually of the same species.

FR banc de poissons

Rassemblement important de poissons, en général de la même espèce; se forme près de la surface, entre deux eaux ou à proximité du fond.

IT banco; banco di pesci

Aggregato di pesci.

NL visschool

Groepering vissen van dezelfde soort die coherent gedrag vertonen.

PT cardume de peixes; banco de peixes

Conjunto de peixes, de uma ou mais espécies, concentrados numa determinada área restrita.

3106

ES ala; banda

Red.

DA vinge; vinge og arm

Trawl, snurrevod: de tilspidsende sektioner i forparten af nettet, som går frem på begge sider. (Er der flere sektioner efter hinanden, betegnes normalt kun den forreste vinge, mens resten kaldes armsektioner). *Den ende, der sættes sidst og hales først (modsat fiskeposen).*

DE Flügel; Netzflügel; Wing

Verlängerung des Scherbrettnetzes an beiden Seiten nach vorn durch den einteiligen Oberflügel und die Unterflügel. Letztere sind aus dem vorderen und hinteren Unterflügel zusammengesetzt.

GR φτερό· μπάντα

Κωνικό τμήμα διχτυού που βρίσκεται στο αντίθετο άκρο από εκείνο το τμήμα που είναι πάνω από το τετράγωνο. *Δίχτυ.*

EN wing

Trawl, seine: tapered net section extending forward from one side of the main body of the net.
Ring net, beach seine: net section at each end of the net.
Purse seine: tapered net section(s) at the opposite end from the bunt.

FR aile

Pièce du filet coupée en forme et située à l'extrémité opposée à la poche.

IT braccio

Rete.

NL vleugel; oor; trawlvleugel; vlerk

Geleidend gedeelte van het net tussen middeling en voortuig. *Bij het vierbladige net vallen hieronder ook de zijperken tussen middeling en voortuig.*

PT asa

Cada uma das secções das redes de arrastar situadas na sua parte oposta ao saco e que limitam lateralmente a boca da rede.

3107

ES copo; saco

Parte de un arte de hilo más fuerte donde se concentra el pescado.
Red de arrastre, red de cerco.

DA pose; fiskepose

Den del af nettet, hvor fisken opsamles.
Her i omkredsende net, f.eks. noter.

DE Bauch; Buk; Tunnel

Netzteil, in welchem sich der Fang beim Einholen des Netzes sammelt.
Lamparanetz; Strandwade; Ringwade; Umschließungsnetz.

GR χαλαρό τμήμα· σάκος

Τμήμα ισχυρότερου δικτυώματος, συχνά χαλαρά αναρτημένο έτσι ώστε να σχηματίζει ένα σάκο ή τσέπη, όπου συναθροίζεται το αλίευμα όταν σύρεται το δίχτυ.
Γρίπος ακτής, πεζότρατα, γρι-γρι, κυκλικό δίχτυ.

EN bunt; bag

Section of stronger netting often loosely hung to form a bag or pocket, where the catch is collected when the net is hauled.
Beach seine; purse seine; encircling net.

FR poche; sac

Partie de l'engin formée d'alèze plus résistante, coupée et montée afin de former une sorte de poche où la capture s'accumule quand on vire le filet.
Filet lamparo, senne de plage, senne coulissante, filet tournant.

IT sacco

Parte della rete di filo più resistente in cui viene raccolto il pesce catturato.
Rete a traino, rete a strascico, rete a circuizione, rete a chiusura.

NL zak

Trechtervormig gedeelte van een vistuig (zegen of trawl), waar de vangst in wordt verzameld tijdens het vangproces.

PT saco; copejada *

Secção de uma arte de pesca com forma de saco ou bolsa, constituída por panos de rede de fio forte e resistente, onde se acumula o peixe capturado.
* Rede de cerco.

3106

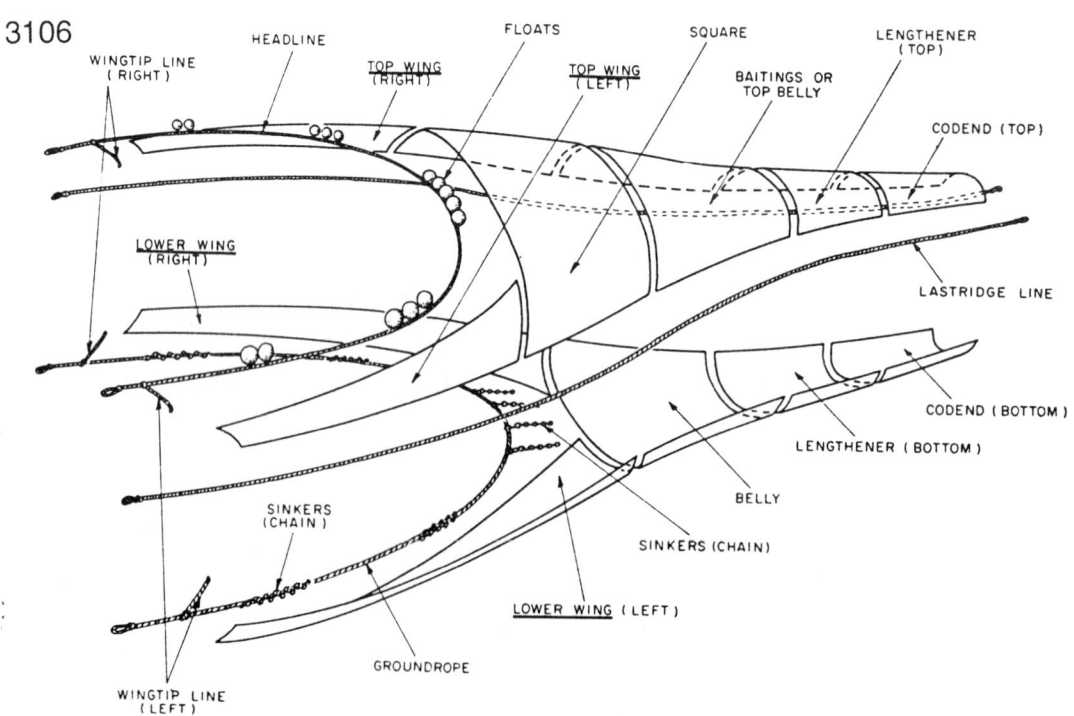

3108

ES **lámparo**

Su diseño particular, con el copo central en forma de cuchara y dos alas laterales, permite retener los cardúmenes de peces cuando se halan las dos alas al mismo tiempo.

Por lo general, las redes del tipo lámparo se maniobran desde un solo barco, a menudo de poco tonelaje.

DA **lampara**

Omkredsende net bestående af en pose i midten og to lange vinger. Sættes ofte fra mindre fartøjer.

DE **Lampara; Lamparanetz**

Halbringwade; Mittelding zwischen Schnur- und Ringwade.

GR **δίχτυ λαμπάρα**

Το χαρακτηριστικό σχήμα αυτού του διχτυού, με κεντρικό χαλαρό τμήμα μορφής κουταλιού και δύο πλευρικά φτερά, του δίνει τη δυνατότητα να συγκρατεί το κοπάδι των ψαριών όταν τα δύο φτερά έλκονται συγχρόνως επάνω.

Ο χειρισμός των διχτυών λαμπάρα γίνεται από ένα σκάφος, συνήθως μικρής χωρητικότητας.

EN **lampara net**

The particular design of this net, with a central bunt in the form of a spoon and two lateral wings, makes it possible to retain the shoal of fish when the two wings are hauled up at the same time.

Lampara nets are generally operated by a single boat, most often of small tonnage.

FR **filet lamparo; lampara; lamparo**

La conception particulière de ce filet, avec une poche centrale en forme de cuiller et deux ailes latérales, permet de retenir le banc de poissons, lorsque les deux ailes sont halées simultanément à bord.

Les filets lamparos sont employés en général à partir d'un seul bateau, le plus souvent de faible tonnage.

IT **lampara**

Tipo di rete a fonte, destinato alla cattura del pesce azzurro. Si tratta di una rete di circuizione manovrata a mano, che possiede nella sua parte centrale una specie di borsa, detta sacco, dove si raduna il pesce durante la manovra di ricupero.

3108

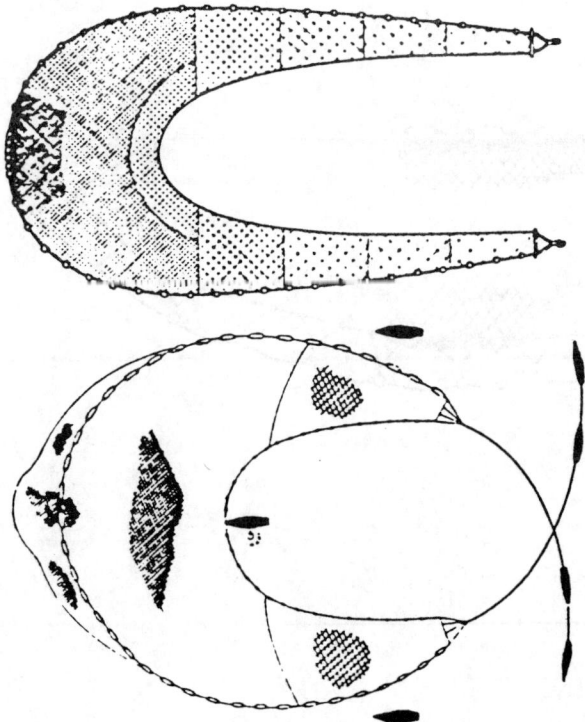 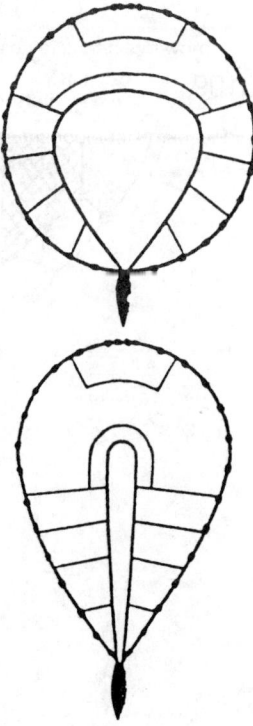

3108

NL lampara

Type zegen, oorspronkelijk gebruikt in het Middellandse-Zeegebied voor de vangst van pelagische vis, met korte grondpees, lange bovenpees en voorzien van vleugels.

PT lâmpara; rede de cerco sem retenida

A sua concepção particular, com saco central em forma de colher e duas asas laterais, permite reter os cardumes quando as alam as duas asas ao mesmo tempo.

Normalmente as lâmparas são utilizadas a partir de uma só embarcação, a maior parte das vezes de reduzida tonelagem.

3109

ES jareta

Cabo que pasa por unas anillas unidas a la relinga inferior por medio de rabizas o pies de gallo. Tirando de él se cierra la parte inferior del arte, evitando de esta manera la evasión de los peces.

DA snurpewire; snurpeline

Den wire eller line, der løber gennem ringe i bunden af en not, og som lukker — snurper — noten sammen i bunden, når stimen er omkredset.

DE Wadenschließleine; Wadenzugleine; Schließleine; Schnürleine; Purseleine

Zum Schließen der Ringwade unten.

GR στίγγος· συστολέας· σχοινί σάκου

Σχοινί, κατασκευασμένο από νήματα ή ατσάλι, το οποίο περνάει μέσα από χαλκάδες που είναι συνδεδεμένοι με το κάτω γραντί ενός γρίπου. Δίνει τη δυνατότητα στο δίχτυ να κλείσει (στιγγάρει) σαν πουγγί και έτσι συγκρατούνται όλα τα συλληφθέντα ψάρια.

EN pursing wire; purse line; purse rope; purse string; draw string

Steel wire running through the pursing rings by means of which the bottom of the net is closed.

3109

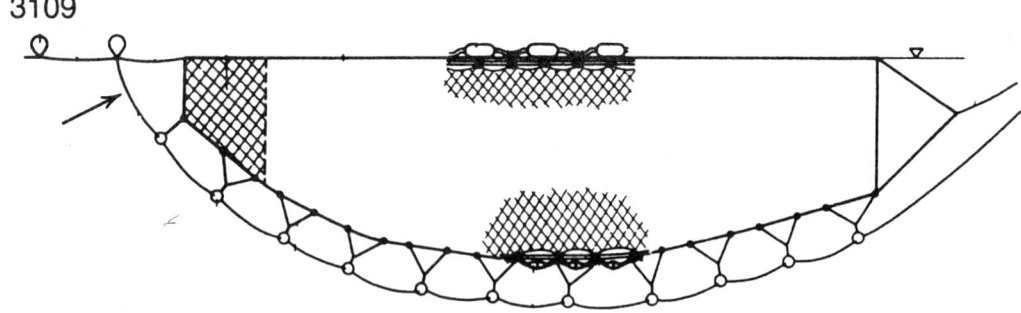

3109

FR coulisse

Cordage en textile ou en acier passant dans des anneaux reliés à la ralingue inférieure de la senne. Assure le boursage du filet qui permet de retenir la totalité du poisson capturé.

IT cavo di chiusura; corda di fondo

Per rete di circuizione.

NL sluitlijn; lijn

Lijn aan de onderkant van een omringend vistuig waarmee dit wordt dichtgetrokken en ingehaald.

PT retenida

Cabo que passa pelo meio das argolas ligadas à tralha dos chumbos das redes de cerco por meio das aranhas, e que ao virar permite fechar a parte inferior da rede evitando, assim, que o peixe se escape.

Em Portugal também se usa este termo para designar um cabo auxiliar de alagem utilizado em certos palangres.

3110

ES pesca con red de cerco de jareta

DA notfiskeri

En fiskerimetode, hvor redskabet, efter at være sat rundt om en stime ved overfladen, snurpes sammen i bunden. Stimen må enten lokaliseres først, f.eks. ved hjælp af sonar, eller lokkes hen til et hjælpefartøj ved lys.

DE Ringwadenfischerei

Fischfangmethode der Küsten- und Hochseefischerei. Die Fischschwärme werden nach dem Feststellen ihrer Entfernung, Richtung und Tiefe vom Fangfahrzeug umfahren, wobei die Ringwade ausgesetzt wird.

Die Ringwadenfischerei läßt dem Fisch so gut wie keine Chance zur Flucht. Sie ist wenig dienlich zur Erhaltung der Bestände.

GR αλιεία με γρι-γρι

EN purse seining

Fishing method using a purse seine, a large, single-panel multi-sectioned net used to encircle pelagic fish, the bottom of which is then drawn together to enclose them.

FR pêche à la senne coulissante

Méthode de pêche qui consiste, dans un premier temps, à localiser le banc de poissons à vue en surface ou au moyen d'appareils de détection acoustique. Parfois aussi, le poisson est concentré sous la surface avant capture en utilisant un procédé d'attraction, par exemple la lumière. Après l'encerclement du banc par la senne, la fermeture du filet par le bas au moyen de la coulisse permet de retenir la totalité du poisson capturé.

IT pesca col cianciolo

NL visserij met ringzegen

PT pesca com rede de cerco com retenida

3111

ES red de cerco con jareta; bolinche*

Red de cerco que se caracteriza por el empleo de una jareta en la parte inferior de la red, que permite cerrarla como una bolsa y retener así todos los peces capturados.

Las redes de cerco con jareta, que pueden ser de gran tamaño, se maniobran desde uno o dos barcos. El caso más común es de una red de cerco de jareta que se maniobra de un solo barco, con o sin embarcación auxiliar.

*Termino local en el País Vasco y en el Cantábrico.

DA not; ringnot; snurpenot

Noten er en netvæg, der sættes rundt om en fiskestime. Når den er sunket ned om stimen, snurpes noten sammen i bunden af snurpewiren. Noter er som regel meget store og sættes almindeligvis fra et enkelt fartøj, evt. med hjælp fra et mindre fartøj. De kan også opereres fra to fartøjer.

DE Ringwade; Beutelnetz; Beutelwade; Ringwadennetz; Drehwade; Schnurwade; Beutelgarn

Netzwand, deren Abmessungen der Größe und Fluchtgeschwindigkeit sowie der Geschwindigkeit der Fangfahrzeuge angepaßt sind. Die Maschenweite entspricht der Fischart und verhindert das Durchschwimmen und Vermaschen der Fische.

GR γρι-γρι

Δίχτυ κυκλωτικό που χαρακτηρίζεται από τη χρήση στίγγου στο κάτω μέρος του διχτυού. Ο στίγγος δίνει τη δυνατότητα στο δίχτυ να κλείσει σαν πουγγί και έτσι συγκρατούνται όλα τα συλληφθέντα ψάρια.

Ο χειρισμός των γρι-γρι, τα οποία μπορεί να είναι πολύ μεγάλα, γίνεται από ένα ή δύο σκάφη. Η πιο συνηθισμένη περίπτωση είναι ο χειρισμός από ένα σκάφος με ή χωρίς βοηθητική βάρκα.

EN purse seine

A large single-panel multi-sectioned net used to encircle pelagic fish, the bottom of which is then drawn together to enclose them.

The purse seines, which may be very large, are operated by one or two boats. The most usual case is a purse seine operated by a single boat, with or without an auxiliary skiff.

FR senne coulissante; senne tournante; bolinche*

Filet tournant caractérisé par l'emploi d'une coulisse à la partie inférieure du filet.

Les sennes coulissantes, qui peuvent atteindre de grandes dimensions, sont manœuvrées par un ou deux bateaux. Le cas le plus courant est la senne coulissante manœuvrée par un seul bateau, avec emploi ou non d'une annexe.

*Terme local.

IT cianciolo; rete da circuizione a chiusura

Rete da circuizione munita di un cavo collegato alla corda dei piombi con anelli per la chiusura della rete nella sua parte inferiore.

NL ringzegen; purse-seine; ringnet; buidelnet; ringzegennet

Omringend vistuig, lang trapeziumvormig, aan de bovenkant voorzien van drijvers en aan de onderkant van lood en ringen waardoor de sluitlijn loopt.

PT rede de cerco com retenida

Rede de cerco caracterizada pelo emprego de uma retenida na parte inferior da rede. A retenida permite fechar a rede como uma bolsa de molde a reter a totalidade do peixe capturado.

As redes de cerco com retenida, que podem atingir grandes dimensões, são manobradas por uma ou duas embarcações. O caso mais vulgar é a rede manobrada por um só barco, com recurso ou não a uma chalandra (embarcação auxiliar).

3111

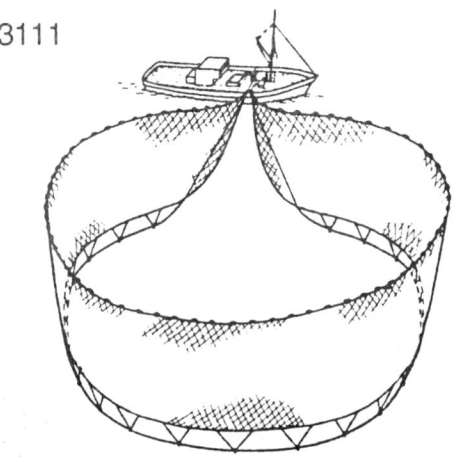

3112

ES relinga superior; relinga alta; relinga de corchos

Cuerda en que van colocados los corchos o flotadores que aseguran la abertura vertical.

DA flydeline; flydetælle

Line monteret med flåd, der sættes på overtællen af et net.
Anvendes på garn og noter.

DE Korkleine

Sperreep mit Korkflotten.

GR σχοινί φελλών

Λεπτό σχοινί όπου τοποθετείται ένας αριθμός πλωτήρων (φελλών) και το οποίο ενώνεται με το επάνω γραντί με ληγαδούρες.
Στάσιμα δίχτυα.

EN float line; float rope

Thin rope on which a number of floats are mounted and then seized to headline.
Set net.

FR ralingue de flotteurs; ralingue de liège; ralingue supérieure

Ralingue supérieure d'un filet (par exemple: filet maillant ou senne) munie de flotteurs de nombre et dimensions variables assurant le déploiement vertical de l'engin.

IT lima da sughero

Lima su cui sono montati i sugheri.

NL drijflijn; drijverspees

Lijn waaraan de drijvers worden bevestigd om de ringzegen of het staande net in de juiste stand to houden.

PT cabo de flutuação; tralha da cortiça; cortiçada; corcho

Cabo onde são colocadas bóias em número variável.

3113

ES red de superficie

DA overfladenet; drivgarn

Drivgarn, der sættes med overtællen helt i overfladen.

DE Oberflächennetz

GR δίχτυ επιφάνειας· αφρόδιχτο

Αλιευτικό δίχτυ του οποίου το σχοινί των φελλών είναι τοποθετημένο στην επιφάνεια της θάλασσας.

EN surface net

Fishing net the upper line of which is set at the sea surface.

FR filet de surface

Filet de pêche dont la ralingue supérieure est maintenue à la surface de la mer.

IT rete da superficie

NL oppervlaktenet

Net dat wordt gebruikt met de bovenpees aan het wateroppervlak.

PT rede de superfície

3112

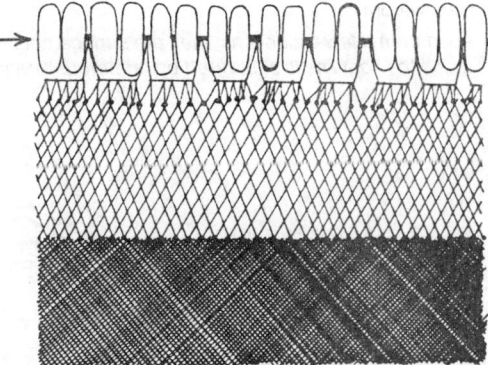

3114

ES red de cerco

Red que captura los peces rodeándolos por los lados y por debajo, evitando de esa manera que, en aguas profundas, bajen a mayor profundidad y escapen.

Con pocas excepciones, se trata de redes de superficie en las que la relinga alta está sostenida por numerosos flotadores.

DA omkredsende net

Samlebetegnelse for net, som fanger fisk ved at indeslutte dem fra siden og fra neden for derved at forhindre dem i at flygte.

Med enkelte undtagelser er det overfladenet, hvor overtællen holdes oppe af talrige flåd, f.eks. not.

DE Umschließungsnetz

Allseitig und am Boden geschlossenes Netz.

Ringwadenartige Konstruktion.

GR κυκλωτικό δίχτυ

Τα δίχτυα αυτά πιάνουν τα ψάρια κυκλώνοντάς τα από τα πλάγια και από κάτω, εμποδίζοντάς τα έτσι να ξεφύγουν σε βαθύτερα νερά, με κατάδυση.

Εκτός από λίγες εξαιρέσεις, είναι δίχτυα επιφάνειας όπου το σχοινί των φελλών υποστηρίζεται από πολυάριθμους πλωτήρες (φελλούς).

EN surrounding net; encircling net

Net which catches the fish by surrounding them both from the sides and from underneath, thus preventing them from escaping in deep waters by diving downwards.

Apart from a few exceptions, they are surface nets in which the float line is supported by numerous floats.

FR filet tournant

Filet qui capture le poisson en l'encerclant à la fois sur les côtés et par en dessous, ce qui l'empêche, en eaux profondes, de s'échapper en plongeant vers le bas.

A part quelques exceptions, ce sont des filets de surface où leur ralingue supérieure est maintenue par de nombreux flotteurs.

IT rete da circuizione

Rete da pesca atta a circondare un tratto di mare nel quale è stato localizzato un banco di pesci che viene catturato con immediata azione di recupero della rete stessa.

NL omringend vistuig; omsluitingsnet; ringnet

Vistuig waarbij het vangen geschiedt door het omsluiten van een watervolume.

PT rede de cercar; rede de cerco

Arte de pesca que captura o peixe envolvendo-o pelos lados e por baixo, o que impede a sua fuga nadando por baixo da rede quando em águas profundas.

São redes de superfície cuja linha de flutuação possui numerosas bóias.

3115

ES **palangrero**

Barco dedicado a la pesca con palangre.

DA **linefartøj**

Fartøj, der er udrustet til at fiske med langliner.

DE **Langleiner; Langleinenfischereiboot**

Fischereifahrzeug, das mit Langleinen fischt.

GR **παραγαδιάρικο**

Σκάφος που αλιεύει με παραγάδι.

EN **long liner**

Vessel fishing with longlines.

FR **palangrier**

Navire pratiquant la pêche à la palangre.

IT **peschereccio armato a palangaro; peschereccio con palangari**

NL **beugschip; beuger; beugvisser**

Schip ingericht voor de visserij met de beug.

PT **palangreiro**

Embarcação destinada à pesca com aparelhos de anzol (palangres).

3116

ES **filástica; hilo de cable**

Producto constituido por uno o varios hilos torcidos o retorcidos para formar una estructura de longitud continua.

Varias filásticas colchadas constituyen un cordón.

DA **garn; kabelgarn**

Et antal spundne tekstilfibre.

Garn anvendes sjældent alene, men samles med et antal andre garn og spindes til en line eller til en kordel. Yderligere samlinger og spindinger frembringer tove og trosser.

DE **Seilgarn; Kabelgarn**

Einfaches Garn oder Zwirn aus textilen Faserstoffen (Natur- oder Chemiefasern), das für die Herstellung von Faserseilen geeignet ist.

Bei der Herstellung von Fasertauwerk wird die Faser zu Garnen (Kabelgarn) gedreht. Mehrere zusammengedreht bilden Kardeele (Litzen), von denen dann drei oder vier zur Trosse und weiter zum Kabel geschlagen werden.

3115

3116

GR κλώσμα σχοινιού· σφιλάτσο

Νήμα ή κλωστή, κατασκευασμένη από έναν αριθμό ινών χαλαρά συστραμένων μεταξύ τους.
Πολλά νήματα σχοινιού συστραμένα μεταξύ τους σχηματίζουν ένα έμβολο.

EN **rope yarn**

A yarn or thread made up of a number of fibres loosely twisted together.
Several rope yarns twisted together form a strand.

FR **fil de cordage; fil de caret; retors**

Élément constitutif d'un toron de cordage, formé de plusieurs filaments assemblés, en général par torsion.
Un cordage est formé de plusieurs torons câblés ensemble, chaque toron étant constitué de fils ou retors assemblés en torsion opposée, chaque fil ou retors comprend un nombre constant de multifilaments (cas des cordages en polyamide).

IT **filo**

Componente di un cavo.

NL **kabelgaren**

Getwijnd samenstel van één of meerdere vezels.
Vezels kunnen eventueel tot strengen van een touw worden geslagen.

PT **filaça; fio de carreta**

Conjunto de fibras de origem vegetal convenientemente torcidas. Um certo número de filaças cochadas constituem um cordão.
No caso de fibras sintéticas denominam-se filamentos; no caso dos aços denominam-se arames.

3117

ES **cordón; cabito; torón**

Producto obtenido por la reunión y torsión de varios hilos o de varios grupos de hilos, retorcidos conjuntamente.

DA **dugt; kordel; part**

En enhed i et tov, der, når den samles med andre dugter og spindes, frembringer et tov. En dugt består af garner, der igen er spundet af naturfibre eller syntetiske fibre.

DE **Litze**

Eine Litze ist das in der ersten Verseilstufe durch Verseilen oder Verflechten von Seilgarnen hergestellte Erzeugnis.

GR έμβολο σχοινιού· κλώνος· έμπουλο

Αριθμός ινωδών νημάτων ή συρμάτων, συστραμένων μεταξύ τους. Τρία ή περισσότερα έμβολα συστραμένα μεταξύ τους σχηματίζουν ένα σχοινί.

EN **strand; primary strand**

One of the component parts which when twisted or laid up together form a rope. Strands are made of yarns or wires and are twisted together in a direction opposite to that in which the yarns are twisted.

FR **toron**

Plusieurs fils ou retors assemblés par torsion forment un toron; plusieurs torons câblés ensemble constituent un cordage.

IT **legnolo**

Prodotto tessile ottenuto dall'unione di più fili o insieme di fili ritorti tra di loro ed unicamente destinato alla successiva operazione di commettitura o trecciatura.

NL **kardeel; streng**

Een uit een aantal garens samengedraaide bundel touwwerk waaruit een lijn of tros wordt geslagen.

PT **cordão**

Conjunto de fios de carreta cochados convenientemente. Três ou quatro cordões enrolados sobre si mesmos formam um cabo ou linha.

3118

ES eslabón aplastado; eslabón plano

DA huggeled; fladled

Ovalt led med et tyndt område, hvor et G-led kan glide ind.
Anvendes, hvor der er behov for en hurtig samling mellem tove, kæder og/eller wirer. Eks.: samlingen af slæbewiren med kæderne på en skovl.

DE Quetschglied

Formschlüssiges Verbindungselement zum lösbaren Verbinden von Ketten und Seilen. Die Verbindung, die unter gleichzeitiger Verwendung von C- oder G-Haken erfolgt, ist bei Entlastung leicht und ohne zusätzliche Werkzeuge zu lösen oder herzustellen.
Quetschglieder werden häufig bei Schleppnetzen zum Herauslösen der Scherbretter aus dem Bereich Kurrleine/Vorgeschirr beim Hieven (Trawler) verwendet.

GR κρίκος εσοχής

Σφυρηλατημένος, ωοειδής κρίκος με εσοχές που επιτρέπει τη σύνδεση με γάντζο-G.

EN recessed link; flat link

Forged oval link with recesses to permit engagement with a G-hook.

FR maille à méplat

Maille ovale forgée comportant une ou deux portions d'épaisseur moindre permettant son assemblage avec un croc en G.

IT maglia per gancio a G

NL kenterpatentschalm; patentschalm; kenterschalm

Losse schalm waarmee kettinglengten aan elkaar gesloten worden.
Deel van een ankerketting, of van een voortuig van een trawl.

PT elo de ligação

Peça de ferro ou aço com forma elíptica, rebaixada nos seus pontos médios, e destinada a ligar a um gato.

3119

ES gancho G

Pieza de hierro o acero en forma de «G» para unirse a un eslabón con rebajo.

DA G-led

G-formet jernled.
Anvendes sammen med et huggeled.

DE G-Haken

G-förmiger Patenthaken.

GR γάντζος-Γ· κρίκος-Γ

Σφυρηλατημένος γάντζος σχήματος Γ για σύνδεση με κρίκο εσοχής.

EN G-hook; G-link

Forged G-shaped hook to connect with a recessed link.

FR croc en G

Croc en forme de G à connecter avec une maille à méplat.

IT gancio a G

NL G-haak

Verbindingshaak in de vorm van een „G", die onbelast op een gemakkelijk uitneembare wijze past in een bijbehorende schalm.
Voor het snel los- of vastmaken bij halen of vieren.

PT elo em G; gato

Peça de ferro ou aço em forma de G que engata num elo de ligação.

3120

ES ayuste; costura

Costura o unión de dos cabos o cables por sus chicotes. Dícese del conjunto de los cabos o cables ayustados.

DA splejsning

Fast forbindelse mellem to tovender dannet ved at flette kordelerne ind i hinanden.

DE Spleiß

Feste Verbindung zweier Tauwerkenden durch Ineinanderflechten der Kardeele.

GR μάτιση

Ένωση σχοινιών, που γίνεται από πλεγμένα άκρα εμβόλων.

EN splice

Join in rope made by intertwining ends of strands.

FR épissure

Jonction de deux cordages par entrelacement de leurs torons.

IT impiombatura; gassa; impalmatura

Attaccatura di due cavi, di canapa o di acciaio, o di due spezzoni di cavo, o dell'estremità di un cavo ripiegata per fare una gassa con il cavo stesso, intrecciando fra loro i legnoli in modo da eliminare il pericolo che i cavi si sfilino.

NL splits

Blijvende verbinding in touw of staaldraad, verkregen door de strengen van de verschillende einden in elkaar te vlechten.

PT costura

Junção de duas extremidades (chicotes) de um cabo por entrelaçamento dos respectivos cordões.

3120

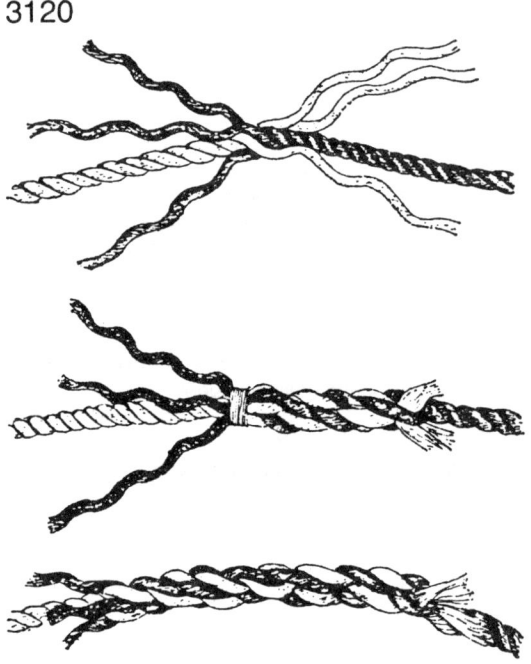

3119   SWIVEL TOWING CHAIN WITH RECESSED LINK

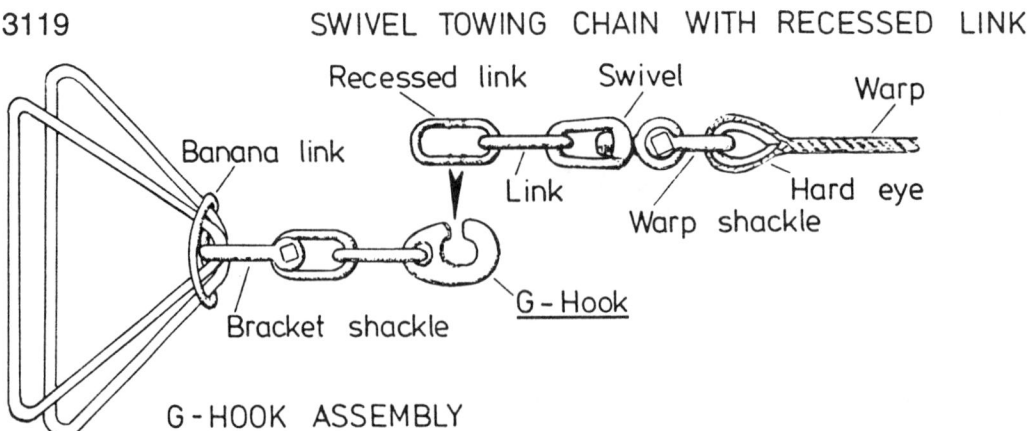

G-HOOK ASSEMBLY

3121

ES rodete

Pieza de red fijada en el interior del copo en su extremo posterior. El rodete podrá replegarse dentro del copo del arte de arrastre.

Está destinado a lograr un cierre más perfecto del copo del arte de arrastre por el rebenque.

DA slør; torquette

Et stykke net fastgjort tæt ved de sidste masker i fangstposen på en trawl.

Stoppes ind i posen, når bindestrikken snøres inden fiskeri, og gør derved posens lukning mere effektiv ved at danne en prop.

DE Einlage

Ein im Innern des Steerts an dessen hinterem Ende angebrachtes Stück Netzwerk. Kann in den Steert zurückgefaltet werden.

Zweck einer Einlage ist es, das Verschließen des Steerts durch die Steertleine zu verbessern.

GR

Κομμάτι δικτυώματος στερεωμένο μέσα στο σάκο, στο πίσω άκρο του. Μπορεί να είναι αναδιπλωμένο μέσα στο σάκο.

Σκοπός του είναι η βελτίωση του κλεισίματος του σάκου από γάιδαρο.

EN torquette

A piece of netting fixed inside the codend at its rear end. It may be folded back into the codend.

Its purpose is to improve the closing of the codend by the codline.

FR torquette

Pièce de filet fixée à l'intérieur du cul à son extrémité arrière. Elle peut être repliée dans le cul du chalut.

Destinée à améliorer la fermeture du cul du chalut par le raban de cul.

IT torquette

Pezza di rete fissata all'interno del sacco della rete, nell'estremità posteriore di quest'ultimo. Essa può essere ripiegata nel sacco della rete.

Serve a migliorare la chiusura del sacco per mezzo della sagola di chiusura.

NL dotje

Stuk netwerk dat binnen de kuil aan het achtereinde daarvan is aangebracht.

Wordt gebruikt om de kuil beter (met de pooklijn) af te sluiten.

PT pano de rede livre

Pano de rede de fio forte fixado no interior da parte final do saco das redes de arrasto. Tem por finalidade tornar mais eficiente o fecho do cu do saco.

Desconhece-se o seu uso em Portugal.

3122

ES guardacabo

Anillo de madera o metálico, acanalado por la superficie exterior para que pueda ajustársele un cabo. Los hay de forma circular y de corazón o alargados, y sirven para pasar otro cabo o cable por dentro sin que roce al anterior o bien para enganchar un aparejo.

DA kovs

En jernring, ofte tilspidset, hvis yderside er konkav. Lægges inden i en øjesplejsning, hvor den forstærker og forhindrer slid.

DE Kausch

Metallschutz in den Augen von Tauwerk. Das Tauwerk wird in die Keep (Rille) der Kausch gelegt und fest verspleißt.

GR ροδάντσα

Μεταλλικός δακτύλιος με κοίλη πλευρά μέσα στην οποία μπορεί να ματίζεται ή να δένεται με ληγαδούρα ένα σχοινί. Συνήθως είναι από γαλβανισμένο σίδερο, ατσάλι, ορείχαλκο ή ερυθρό ορείχαλκο.

EN thimble

Metal ring, with concave side into which a rope may be spliced, or seized. Usually of galvanized iron, steel, brass, or gunmetal.

FR cosse

Pièce métallique, généralement en forme de cœur ou d'anneau, comportant à son pourtour une gorge destinée à recevoir le cordage.
Sert à protéger un œil épissé quand on le fixe avec une manille.

IT redancia; radancia

Rinforzo metallico delle gasse.
Serve per inserire nella gassa un gancio o un grillo.

NL kous

Metalen beslag of voering in een oogsplits of gebindseld oog, dienende om het oog van binnen tegen schavieling te beveiligen en het in de gewenste vorm te houden.

PT sapatilho

Artefacto constituído por um aro metálico fechado e com forma oval ou circular, tendo meia cana na periferia.
Destina-se a proteger internamente uma mãozinha.

3122

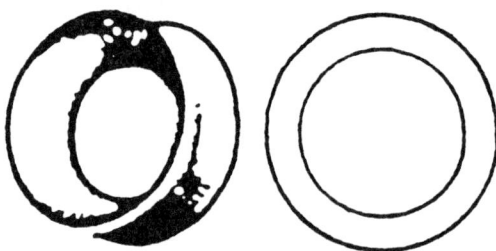

3123

ES grillete

Pieza de acero forjado, en forma de U, atravesada por un perno, que sirve para unir dos trozos de cadena o dos elementos de jarcia.

DA sjækkel

En bøjle af metal, ofte galvaniseret jern, der kan lukkes med en bolt. Udstrakt anvendelse ved alle former for samlinger af kæder, tove og wirer.

DE Schäkel

Universell einsetzbares, U-förmiges Verbindungs- und Befestigungsglied aus Metall (meist feuerverzinkter Stahl) für Tauwerk und Ketten.

GR κλειδί· αγκύλιο

Κρίκος σχήματος αγκύλης ο οποίος είναι κλειστός με έναν πείρο (καβίλια). Χρησιμοποιείται για να ενώνει δύο κομμάτια σχοινιού ή αλυσίδας.

EN shackle

'U' shaped steel forging with a pin through an eye on each end of the 'U' which serves as connecting links for rigging components.

FR manille

Pièce d'acier en forme d'étrier, munie d'un axe à vis, servant à la liaison de deux chaînes ou cordages ou à leur fixation sur une pièce fixe ou mobile.

IT grillo; maniglia; gambetto; maniglione

Semianello chiuso da uno spinotto generalmente avvitato.

Usato per congiungere tra di loro le varie parti dell'attrezzatura di pesca.

NL sluiting; harp

U-vormige smeedstalen schalm of hoefijzervormige beugel met inschroefbare of ingestoken sluitbout om einden aan elkaar of aan het scheepsverband te verbinden.

PT manilha

Peça com forma de U, D ou em raco, de ferro forjado ou aço, tendo um cavirão através da garganta, usada para ligar as quarteladas de uma amarra, ou para prender outros artefactos.

O cavirão pode ser de rosca, travão ou cavilha.

3123

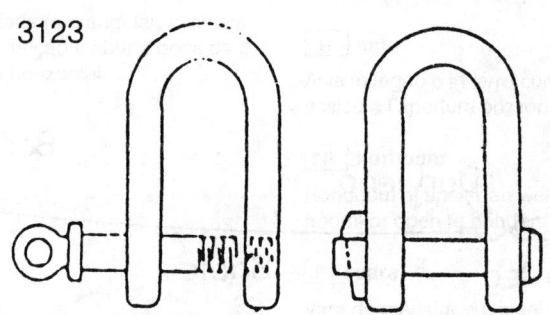

3124

ES ocho; ocho deslizante

Pieza de hierro con dos orificios, que se coloca en el extremo anterior de la malleta, allí donde está en contacto con la puerta, para servir de unión entre ésta y aquélla.

DA brille; kameløje; samleled

8-tal-formet metalled. Det lille øje samler bagstropperne fra skovlen, det store tillader mellemlinen at glide igennem indtil stopperen, der er mellemlinens samleled med indhalertampen.

DE Ochsenauge

Durchzugring am Scherbrett.

GR οκτώ

Σφυρήλατο ατσάλινο κλειδί σχήματος 8, του οποίου ο μικρότερος δακτύλιος συνδέεται με τον οπίσθιο στρόφο και ο μεγαλύτερος, μέσω του οποίου περνά το χοντρό συρματόσχοινο, σταματά τον εμπλοκέα στο μπροστινό μέρος του χοντρού συρματόσχοινου.

EN Kelly's eye

8-shaped steel forging, the smaller ring for attachment to the backstrop, the larger through which passes the bridle, to arrest the stopper at the fore end of the bridle.

FR huit; craquelin

Pièce forgée en forme de huit, fixée à l'extrémité des pattes de panneau; le bras coulisse dans l'un de ses trous où il vient se bloquer en traction par une autre pièce forgée appelée stoppeur.

IT anello ad otto

NL bril; oog

Oog dat past in een speciale stopper en dat het doorvoeren van kabels en kabelverbindingen mogelijk maakt, zodat ontkoppelen niet nodig is.
Wordt aangebracht aan de verbindingsdraden; wordt ook gebruikt voor het doorwinden van de voorlopers.

PT argolão; oito

Peça de ferro com dois orifícios de diferente diâmetro; o orifício menor para ligação aos brincos da porta, e o maior, através do qual passa a malheta, para servir de batente ao elo em oito que faz a ligação entre o brinco da malheta e a malheta.

3124

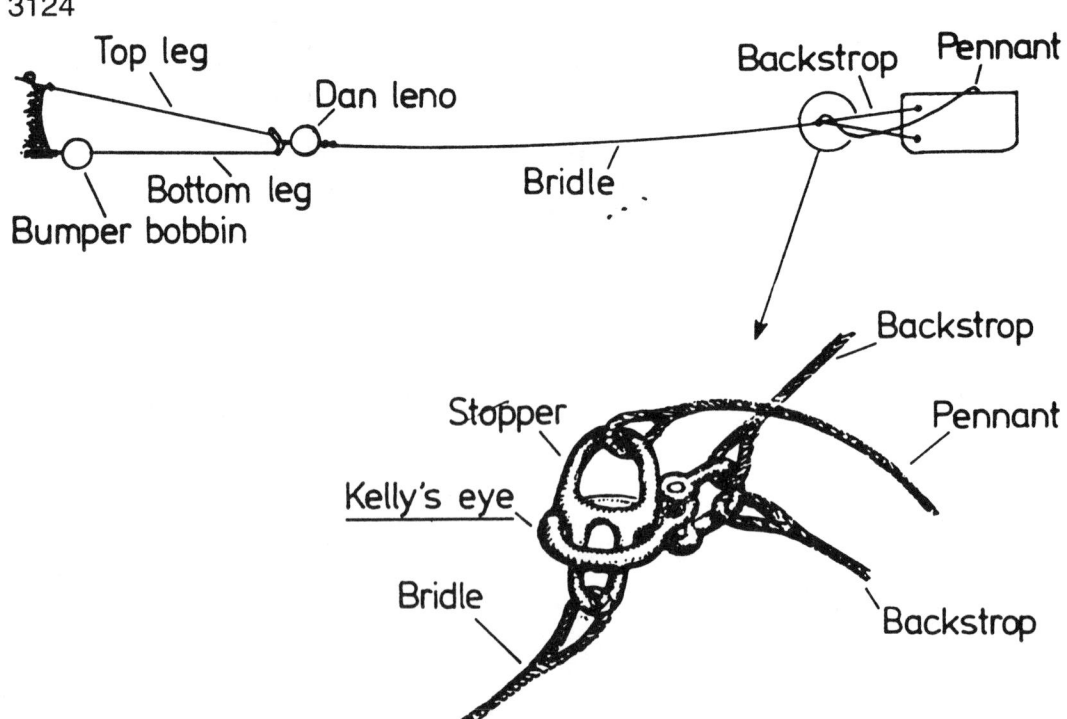

3125

ES calón de vara; calón

Palo de aproximadamente un metro de largo, con el que se mantienen extendidas las redes de pesca.

DA stav

Kort jern- eller træstav sat ind mellem stjerterne for at give afstand og begrænse rivninger i vinger og arme.
Trawl, snurrevod, strandvod.

DE Knüppel

An den Flügelenden senkrecht angebrachtes, unten beschwertes Holz, um das Zusammenfallen des Flügels zu verhindern.
Bei Netzen mit Rollergrundtau ist der Knüppel durch ein kleineres Scherbrett, das Pony, ersetzt.

GR ματσόξυλο· σταλίκι

Ξύλινο ερματισμένο κοντάρι με κοντά σχοινιά συναρμολογημένα σ'αυτό.
Γρίπος ακτής· δανέζικη τράτα.

EN Dan leno stick; spreader

Dan leno in form of ballasted wood pole with short rigging ropes attached.
Trawl; beach seine; Danish seine.

FR guindineau

Pièce de forme variable, généralement en fer forgé, placée à la liaison entre le bras et les entremises.
Gréement de chalut de fond.

IT asta della mazzetta; asta della stazza

Parte della mazzetta della rete a strascico in legno o ferro che mantiene distanziata la linea da sughero da quella da piombo.

NL knuppel

Bij een trawlnet gebruikte houten of ijzeren staaf, die dient om de stroppen aan de vlerken van het net verticaal op afstand te houden.
Wordt ook wel rechtstreeks aan de vlerken bevestigd en dient dan om het voorste uiteinde van het net verticaal gestrekt te houden.

PT calão

Peça de aço ou madeira com forma variável situada entre os cabos de alagem de uma rede de arrasto e os cabos superior e inferior de cada uma das asas da rede e que tem por finalidade manter estes últimos afastados um do outro.
Redes envolventes-arrastantes e redes de arrasto pelo fundo.

3125

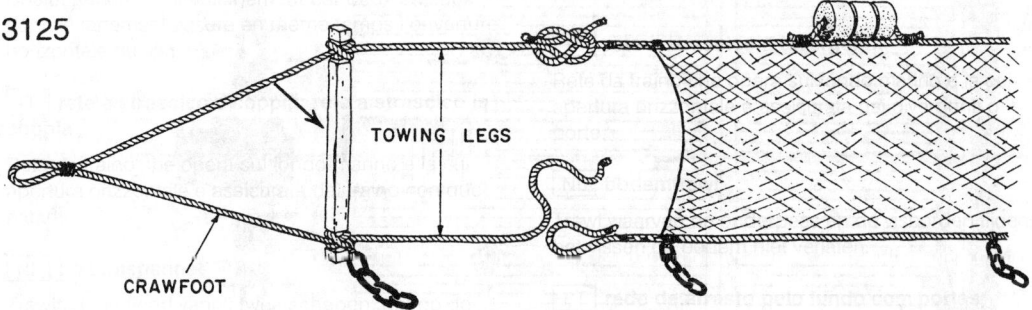

TOWING LEGS

CRAWFOOT

3126

ES carretel de la maquinilla; tambor de la maquinilla

DA spiltromle; wiretromle

Tromle, hvorpå trawlwire e.l. rulles op.

DE **Kurrleinentrommel**

Kuppelbare Speichertrommel, begrenzt durch Trommelschilder zur Aufnahme der rechten bzw. linken Kurrleine. Die mechanisch angetriebene Aufleitvorrichtung gewährt ein geordnetes Aufwickeln der Leine.

GR τύμπανο βαρούλκου· τύμπανο βιντσιού· κύλινδρος βαρούλκου· κύλινδρος βιντσιού

EN **winch barrel; warp drum; barrel; drum**

Barrel of winch on to which the warp is wound.

The trawl winch is provided with two large barrels, one for each warp, which can be worked together or separately by the action of clutches.

FR **bobine de fune; bobine de treuil du chalut; tambour de la fune; tambour de treuil**

Bobine sur laquelle est enroulée la fune.

Un treuil de chalutier comporte en général deux bobines qui peuvent être accouplées ou scindées; pour certains types de chaluts (gréement double, chaluts jumeaux), le treuil peut comporter trois bobines ou davantage.

IT **tamburo avvolgicavo**

NL **vislijntrommel; liertrommel**

Onderdeel van een lier waarop de vislijnen, breidels en andere kabels worden gewonden.

PT **tambor do cabo real**

Trata-se dos tambores dos guinchos dos cabos reais e onde estes são enrolados.

3126

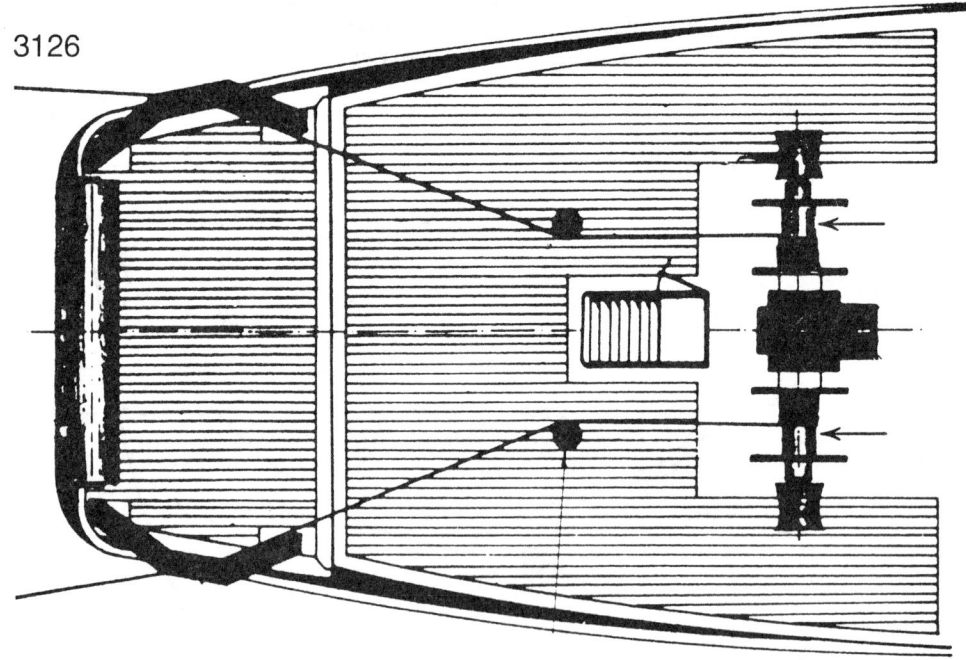

3127

ES | brazo; triángulo

Las puertas en su cara interna llevan unos elementos, generalmente de tubo metálico y de forma triangular. Estos brazos o triángulos se unen a la puerta por su base y de forma que puedan girar libremente sobre la misma.

DA | bøjle

Trekantet jernbøjle, hængslet til skovlens forside. En eller to af disse tjener til fastgørelse af slæbewiren.
Kan anvendes i stedet for jernkæder.

DE | Bügel

Befestigungsstück für die Kurrleine, das an der Scherbrettinnenseite angebracht ist. Der große Bügel ist auf halber Brettlänge befestigt, der kleine Bügel in Zugrichtung der Kurrleine vor dem großen Bügel.

GR | μπράτσο· ορθοστάτης

Ένας ή δύο ατσάλινοι σκελετοί τριγωνικού σχήματος που αναρτώνται στο μπροστινό μέρος της πόρτας, όπου είναι συνδεδεμένο το χοντρό συρματόσχοινο.

EN | bracket; triangular bracket

One, or a pair of, triangular shaped steel frames hinged to the front face of otter board, to which the warp is attached.

FR | branchon; braguet, braquant

Pièce métallique, en général de forme triangulaire, fixée au panneau de chalut et sur laquelle se fixe la fune.

IT | triangolo del divergente

NL | beugel

Scharnierend driehoekig ijzer aan de drukzijde (binnenzijde) van het visbord ter bevestiging van de vislijn.
Bij platte houten borden worden er twee gebruikt, parallel aan elkaar, met de punten tegen elkaar.

PT | triângulo da porta de arrasto; triângulo

Cada uma das peças metálicas com forma de triângulo articuladas na face anterior da porta de arrasto e às quais se liga o cabo real.

3127

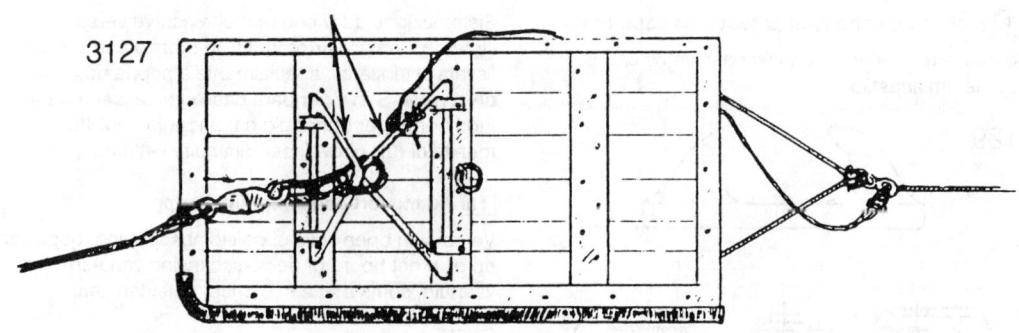

3128

pasteca del pescante

Pasteca a través de la cual pasa el cable en el pescante de un buque de arrastre.

DA **trawlblok**

Den specielt udformede blok, der er placeret i trawlgalgen og fører slæbewiren ud over fartøjets side.

DE **Scherblock; Kurrleinenblock**

Stahlkonstruktion mit wesentlich längeren Backen als ein normaler Block. Die Blockscheibe liegt bei schweren Ausführungen in Wälzlagern gelagert. Die Kiep ist dem Leinenprofil angepaßt. Aufhängung mit Wirbel im Galgenauge.

GR **τροχαλία· μακαράς· μπαστέκα**

EN **warp block; warping block**

A block used to guide the fishing line.

FR **poulie de fune; poulie de potence**

Poulie fixée à la potence de chalutage, dans laquelle passe la fune.

IT **bozzello dell'archetto; pastecca dell'archetto**

NL **galgblok**

Blok ter geleiding van de vislijn.

PT **moitão do cabo real; patesca do cabo real**

Patescas ou moitões nos quais passam os cabos reais de um arrastão.

3129

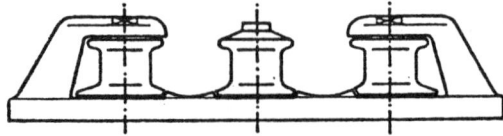

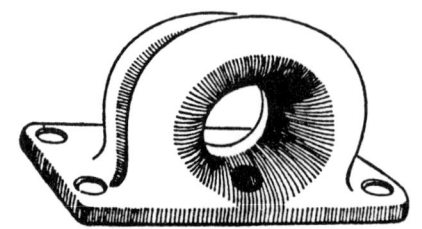

3129

ES **galápago; guía; alabante**

Pieza de madera o metálica que, bien afirmada, sirve de guía.

DA **klyds; fodblok**

Indretning af metal til at lede en wire eller et tov. To bøjler (klyds) eller blokke (fodblok) fører tovet eller wiren.

DE **Lippklampe; Verholklampe**

Vorrichtung zur besseren Führung der Leinen, häufig mit Rollen versehen.

GR **οδηγός συρματόσχοινου· τονοδηγός**

Βαρύ ξύλινο ή μεταλλικό εξάρτημα στερεωμένο σε κατάστρωμα ή αποβάθρα, το οποίο έχει σιαγόνες απ' όπου περνά σχοινί ή καλώδιο και χρησιμοποιείται σαν οδηγός συρματόσχοινου.

EN **chock; cable chock; fairlead**

A heavy wooden or metal fitting secured on a deck or on a dock, having jaws through which line or cable passes, and which can serve as a fairlead.

FR **chaumard**

Pièce de guidage pour les amarres, dont toutes les parties présentent des arrondis pour éviter d'user ou de couper les cordages.

IT **passacavo; bocca di rancio**

Sistemazione ad U con branche ricurve verso l'interno e spigoli arrotondati, in bronzo o in acciaio, fissata ai trincarini, specialmente a poppa ed a prua della nave. Serve per dare passaggio ai cavi da dare fuori bordo per ormeggio o tonneggio; talvolta è munita di rulli girevoli per diminuire l'attrito.

NL **kam; verhaalklamp; lipklamp**

Van boven open kluis of geleidingsopening, geplaatst op of in het boord of de verschansing van een vaartuig, soms met een of meer geleiderollen.

PT **castanha**

Peça metálica, segura no convés, de boca aberta ou fechada, servindo para guia de cabos. Pode ter um rolete vertical, a meio, para diminuição do atrito nos cabos.

3130

ES cable; cable de arrastre

DA trawlwire; wire; slæbewire

Stålwire, som trawlen slæbes i.

DE Kurrleine

Trosse aus Stahldraht oder Fasermaterial zum Aussetzen, Schleppen und Einholen eines geschleppten Fanggeräts (Schleppnetz).

GR συρματόσχοινο τράτας

Μακρύ και ευλύγιστο συρματόσχοινο που συνδέει το σκάφος με την τράτα.

EN warp; trawl warp

Long flexible steel rope connecting vessel to the trawl gear.

FR fune

Câble d'acier servant à remorquer le chalut.

IT cavo di traino

Cavo di acciaio per il traino della rete.

NL trawllijn; vislijn

Staaldraad in gebruik op trawlers en kotters om het vistuig, bestaande uit visborden, voorlopers, breidels en sleepnet, door het water te slepen.
Wordt ook bij boomkorren gebruikt, vaak dubbel ingeschoren.

PT cabo real

Cabo de aço longo e flexível que liga a embarcação (arrastão) às portas de arrasto.

3131

ES gancho de escape; gavilán

DA slipkrog

En hængslet krog for enden af en kæde eller et tov. Tjener som samleled til en anden kæde. En sikkerhedsring låser krogen fast, så den først, når ringen tages, bryder samlingen.
Anvendes ved udsætningen af trawlen, hvor den fastgøres, når den er sat ud og skal strækkes op. Slipkrogen sidder da på en kæde på fartøjet og udløses, når opstrækningen er tilendebragt.

DE Sliphaken; Pelikanhaken

Haken, der nach Öffnung einer Sicherung unter Belastung aushakt.

GR κόρακας εκφυγής· σκύλα· άγκιστρο ολίσθησης

Αρθρωτός κρίκος που επιτρέπει την εύκολη αποσύνδεση του συρματόσχοινου ή της αλυσίδας που περνά από μέσα.

EN slip hook; pelican hook; tripping link*; quick release hook

A long-shank, hinged hook with a link over the beak. When the link is knocked off, the hook collapses. It is used for chain plates, cable stoppers, boat gripes, clear-hawse pendants, and so on.
* Term used in connection with 'tripping line'.

3130

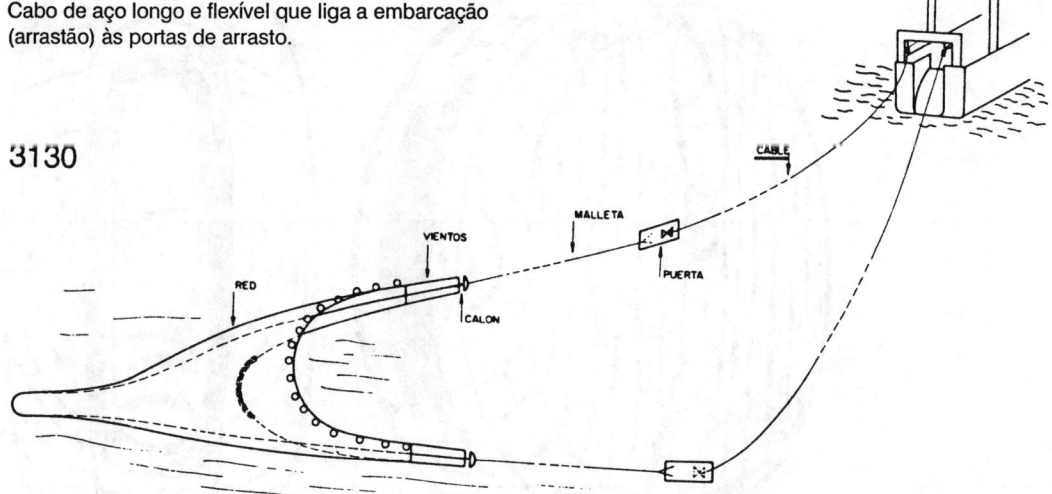

109

3131

FR croc à échappement; croc à largage rapide

Croc articulé permettant de dégager aisément le cordage ou la chaîne qui s'y trouve logé.

IT gancio a scocco

NL sliphaak

Haak met speciale constructie die het mogelijk maakt de haak te openen wanneer er spanning op staat.

PT gato de escape

Gato articulado que permite desengatar com facilidade cabos ou correntes que liga entre si.

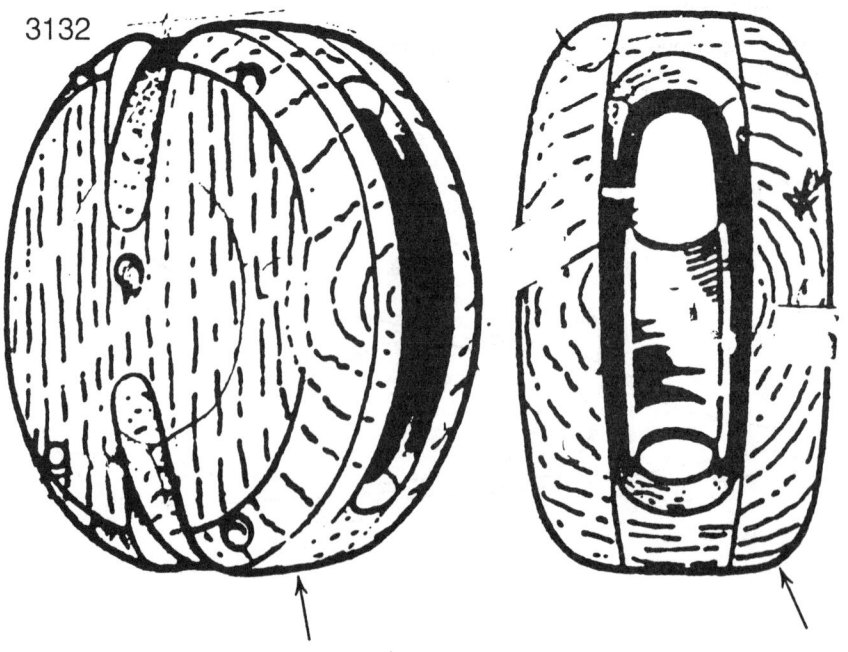

3131

3132

ES quijada

Cada uno de los dos lados entre los cuales median la cajera de un motón o las de un cuadernal.

DA

Siden på et blokhus.

DE Backe

Tragendes Blockteil für den Schäkelbügel und die Seilscheibe.

GR πλαϊνό τοίχωμα· μάγουλο· παρειά

Ένα από τα δύο πλαϊνά μιας τροχαλίας.

EN cheek

One of the two sides or framing of a block.

FR joue

Face extérieure de la caisse d'une poulie.

IT maschetta; guancia

Ognuna delle due facce di un bozzello fra le quali gira la puleggia.

NL wang

Houten of kunststof zijplaat van een blok die om het geraamte heen is bevestigd.

PT face do moitão

Lado exterior da caixa de um moitão.

3132

3133

ES pescante; potencia

Pesada pieza de hierro en forma de uve invertida y de ángulo redondeado, del que pende una roldana.

DA trawlgalge

David eller galge til at løfte trawlskovlene ud af vandet. I toppen af galgen hænger trawlblokken. Der forefindes normalt to galger på en trawler, placeret i den ene side (sidetrawlere) eller på hver side agter (hæktrawlere).

DE Galgen

Bogen- oder T-förmiger Träger auf Trawlern, an denen der Kurrleinenblock hängt. Bei Seitenfängern ist am Fußpunkt eine Umlenkrolle angebracht.

GR καπόνια

Τέσσερα ισχυρά εξαρτήματα τοποθετημένα στις δύο πλευρές μιας μηχανότρατας, τα οποία σχηματίζονται από μία δοκό σε σχήμα διατομής η οποία προσαρτάται με μπρατσόλια στο κατάστρωμα. Τα καπόνια χρησιμοποιούνται για να ανασηκώνουν τις πόρτες όταν η τράτα βρίσκεται σε λειτουργία. Κάθε καπόνι είναι εξοπλισμένο με ένα βαρύ ατσάλινο μακαρά στην κορυφή του, απ' όπου περνά το συρματόσχοινο της τράτας.

EN trawl gallows

Four strong fittings placed on both sides of a trawler, generally formed by an H bar with bracket attachments to the deck. The purpose of the gallows is to raise the otter boards when working the trawl. Each gallow is provided with a heavy steel pulley at the top, through which the trawl warp is rove.

FR potence de chalut

Désigne les deux supports placés de chaque côté en abord à l'arrière du chalutier. Ils soutiennent les poulies dans lesquelles passent les funes remorquant le chalut.

IT archetto del divergente

NL galg

Davit voor de ophanging van het trawlblok waarover de vislijn naar het visbord loopt.

PT aro de pesca de arrasto

Pesada peça de ferro em forma de U aberto e invertido da qual pende o moitão por onde passa o cabo real.

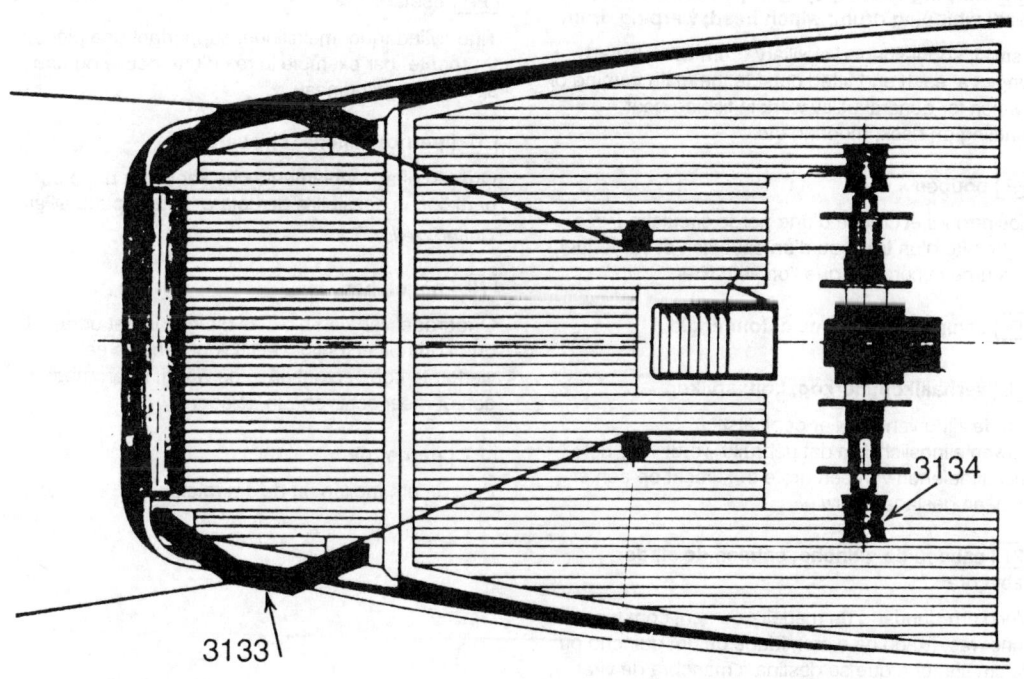

3134

3133

3134

ES capirón; muñón de la maquinilla; virador

Pieza de acero, en el extremo del eje de una maquinilla, para tomar vuelta a los cabos o cables y poder virarlos.

DA spilkop

Kort spole anbragt på ydersiden af et spil. Talrige anvendelsesmuligheder ved indhaling af liner, løft af last og lignende.

DE Spillkopf; Windenkopf; Verholkopf

Angebracht am äußeren Wellenende einer Winde. Vielseitige Verwendung zum Einholen von Leinen, Heben von Lasten über einen Block oder zum Verholen des Schiffes.

GR εργατόκρανο· κεφαλάρι· ράουλο βιντσιού· αλυσέλικτρο· κεφαλή αλυσέλικτρου· κεφαλάρι βαρούλκου

Μικρό βοηθητικό τύμπανο σχήματος καρουλιού με ταινιοειδείς φλάντζες σε κάθε άκρο, εφαρμοσμένο εξωτερικά του κύριου σκελετού ενός βαρούλκου, βαρούλκου αγκυρών ή βαρούλκου πόντισης που χρησιμοποιείται κατά το χειρισμό συστημάτων ανύψωσης και γενικά σε χρήσιμες εργασίες κατά την πόντιση και άλλες παρεμφερείς λειτουργίες.

EN warping end; gipsy; gipsy head; warping head; whipping drum; winch head; warping drum

A small, spool-shaped auxiliary drum with filletted flanges at each end fitted outside the main framing of a winch for general utility work in fishing gear handling on deck, mooring etc.

FR poupée

Bloc arrondi et creusé d'une gorge circulaire fixé à l'extrémité d'un treuil ou d'un cabestan et sur lequel on tourne le cordage que l'on veut virer.

IT campana; campana di tonneggio

NL verhaalkop; lierkop; kop; spilkop

Aan de zijde van een lier geplaatst omwentelingslichaam dat gebruikt wordt voor het snel uitoefenen van een grote trekkracht op een er omheen geslagen touw.

PT cabeço do guincho; cabeço do alador; cabeçote

Bloco arredondado de madeira ou aço, com faces côncavas, fixado na extremidade de um guincho ou de um alador e que se destina à manobra de virar cabos.

3135

ES perno de motón; paja

Especie de clavo largo y grueso de hierro, acero, latón, u otro material. Por un lado los pernos tienen cabeza y por el otro tuerca o chaveta.

DA bloknagle; skivebolt

Metalaksel, hvorom skiven i en blok drejer.

DE Blockbolzen

Runder Metallbolzen mit einem als Sicherung in die Backe eingelassenen Vierkantkopf an einem Ende sowie einem Gewinde mit Mutter am anderen Ende. Der Bolzen trägt die Seilscheibe.

GR πείρος

Μεταλλικός άξονας πάνω στον οποίο περιστρέφεται το ράουλο μιας τροχαλίας. Συνήθως είναι κατασκευασμένος από μαλακό χάλυβα και έχει τέτοιο μέγεθος ώστε να αντέχει σε διατμητική τάση ίση τουλάχιστον με την τάση θραύσης του σχοινιού.

EN pin

The metal axle upon which the sheave of a block revolves. It is usually made of mild steel and of such size as to withstand a shearing stress at least equal to the breaking load of the rope.

FR essieu; axe

Tige cylindrique, métallique, supportant une pièce tournante, par exemple le réa d'une poulie ou une sphère de guindineau.

IT perno; bullone; chiavarda

Organo di accoppiamento che serve per unire due parti con la possibilità di ruotare l'una rispetto all'altra. *Del bozzello.*

NL nagel; pen

Onderdeel van een blok; as die rust in het geraamte en/of omhulsel (huis), veelal voorzien van een vierkante kop verzonken in de wang, waaromheen een of meer schijven zijn gelagerd.

PT perno; pino

Eixo de um moitão ou de um rolete.

3136

ES roldana

Rueda de madera o metal sobre cuya periferia gira el cabo de los cuadernales o motones y en cualquiera otra cajera destinada al laboreo de un cabo.

DA skive; blokskive

Skive med konkav kant, hvorover tovet løber i en blok.

DE Scheibe; Blockscheibe

GR ράουλο

Αυλακωτός τροχός σε μία τροχαλία, άρμουρο, αντένα κλπ., πάνω στον οποίο περνά ένα σχοινί. Το ράουλο, το οποίο έχει επενδυθεί με αντιτριβικό δακτύλιο, περιστρέφεται πάνω σε πείρο. Τα ράουλα είναι κατασκευασμένα από ξύλο, μπρούντζο, γαλβανισμένο χυτοσίδηρο ή ατσάλι. Στην περίπτωση των εκφορτωτήρων και των συστημάτων ανύψωσης, χρησιμοποιούνται μπρούντζινα ράουλα διότι αναμένονται βίαιες ή ισχυρές τάσεις κατά διαστήματα.

EN sheave

A grooved wheel in a block, mast, yard, and so on, over which a rope passes. The sheave, which is bushed, rotates upon the pin. Sheaves are made of wood, bronze, or galvanized cast-iron or steel. For running rigging where severe or heavy intermittent strains are expected, as in the case of runners and topping lifts, brass sheaves are used.

FR réa

Pièce tournante des poulies, palans, clans et chaumards; son pourtour est creusé pour recevoir le cordage.

Très souvent, le terme désigne la poulie de potence.

IT puleggia

Rotella in legno duro, bronzo o ferro che intorno alla circonferenza ha un incavo per alloggio della cima che vi deve lavorare ed al centro un foro per il passaggio di un perno.

Parte del bozzello.

NL schijf

Al of niet gegroefd wiel van een blok waarover de lijn loopt.

PT roda do moitão

Roda escavada de um moitão na qual corre um cabo que muda de direcção. Pode ser de madeira, ferro ou bronze.

3137

ES cáncamo

Sirve para afirmar el motón o aparejo.

DA fæstepunkt for en blok

DE Blockaufhängung

Vorrichtung zur Aufhängung eines Blocks mittels Schäkel oder Wirbel.

GR ανάρτηση τροχαλίας

EN block suspension

Device on which the block is hung.

FR suspension de poulie

IT punto di aggancio della pastecca; sospensione del bozzello

NL galgblokhanger

Deel van een blok waaraan dit wordt opgehangen.

PT alça de suspensão do moitão

Aro metálico ou alça de cabo de fibra ou aço que se destina à fixação de moitão a um ponto de apoio.

3138

ES motón; cuadernal; polea; garrucha

Denominación marinera de las poleas por donde pasan los cabos y que sirven para cambiar la dirección del movimiento de éstos.

Los motones pueden ser de madera, metal o plástico, y reciben este nombre cuando tienen una sola roldana, porque los de dos o tres se denominan cuadernales.

DA blok

Består af et hus og en eller flere skiver. Anvendes, hvor der er behov for at ændre retningen af et tov eller en wire, eller i et taljesystem til at hale eller løfte.

DE Block

Gehäuse mit einem oder mehreren drehbar gelagerten Seilscheiben zur Führung von Tauwerk oder Ketten.

Wird für Taljen sowie für Lade- und Fischereieinrichtungen verwendet.

GR τροχαλία· μακαράς

Αυλακωτό ράουλο που δουλεύει μέσα σ' ένα σκελετό ή θήκη. Χρησιμοποιείται για να αλλάζει τη διεύθυνση ενός σχοινιού ή αλυσίδας, ή να κερδίζει μηχανικό όφελος όταν περνάει το σχοινί από ένα σύσπαστο (παλάγκο).

EN block; pulley

Grooved sheave working in a frame or shell. Used to alter direction of a rope or chain, or to gain a mechanical advantage by reeving a purchase.

FR poulie

Corps généralement en métal, supportant un ou plusieurs réas qui tournent autour d'un axe ou essieu passant par le milieu du corps. Servant à modifier la direction d'un cordage.

IT bozzello; pastecca; carrucola

Una cassa in legno od in ferro, con una o più cavatoie, in ognuna delle quali è sistemata con un perno la puleggia.

NL blok

Werktuig ter geleiding van touwwerk, staaldraad of ketting (loper) met een of meer schijven.

Hiermee kunnen grotere krachten worden overgebracht dan met een enkelvoudige lijn, of kan de richting waarin een lijn loopt worden gewijzigd.

PT moitão; moutão; roldana; polia

Peça de poleame de laborar constituída por uma caixa de forma elíptica, de madeira ou de ferro, que abriga, na abertura ou gorne, a roda e o seu eixo ou perno, tendo na cavidade ou goivado que a abraça a alça. É empregue para mudar a direcção de movimento de um cabo.

Quando tem mais de uma roda designa-se por cadernal.

3139

ES falsa boza

Trozo de cable que une la malleta con el cable de arrastre.
Red de arrastre.

DA indhaler

Line af tov eller wire, der forbinder trawlwiren med mellemlinen eller stjerterne, forbi skovlene. Bruges under udsætning og haling af trawlen ved til- og frakobling af skovlene.

DE Umgehungsstander; Zwischenstander

Zwischen dem Stoppereisen des Jagers und dem Verbindungsglied des Patenthakens befindlicher Hilfsstander, mit welchem nach der Abhängung der Scherbretter weitergehievt wird.

GR ανεξάρτητο σύρμα

Σύρμα χειρισμού που συνδέει το συρματόσχοινο της τράτας με το χοντρό συρματόσχοινο. Επιτρέπει στο χοντρό συρματόσχοινο να παρακάμψει την πόρτα όταν ρίχνεται (μαϊνάρεται) ή σηκώνεται (βιράρεται) το εργαλείο.
Τράτα.

EN pennant

Handling wire connecting warp to bridle and allowing the bridle to by-pass the otter board when shooting or hauling the gear.
Trawl.

FR rapporteur; faux bras

Cordage de manœuvre reliant la fune au bras; permet de filer ou de virer le bras quand le panneau est déconnecté de la fune.

IT penzolo; tira su

Cavo di manovra per recuperare i calamenti.
Rete da traino.

NL verbindingsdraad

Lijn of ketting (staaldraad of kabel) die de vislijn van een trawlnet verbindt met de voorlopers, kabels of breidels zodat deze kunnen worden doorgewonden na het ophangen van de visborden.

PT brinco da malheta

Cabo de manuseamento que liga o cabo real à malheta e que permite efectuar um «by-pass» à porta de arrasto quer ao largar quer ao virar a rede.

3139

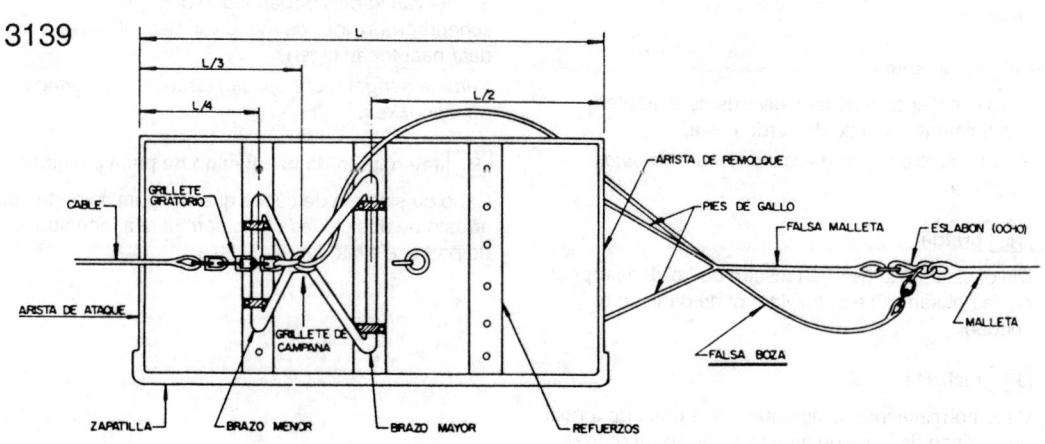

3140

ES malleta

Cabo de tiro de una red de pesca, especialmente el del arte de arrastre. Suele ser reforzada y a veces con alma de acero.

DA mellemline

Tov af stål eller en kombination af stål og kunststof. Forbinder skovlenes bagstrop med stjerterne i en trawl.

DE Jager

Stander des Vorgeschirrs eines Grundschleppnetzes zwischen Scherbrett und Joch bzw. Ponybrett oder nachfolgenden Standern, die zu den Maulleinen des Schleppnetzes führen.

Gelegentlich werden mit Jager auch andere Stander des Vorgeschirrs bzw. vorderen Netzteils bezeichnet.

GR χοντρό συρματόσχοινο

Σχοινί, συνήθως από σύρμα, που βρίσκεται μεταξύ πόρτας ή οπισθίου στρόφου και διχτυού ή dan leno ή σκελών.

Τράτα με πόρτες.

EN bridle; sweep *

The rope usually of wire, between otter board and net or dan leno or legs.

* *Scottish term.*

FR bras de chalut; bras

Cordage reliant le panneau au guindineau ou aux entremises.

IT calamento

Cavo di collegamento tra il divergente e la rete, generalmente formato da corda mista.

Se è formato da cavo d'acciaio viene chiamato maglietta.

NL breidel

Lijn of staaldraad gelegen tussen de bordenstroppen en de nokken van een trawlnet of de danleno of knuppel.

PT malheta

Cabo normalmente de aço que liga a porta de arrasto ou o brinco da porta de arrasto ao calão da rede de arrasto.

3141

ES vientos de la puerta elevadora

Sistema de cables unidos a las puertas o calones y al centro de la relinga de corchos, para el montaje de las puertas elevadoras.

DA forlængerwire

En rigning af liner eller tove for montering af et eller flere skærebrædder over eller evt. foran overtællen.

DE falsche Headleine; falsches Kopftau

Verbindung der Kopfscherbretter mit den Kurrleinen.

GR ψευδές επάνω γραντί

Σύστημα από σύρματα που είναι προσαρτημένα στις δύο πόρτες ή στα δύο dan leno για την εξάρτηση των αετών.

Τράτα.

EN false headline

System of wires attached to either the otter boards or the dan lenos for the rigging of kites.

Trawl.

FR petit bras

Cordage reliant un plateau élévateur au guindineau ou au panneau.

IT finta lima da sugheri

NL vrije jaaglijn; headlijn

Stelsel van kabels opgehouden door de scheerborden voor de mond van het net om de vis daar naar toe te jagen.

Kunnen aangebracht zijn aan visborden, danleno's of bovennokken.

PT falso cabo da pana; cabo da pana auxiliar

Cabo ou sistema de cabos que se ligam às portas de arrasto ou aos calões e que permitem a montagem de portas elevatórias.

3141

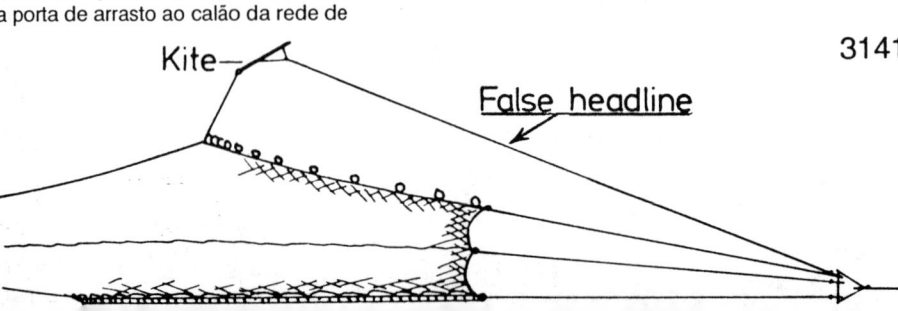

Kite

False headline

3142

ES abertura de la malla; luz de malla

a) Para la red anudada, distancia interior existente entre dos nudos opuestos en una misma malla completamente tensa en el sentido N.
b) Para la red sin nudos, distancia existente entre dos cruces opuestos en una misma malla completamente tensa en el sentido que da a la dimensión su valor máximo.

DA indvendig maskevidde

Angivelse (i mm) for den indvendige åbning mellem to modstående knuder i en strakt maske.

På dansk anvendes maskevidde og maskestørrelse ofte synonymt, og man skelner mellem:
a) maskestørrelse eller halvmaske: afstanden mellem centrum af to knuder i siden af en maske, se 3310
b) strakt maskestørrelse eller helmaske: afstanden mellem centrum af to modstående knuder i en maske, se 3173
c) indvendig maskevidde (denne);
se tillige 3008, 3013 og 3315.

DE Maschenöffnung

a) Bei geknoteten Netztuchen: lichter Abstand zwischen zwei einander gegenüberliegenden Knoten einer zum Zwecke der Messung in N-Richtung gestreckten Masche;
b) bei knotenlosen Netztuchen: lichter Abstand zwischen zwei einander gegenüberliegenden Verbindungsstellen einer Masche, die im Sinne der N-Richtung zum Zwecke der Messung gestreckt ist.

GR άνοιγμα ματιού· άνοιγμα βρόχου

α) Δικτύωμα με κόμπους: η εσωτερική απόσταση μεταξύ δύο απέναντι κόμπων στο ίδιο μάτι, όταν αυτό είναι τελείως τεντωμένο κατά τη διεύθυνση N.
β) Δικτύωμα χωρίς κόμπους: η εσωτερική απόσταση μεταξύ δύο απέναντι ενώσεων στο ίδιο μάτι όταν αυτό εκτείνεται τελείως κατά το μεγαλύτερό του άξονα.

EN opening of mesh

(a) For knotted netting, the inside distance between two opposite knots in the same mesh when fully extended in the N-direction;
(b) for knotless netting, the inside distance between two opposite joints in the same mesh when fully extended along its longest possible axis.

FR ouverture de maille; maillage

— Pour la nappe nouée, distance intérieure existant entre deux nœuds opposés dans une même maille complètement tendue dans le sens N.
— Pour la nappe sans nœuds, distance intérieure existant entre deux croisements opposés dans une même maille complètement tendue dans le sens qui donne à la dimension sa valeur maximale.

IT apertura di maglia

a) Per la rete annodata rappresenta la distanza interna esistente tra due nodi opposti di una stessa maglia completamente tesa nella direzione N;
b) Per la rete senza nodi rappresenta la distanza interna estistente tra due incroci opposti di una stessa maglia, completamente tesa, nella direzione che fornisce alla dimensione il valore più alto.

NL maaswijdte; maasopening

Binnenafstand tussen twee tegenover elkaar liggende knopen of verbindingen van dezelfde maas, terwijl het netwerk in de lengterichting is gestrekt.

PT abertura da malha; vazio da malha

Corresponde à distância interna entre dois nós opostos da malha quando esta se encontra completamente estirada na direcção normal.

Expressa normalmente em milímetros.

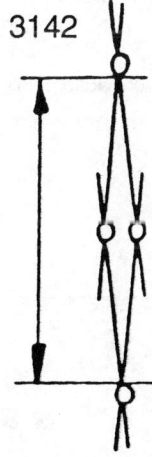

3142

3143

ES malla

Abertura formada especialmente y limitada por la materia constitutiva de la red.

DA maske

DE Masche

Eine bestimmt geformte Öffnung, die von Netzgarn oder anderem Netztuchmaterial umgeben ist.

GR μάτι· βρόχος

Διάκενο με χαρακτηριστική μορφή το οποίο περιβάλλεται από υλικό δικτυώματος.

EN mesh

The openings in a piece of netting bounded by the material from which the netting is made.

FR maille

Ouverture formée à dessein et limitée par la matière constitutive de la nappe de filet.

IT maglia

Apertura di forma geometrica, limitata dal materiale costituente la pezza di rete.

NL maas

Elk der ogen (openingen) gevormd door de elkaar kruisende draden van een netwerk.

PT malha

Cada uma das aberturas formadas pelos fios de um pano de rede de pesca.

3144

ES viento

Uno de los cables que unen la red al calón.

DA stjert

To eller flere stjerter forbinder trawlen med mellemlinen eller med en danleno. Består af tov af stål eller en kombination af stål og kunststof.
Lokalt anvendes tillige termerne forlænger og hanefod.

DE Stander

Tau des Vorgeschirrs des Schleppnetzes.

GR σκέλος

Ένα από τα σύρματα ή αλυσίδες που συνδέουν δίχτυ με dan leno ή πόρτα ή χοντρό συρματόσχοινο.

EN leg ; spreader; spreading wire *

One of the wires or chains connecting net to dan leno or bridle or otter board.
* Scottish term.

FR entremise; patte de chalut

Un des cordages reliant l'aile du chalut au guindineau, au bras ou au panneau.

IT braccio finto

NL voorloper

Verbindingsdraad tussen danleno en nok, of tussen visbord of bordstrop en nok, of tussen breidel en nok.

PT pernada; tirante; forcada

Cada um dos cabos (ou correntes) que ligam a asa da rede de arrasto ao calão, à malheta ou à porta de arrasto.

Podem existir, de acordo com a tipo de rede de arrasto e respectivo armamento, três tipos de pernadas:
a) pernada superior ou do cabo pana
b) pernada do meio ou do porfio
c) pernada inferior ou do arraçal.

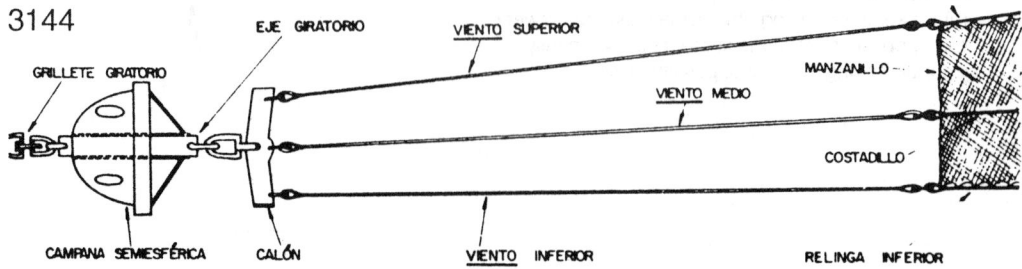

3144

UNIÓN DEL CALÓN CON EL ALA POR MEDIO DE VIENTOS

3145

ES conjunto del calón «danleno»

Se refiere al más complejo sistema de calón usado y que consiste en un diábolo montado con eje de acero y el calón propiamente dicho, en forma de bumerán.

DA danleno

Et komplekst arrangement af en kraftig bobbins og en trekant eller sommerfugl. Monteres foran undervingen på en trawl eller ved overgangen mellem stjerter og mellemline. Udgør dels en beskyttelse for trawlen på hård bund, dels en vægt, der holder det forreste af trawlen nede.

DE Vorgeschirr

Bereich des Schleppnetzfanggeschirrs zwischen den Flügeln und Kurrleinen. Die Aufgabe des Vorgeschirrs ist, im Zusammenwirken mit den Kurrleinen, die Bildung der horizontalen und vertikalen Netzmaulöffnung des Schleppnetzes und die Bestimmung seiner Lage im Wasser.

GR σύστημα dan leno

Συνήθως αναφέρεται στην πιο σύνθετη μορφή dan leno που χρησιμοποιείται για το καλάρισμα της τράτας σε βαθιά νερά και που αποτελείται από ατσάλινο καρούλι προσαρτημένο σε άξονα και πεταλούδα.

EN Dan leno assembly

Usually refers to the more complex form of Dan leno used for deep sea trawling and consisting of spindle-mounted steel bobbin and butterfly.

FR ensemble de guindineau

Dans un ensemble de ce type, employé habituellement sur les chalutiers de grande pêche, le guindineau est protégé sur l'avant par une sphère, un cône ou une «casserole» (disque renforcé) sur un axe.

IT mazzetta a Dan leno

Non è usata sulle reti italiane mediterranee.

NL danleno

Constructie ter verbinding van breidels of voorlopers, bedoeld om het net meer verticale opening te geven; veelal bestaand uit een danlenokogel en een zwaluwstaartvormige plaat (spreider, butterfly).

PT conjunto de calão; calão

Usualmente refere-se à mais complicada forma de calão utilizado em redes de arrasto pelo fundo a grandes profundidades e que é constituída pela borboleta, esfera do calão e respectivo eixo, manilha e tornel.

3145

Dan leno assembly

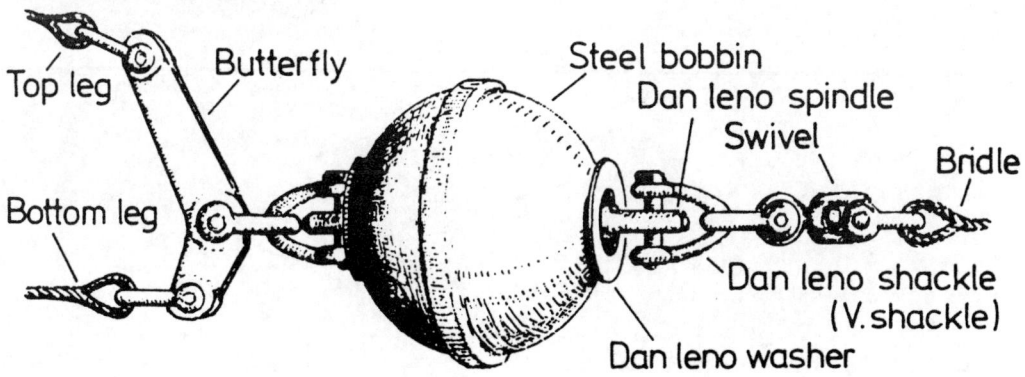

Top leg — Butterfly — Steel bobbin — Dan leno spindle — Swivel — Bridle

Bottom leg — Dan leno shackle (V. shackle) — Dan leno washer

3146

ES cordón de conexión; trencilla

DA bændsel

En sammenføjning fremstillet ved omvikling med garn eller line. Bruges ved sammenføjning af to tove, ved fastgørelsen af nettet til tællerne i et fiskeredskab osv.

DE Bändsel

Verbindung von zwei parallel nebeneinanderliegenden Leinen durch Umwickeln mit dünnem Faser- oder Drahttauwerk.

GR ληγαδούρα

Σπάγγος μικρού μήκους για να ενώνει (δένει) δύο τμήματα ενός εργαλείου μαζί.

EN setting

Short length of twine used to seize two parts of a gear together.

FR amarrage

Fil utilisé pour relier deux parties d'un filet ou d'un engin.

IT cavetto d'unione

NL bindsel; bendel

Stuk touw of garen gebruikt om onderdelen van een vistuig te verbinden.
Men gebruikt hierbij vooral boet- of breinaalden.

PT fio para pegamento; fio para pegar

Fio utilizado para ligar dois panos de rede ou duas secções de uma rede de arrasto entre si.

3147

ES burlón; relinga inferior; relinga de plomo; bolera

Cabo sobre el que se colocan piezas de plomo para lastrar el arte.

DA rup; gear

Den del af trawlen, som beskytter nettet mod slid fra bunden, og som fæstnes til undertællen med korte kæder eller stropper. Den monteres, så den løber under eller lidt foran undertællen. Udformningen varierer meget og afhænger af bundtypen. Består i reglen af en wire eller kæde beskyttet med:
— omviklet tov eller trukket over med gummipropper (glat bund)
— gummiskiver og -propper (hårdere bund)
— stålbobbins (eller gummiruller) og afstandstykker (hård bund, sten og klipper)
— plastikkugler og gummipropper (blød bund).
For at sikre en god bundkontakt monteres ruppen efter behov med blylodder, kæder eller andet tungt.

DE Grundtau

Teil des Grundschleppnetzgeschirrs, das in verschiedenen Ausführungen existiert:
a) als Rollergrundtau: besteht aus Rollenstander, großen Rollen, Zwischenrollen, Rollenstropps, Vorstander und Bülschleinen. Wird zur Befischung auf rauhem Grund eingesetzt;
b) als Gummihopser-Grundtau: Drahtgrundtau, bestückt mit aus alten Autoreifen geschnittenen Scheiben div. Durchmesser. Eingesetzt z. B. in der Aalfischerei (Aalhopser) oder in der Kabeljaufischerei;
c) als mit Ketten bewickeltes Grundtau.

3147

GR κάτω γραντί

Ενωμένα τμήματα σχοινιού, συνήθως συρμάτινου, περιτυλιγμένου με σχοινί για προστασία ή με λαστιχένιους δίσκους ή με διάφορους τύπους καρουλιών (κουτρουμπούκια), προσαρτημένου στη γραμμή αλιείας και μπροστά απ' αυτή, για να προστατεύει το κατώτερο προπορευόμενο χείλος μιας τράτας βυθού από καταστροφή εξαιτίας του πυθμένα, ενώ η τράτα εξακολουθεί να παραμένει σε επαφή μ' αυτόν.

EN groundrope; foot rope

Connected sections of rope, usually of wire, protected with rope rounding or rubber discs or various types of bobbins, attached to and in front of the fishing line, to shield the lower leading margin of a bottom trawl from ground damage whilst maintaining ground contact.

FR bourrelet

Ralingue inférieure d'un chalut de fond, habituellement garnie de cordages ou de disques en caoutchouc. Formée d'une seule pièce ou en plusieurs morceaux reliés entre eux.

IT lima da piombo; bullone

Lima su cui sono montati i piombi o i pesi.
– Più in generale lima inferiore della rete.
– Reti atlantiche.

NL onderpees; grondpees

Touw, staaldraad of ketting, enkelvoudig dan wel omwoeld met touw, voorzien van rubber schijven, welke de voorste begrenzing vormt van de onderzijde van een vistuig (behalve indien de klossenpees wordt gebruikt).

PT arraçal

Cabo inferior da boca das redes de arrasto. Normalmente cabo de aço forrado, é no caso das redes de arrasto pelo fundo algumas vezes dotado de rodelas de borracha (bolachas), roletes, diábolos, esferas de protecção contra fundos acidentados.

3148

ES esfuerzo pesquero

La relación entre el volumen de capturas de una especie y el esfuerzo empleado en obtenerlas se expresa como captura por unidad de esfuerzo: CPUE. Éste es un índice de rendimiento pesquero que, junto con un conocimiento de la biología y las reservas de la especie pertinente, puede ayudarnos a encontrar el óptimo esfuerzo pesquero.

DA fiskeriindsats; effort

Mål for fiskeraktiviteten. Henviser normalt til antallet af timer, der bliver fisket, men kan også angives ved antallet af redskaber i brug, antallet af aktive fartøjer, antallet af sæt eller slæb, mængden af motorkraft, der sættes ind osv.
Fangsten pr. fiskerindsatsenhed anvendes i forbindelse med den biologiske rådgivning som et indeks for udviklingen i et bestemt fiskeri.

DE Fischereiaufwand

GR αλιευτική προσπάθεια

EN fishing effort

A measure of the activity of fishing boats. Fishing effort is strictly defined in terms of 'total standard hours fishing per year' but is often described less rigorously in terms of numbers of vessels, fishing time or fishing power for instance.
Fishing intensity.

FR effort de pêche

Activité des navires de pêche dans une région donnée. La mesure de l'effort de pêche tient compte du temps de pêche, de la puissance de pêche et du nombre de navires exerçant cette activité.

IT sforzo di pesca

Attività di pesca in una data zona, tenendo conto del numero dei pescherecci, potenza e durata della pesca.

NL visserij-inspanning

Een maat voor de operationele inbreng van een vissersvloot, uitgedrukt in aantal schepen, motorvermogen van de schepen en afmetingen van de netten en eventueel de duur van het vissen.

PT esforço de pesca

Trata-se de uma medida da actividade exercida pelos navios de pesca numa dada área.

3149

ES diábolo del calón; campana semiesférica

DA danlenokugle

Stor stålkugle foran undervingerne på en trawl; den skal forhindre trawlen i at gå i hold på havbunden. Vægten af danlenoen tjener samtidig til at holde undervingerne på bunden.

DE Kugel; Dan-Leno-Rolle

Stahlkugel als Vorgewicht für das Joch.

GR μπομπίνα Dan leno· καρούλι Dan leno· μπομπίνα· καρούλι

Μεγάλη ατσάλινη κατασκευή που εμποδίζει τα φτερά του διχτυού της τράτας να σκαλώσουν σε μικρά αντικείμενα.

EN Dan leno bobbin; bobbin

Large hollow steel sphere that prevents the wings of the trawl net from becoming caught up on small obstacles.

FR sphère de guindineau; casserole; cône

Pièce tournante en acier qui empêche les ailes du chalut de s'accrocher aux obstacles du fond.

IT diavolone; sfera; sfera della mazzetta

NL danlenokogel; kogel

Een aan de ondernokken of de danleno zelf bevestigd stalen bolvormig lichaam ter beperking van netschade bij het vissen op harde gronden.

PT esfera do calão

Cada uma das grandes esferas de aço destinadas a evitar que as asas das redes de arrasto pelo fundo se prendam em pequenos obstáculos do fundo (peguilhos).

3150

ES diábolo; esférico; bolo;

En la parte central de la relinga inferior central (burlón), que es la que corresponde al vientre, puede acoplarse un rosario de ruedas o esferas de goma (diábolos) para evitar las enganchadas cuando se trabaja en fondos rocosos.

DA bobbin

En af de store stål- eller gummikugler, der er monteret på ruppen af en trawl. Anvendes på trawl, der skal fiske på bund med sten og klipper.

DE Roller; Grundtaukugel; Bomber

Stahlkugel, die auf dem Rollergrundtau befestigt ist, bzw. Roller in Scheibenform aus Holz oder Stahl, letzterer hartgummibereift.

GR καρούλι· μπομπίνα· τροχός· κουτρουμπούκι

Κύλινδροι που έχουν περασθεί σε σύρμα συγκεκριμένου μήκους έτσι ώστε να αποτελούν τμήμα του κάτω γραντιού.

Τα καρούλια μπορεί να είναι από ξύλο, ατσάλι, λάστιχο και σε σχήμα σφαιρικό, κυλινδρικό, ημισφαιρικό ή ελλειπτικό.

EN bobbin

Roller, numbers of which are threaded on wire of specified length to form part of a groundrope.

Bobbins may be made of wood, steel, composition rubber, and may be cylindrical, spherical, hemispherical or elliptical in shape.

3149
Dan leno assembly

Top leg · Butterfly · Bottom leg · Steel bobbin · Dan leno spindle · Swivel · Bridle · Dan leno shackle (V. shackle) · Dan leno washer

3150

FR sphère; diabolo

Élément d'un chapelet de sphères ou de disques (en métal, plastique, caoutchouc ou bois) garnissant le bourrelet d'un chalut.

Il empêche les avaries ou déchirures du filet sur des fonds durs ou irréguliers.

IT diavolone; sfera

Una delle sfere, molte delle quali sono infilate su un filo metallico di una determinata lunghezza e costituiscono parte integrante di una lima da piombi.

Rete a strascico.

NL klos

Vrij om de pees rollend omwentelingslichaam dat het mogelijk maakt op moeilijke grond te slepen zonder noemenswaardige schade aan het net te veroorzaken.

Kan de vorm hebben van een bol, cilinder of ellipsoïde.

PT esfera; diábolo; bolacha; rolete; meia-esfera; meio-melão

Cada um dos elementos rolantes com forma esférica, cilíndrica, elipsoidal ou semiesférica, de ferro, aço, madeira ou borracha, colocados no arraçal das redes de arrasto pelo fundo, formando um rosário (rosário do arraçal).

Têm por finalidade evitar peguilhos quando a rede arrasta em fundos rochosos.

3151

ES intercalador con cadenilla

DA afstandsstykke med kædeophæng

Afstandsstykker er kraftige jernrør, der monteres mellem bobbins på en rup. Nogle eller alle er forsynet med kædeophæng, som fæstner ruppen til undertællen.

DE Klotje mit Kette

GR διαχωριστικό με αλυσίδα

Σιδερένιο διαχωριστικό τμήμα με αλυσίδα σύνδεσης που χρησιμοποιείται στο κάτω γραντί.

EN lancaster

Iron spacer between bobbins with connecting chains to the fishing line, used in groundropes.

FR intermédiaire avec chaîne

Intermédiaire en acier muni d'une chaîne de liaison pour le bourrelet.

IT catena con distanziatore per bullone; catena con yoyo

Reti atlantiche.

NL klosjes met ketting

Kleine ijzeren klossen met kettingeinden ter bevestiging van klossenpees aan grondpees.

PT carrinho do arraçal com bichanas

Pequenos roletes colocados no arraçal entre os roletes normais e aos quais se fixam as mãozinhas (bichanas), normalmente de aço, que têm por finalidade ligá-los ao cabo de entralhe das asas e barrigas inferiores.

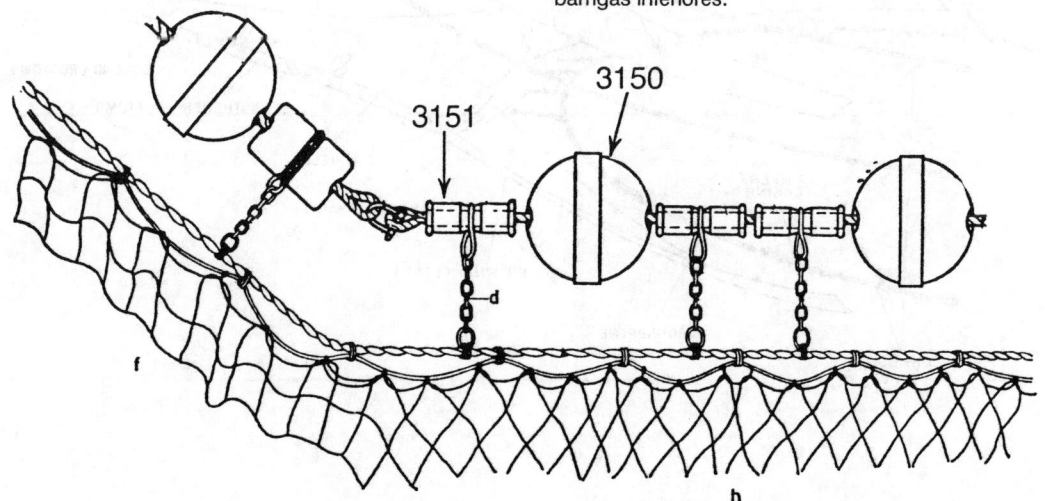

3152

ES relinga superior; relinga de corchos

Cabo mixto o cable de nailon, en el que se colocan los flotadores de la red.

En España los corchos van colocados directamente en la relinga superior.

DA overtælle; overlig

Tov, som nettet er fastgjort til langs den øverste kant af et fiskeredskab, såsom trawl, garn, not osv.

DE Headleine; Kopftau

Tau, das mit den Schwimmkugeln bestückt und mittels Bändseln und Bülschleine mit dem Netz verbunden ist.

GR επάνω γραντί· επάνω καζίλι

Το κύριο σχοινί στο επάνω τμήμα του πλαισίου.

Όρος χρησιμοποιούμενος για τράτες κυρίως· όρος χρησιμοποιούμενος για απλάδια κλπ.

EN headline; headrope

The principal upper frame rope of a net (such as gill net or trawl) to which the netting is attached.

FR ralingue supérieure; corde de dos *

Ralingue bordant la partie supérieure de l'ouverture d'un filet.

* Terme utilisé pour le chalut.

IT lima superiore della rete; lima da sughero

Lima su cui sono montati i sugheri.

In Italia i sugheri sono montati direttamente sulla lima superiore.

NL bovenpees

Staalkabel of touwwerk waarmee een trawlnet langs de bovenzijde rechtstreeks is afgeboord.

PT cabo da pana

Cabo superior da boca das redes de arrasto.

3152

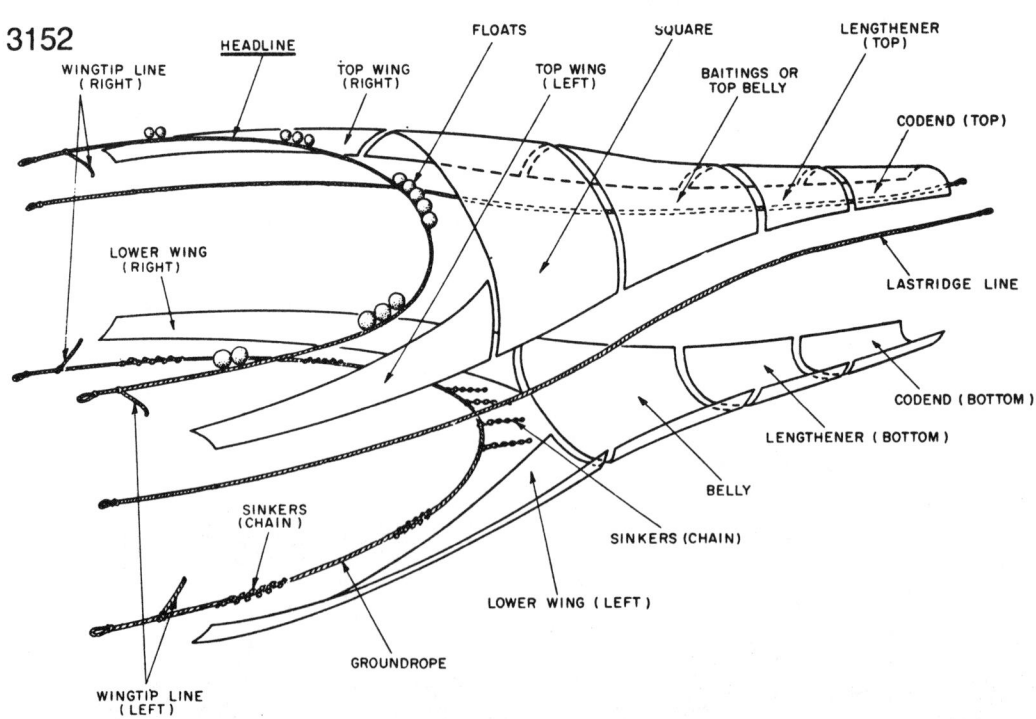

WINGTIP LINE (RIGHT)
HEADLINE
TOP WING (RIGHT)
FLOATS
TOP WING (LEFT)
SQUARE
LENGTHENER (TOP)
BAITINGS OR TOP BELLY
CODEND (TOP)
LOWER WING (RIGHT)
LASTRIDGE LINE
CODEND (BOTTOM)
LENGTHENER (BOTTOM)
BELLY
SINKERS (CHAIN)
SINKERS (CHAIN)
LOWER WING (LEFT)
GROUNDROPE
WINGTIP LINE (LEFT)

3153

ES visera; cielo

Unido por su parte anterior a las alas y relinga de corchos y por la posterior a la espalda.

DA tag

Den del af overpanelet i en bundtrawl, som er trukket frem over og foran underpanelet. Forhindrer fiskene i at flygte opad.

DE Square; Dachstück

Es erzeugt durch seine Anstellung zur Strömung beim Schleppen des Schleppnetzes eine vertikale hydrodynamische Kraft, die zusammen mit den Auftriebskörpern die für die Fängigkeit des Schleppnetzes bedeutende vertikale Netzmaulöffnung erzeugt.

GR τετράγωνο

Περιοχή του κορυφαίου διχτυωτού μεταξύ επάνω γραντιού και γούλας.
Τράτα.

EN square

Section of top panel between headline and batings.
Trawl.

FR grand dos; carré

Pièce de la face supérieure d'un chalut située entre les ailes supérieures et le petit dos.

IT pezza di copertura

Prima parte del cielo di una rete a strascico o sfogliara.
È la parte che «coprendo» impedisce la fuga del pesce sollevato dalla lima da piombo.

NL bovenboel; kap

Gedeelte van de bovenkant van een trawl vlak achter de middeling van de bovenpees en begrensd door de bovenste zijnaden.
Bevindt zich ook boven de middeling van de onderpees, indien deze langer is dan de bovenpees.

PT quadrado

Secção da face superior das redes de arrasto situada entre o cabo de pana e a primeira barriga.

3153

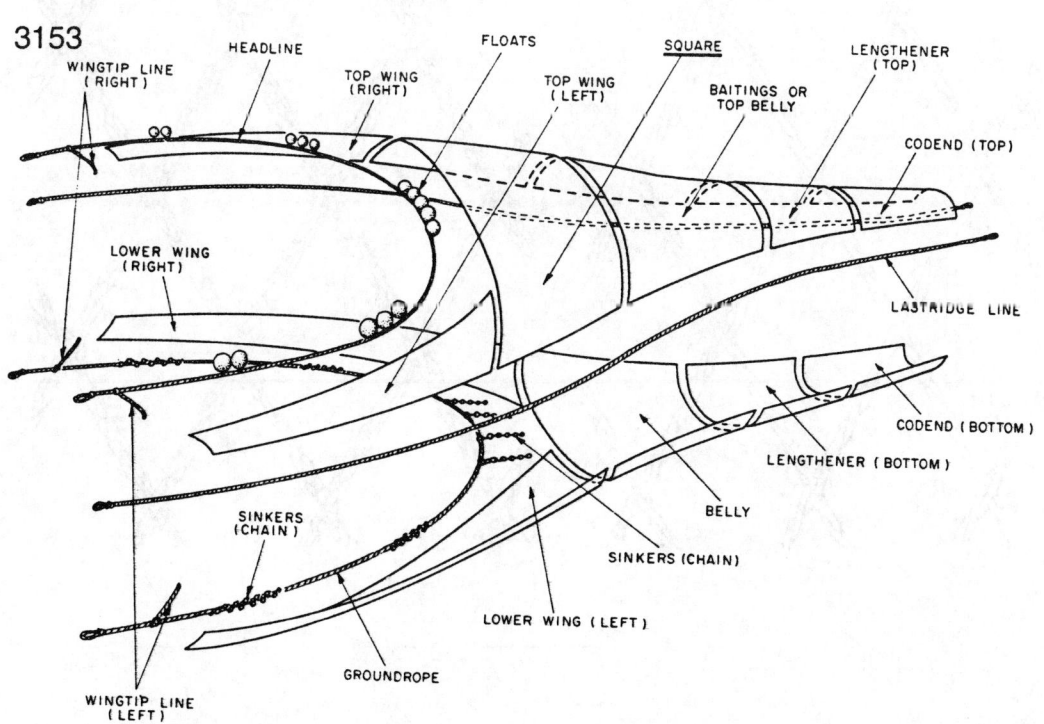

3154

ES red de pesca sin nudos

DA knudeløst net

Net fremstillet ved en metode, hvor maskerne ikke bindes med knuder, men flettes.

DE knotenloses Fischnetz; Fischnetz ohne Knoten

GR αλιευτικό δίχτυ χωρίς κόμπους

EN knotless fishing net

Fishing net made of knotless netting.

FR filet de pêche sans nœuds

Filet de pêche dont les jonctions des côtés de maille sont formées par tressage ou entrecroisement des filaments constitutifs.

IT rete da pesca senza nodo

Rete da pesca armata con pezze di rete senza nodo.

NL visnet zonder knopen; knooploos visnet

Visnet gemaakt van netwerk, dat zodanig machinaal wordt gebreid, dat de knopen ontbreken.
Wordt vooral in Japan gemaakt en in Nederland weinig gebruikt.

PT rede sem nós

3154

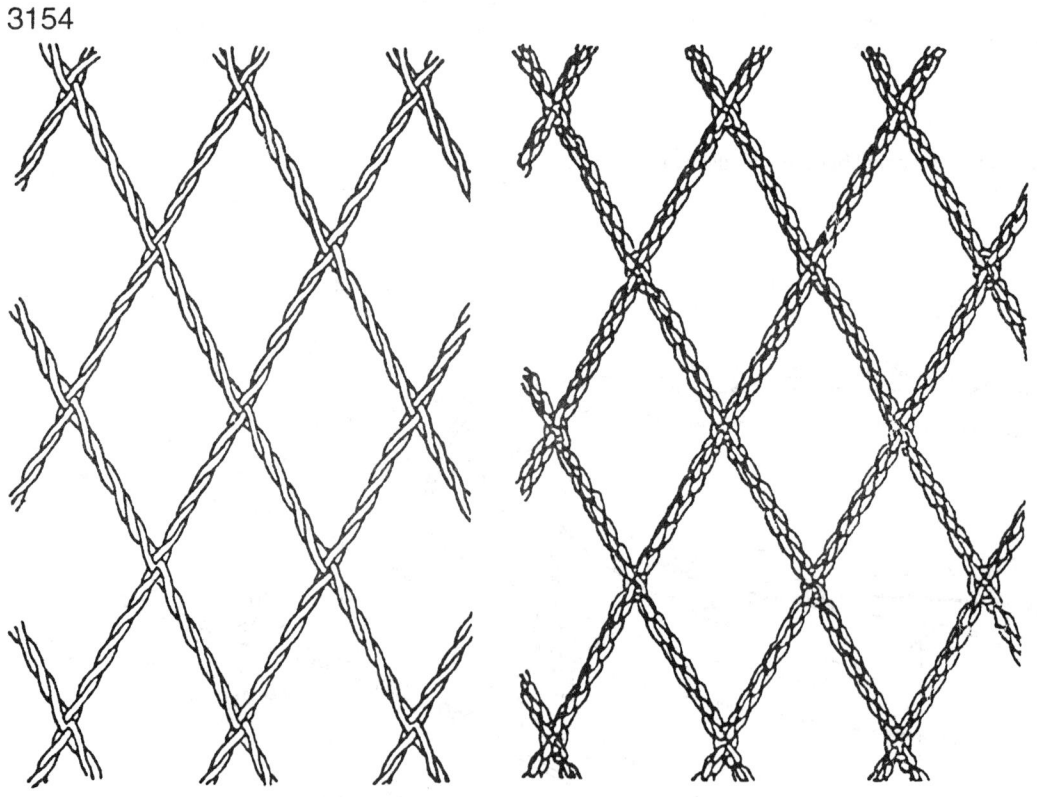

3155

ES pie de gallo; pata de gallo

Cables que salen de la cara exterior de la puerta o cadenas y que van unidas a una pieza de hierro con dos orificios, llamada «el ocho», por cuyo agujero libre pasa la malleta de la red.

DA bagstrop; agtertræk

System af to til fire kæder på bagsiden af en trawlskovl. Danner forbindelsen til mellemlinen.

DE Hahnepot

Verbindungsstück zwischen Scherbrett und Ochsenauge zur Kurrleine.

GR οπίσθιος στρόφος

Σχοινί, συνήθως συρματόσχοινο, που βρίσκεται μεταξύ πόρτας και διχτυού ή Dan leno ή σκελών.

EN backstrop

The short rope system, usually of wire or chain, between the otter board and bridle.

Otter trawl.

FR patte de panneau

L'un des deux câbles en fil d'acier qui constituent la patte d'oie reliant le bras au panneau de chalut.

IT braga del divergente; patta d'oca

Uno dei due cavi d'acciaio che formano il collegamento tra calamento (o maglietta) e divergente.

NL bordstrop

Staaldraad of lijn tussen visbord en breidels van een trawl.

Deze kan uitgevoerd zijn in twee aparte parallelle lijnen (onder- en bovenbordstrop), of in één hanepoot.

PT brinco da porta

Conjunto de cabos de aço, correntes de ferro ou ambos ligados à face posterior das portas de arrasto e ao argolão ou à malheta.

3155

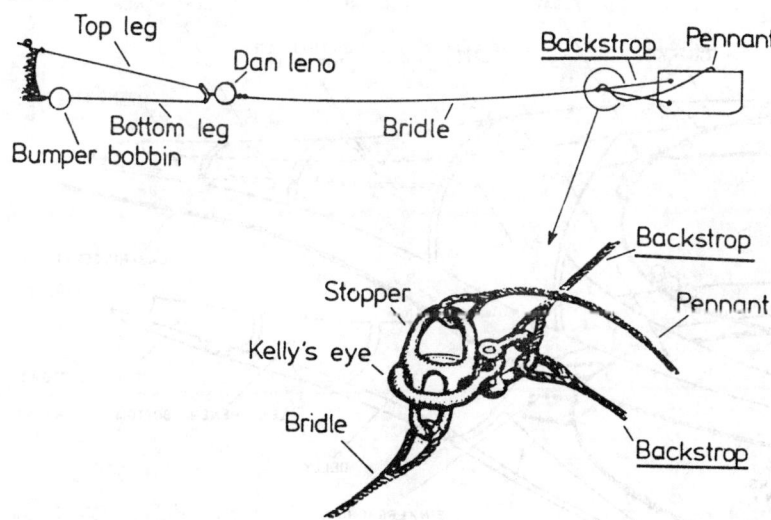

3156

ES banda; ala superior; banda de corcho

Unida a la relinga de corchos en su parte alta, a las alas inferiores por su parte baja y al cielo en la parte posterior.

DA overvinge

Den forreste netsektion i overpanelet på en trawl.

DE Oberflügel

Teil der Maulöffnung des Vornetzes, abgehend vom Knüppeltau, bzw. Fischleine bis zum Dachstück.

GR άνω φτερό· άνω μπάντα

Τμήμα διχτυού που εκτείνεται προς τα εμπρός από τη μία πλευρά του τετράγωνου και που συνήθως είναι συνδεδεμένο με το προσκείμενο κατώτερο φτερό (τράτα δύο φύλλων) ή με το προσκείμενο πλαϊνό φτερό (τράτα τεσσάρων φύλλων).

EN top wing; upper wing

Net section extending forward from one side of the square and usually joined to the adjacent lower wing (two panel trawls) or adjacent side wing (four panel trawls).

FR aile supérieure

Pièce de la face supérieure du chalut située en avant du grand dos.

IT braccio superiore

NL bovenvleugel; bovenvlerk

Taps toelopend netgedeelte tussen de kap en de boven- of zijnaad.

PT asa superior

Secção das redes de arrasto que se estende para a parte anterior da rede a partir de cada um dos lados do quadrado e que se liga à asa inferior adjacente (redes com duas faces) ou à face lateral adjacente (rede com quatro faces).

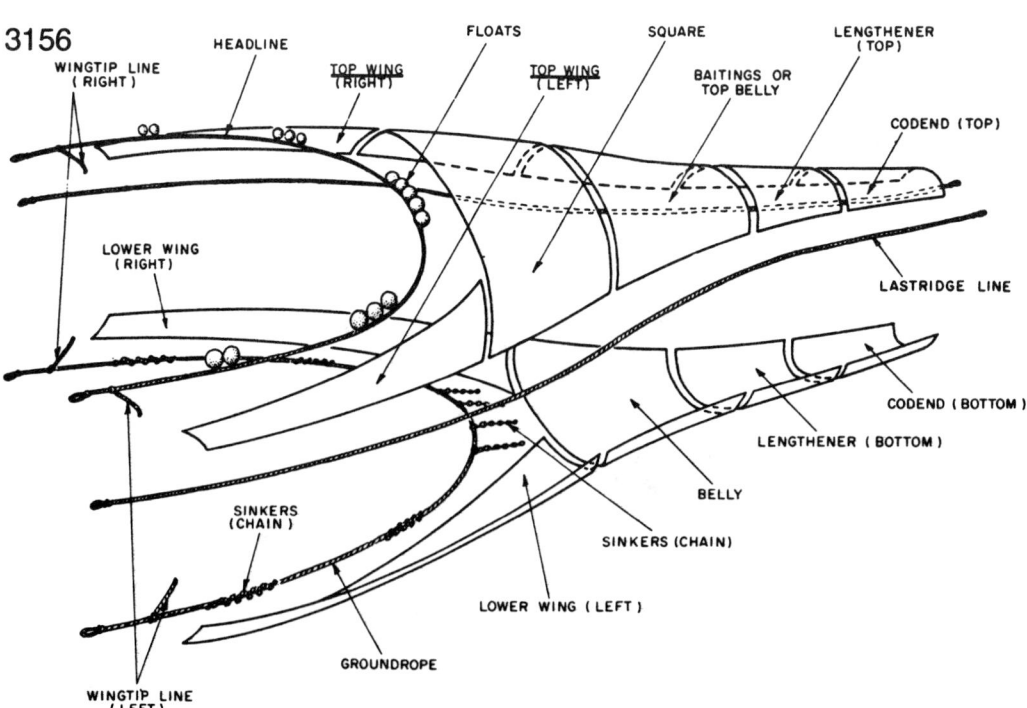

3156

3157

ES piel de vaca

Piel de vaca que protege la cara inferior del saco de la red.

DA hud

Dyreskind, der anvendes som undersideslidlag på en trawlpose.

DE Ochsenhaut; Ochsenfell

Scheuerschutz für den Steert von Grundschleppnetzen.

GR προστατευτικά δέρματα· πετσάλια

Δέρματα βοδιών προσαρτημένα στο κάτω μέρος του σάκου, για προστασία του δικτυώματος από τη φθορά λόγω τριβής.

EN hides

Cowhides attached to underside of codend to prevent chafing of the netting.

FR peau de vache

Tablier en cuir pour la protection de la partie inférieure du cul de chalut.

IT pelle di vacca

NL huid; lap; beschermlap

Slijtlap bevestigd aan de onderzijde van de kuil van een trawl ter voorkoming van netschade of slijtage.

PT couro do saco

Forra de couro ou cabedal que protege a face inferior do saco das redes de arrasto pelo fundo.

3158

ES rosario de diábolos; burlón con diábolos; relinga de plomo con diábolos

Relinga de plomo protegida por esferas de goma o acero para evitar las enganchadas cuando se trabaja en fondos rocosos.

DA bobbinsgear

Kraftig rup eller gear med stål- eller gummibobbins. Til trawl, der anvendes på hård bund. Se 3147.

DE Rollengeschirr

Bauteil des Grundschleppnetzes und des semipelagischen Schleppnetzes. Es dient zur Beschwerung und zum Schutz der Unterseite des Netzes vor Beschädigungen durch Grundberührung.

GR κάτω γραντί με καρούλια

EN bobbin wire

Assembly of spheres or discs attached along the forward edge of the lower panel of a trawl to protect it on hard or uneven ground.

3158

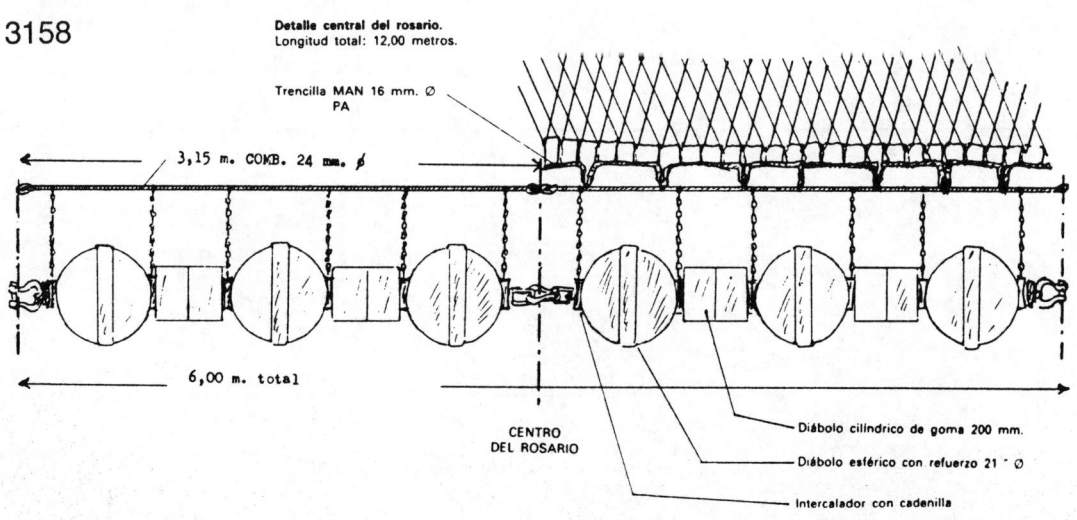

Detalle central del rosario.
Longitud total: 12,00 metros.

Trencilla MAN 16 mm. Ø PA

3,15 m. COKB. 24 mm. Ø

6,00 m. total

CENTRO DEL ROSARIO

Diábolo cilíndrico de goma 200 mm.

Diábolo esférico con refuerzo 21 " Ø

Intercalador con cadenilla

3158

FR **ligne de sphères; diabolos**

Chapelet de sphères ou de disques servant à la protection du dessous du chalut sur les fonds durs et irréguliers.

IT **bullone con diavoloni**

Lima da piombo particolare con sfere di acciaio e dischi di gomma.

NL **klossenpees**

Snoer van klossen dat met kettingeinden aan de grondpees van een trawl wordt bevestigd om zonder netschade op steenachtige of scherpe grond met de trawl te kunnen vissen.

PT **rosário do arraçal**

Conjunto dos elementos rolantes do arraçal das redes de arrasto pelo fundo que têm por finalidade evitar peguilhos quando se arrasta em fundos rochosos.

3159

ES **costadillo**

DA **sømline; sømstræber**

Forstærkningstov, der bændsles til sømmen på en trawl.

DE **Bortenleine; Nahtleine; Rebenleine**

Zur Verstärkung über den seitlichen Nähten, z. B. von Ober- und Untertrawlnetz.

GR **σχοινί πλάγιας ενίσχυσης**

Σχοινί μεγάλης αντοχής, τοποθετημένο κατά μήκος του κοψαδούρου.

EN **boltrope; selvage rope**

Load-bearing rope usually fixed to the seam of a net. *Load-bearing rope fixed to length of lestridge.*

FR **ralingue de côté; ralingue latérale**

Cordage de renfort fixé le long de la couture latérale d'un chalut.

IT **relinga laterale**

NL **naadlijn**

Touw of lijn ter versteviging van de naden.

PT **cabo de porfio lateral**

Cada um dos cabos a que se apontoa ou porfia os porfios de união das faces superior e inferior entre si ou das faces superiores e inferiores às faces laterais das redes de arrasto.

3160

ES las

Cabo que sirve para acercar el saco al costado y que va unido al estrobo del mismo. Su extremo anterior va amarrado a la relinga de corchos.

DA frelserline

Tov anvendt til at hale trawlposen hen til skibssiden under indhaling af trawlen. Fastgøres i den ene ende til delestroppen eller takkelstroppen. Under fiskeri fastgøres den anden ende til en af vingerne.

DE Leit; Pokleine

Das Leit dient als Hievstropp und zum Holen des Teilstropps.

GR

Πλεκτό σχοινί που χρησιμοποιείται για να έλκει το σάκο προς την πλευρά του σκάφους. Το μπροστινό άκρο του συνήθως είναι δεμένο στο επάνω γραντί.

EN lazy deckie; lazy decky; lazyline; poke line; recovery rope; messenger

Fibre rope used for hauling codend to ship's side. It may be secured to the halving becket, or by two tails to both codend lestridges. The fore end is usually hitched to the headline.

FR baîllon; hale-à-bord; ballant

Cordage en textile utilisé pour le halage de la poche ou du cul de chalut. Il est relié soit à l'erse de cul, soit aux deux ralingues de côté par une patte d'oie. Son extrémité antérieure est habituellement amarrée à la corde de dos.

IT baione

Cavo che va dalla parte della rete allo strozzatoio del sacco e che serve per ricuperare la saccata.

NL kuilstrop; kuiltouw

Touw met één einde bevestigd aan de verdeelstrop of aan de zijnaden van de kuil, en met het andere einde aan de slof (bij boomkor) of aan de nok of de bovenpees (bij trawl), ten einde de kuil bij het halen naar het schip toe te kunnen trekken.

PT laracho

Cabo de fibra cuja extremidade posterior liga ao estropo do laracho e se estende até ao cabo de pana onde a respectiva extremidade anterior se prende. Tem por finalidade virar o saco das redes de arrasto.

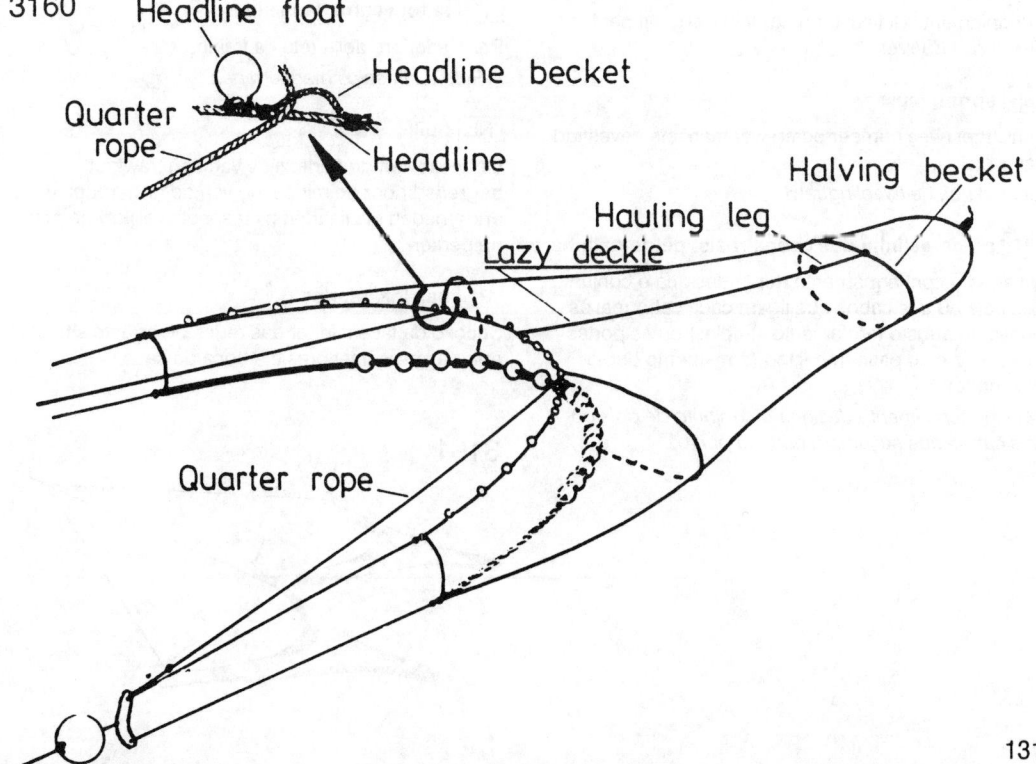

3160

Headline float

Headline becket

Quarter rope

Headline

Halving becket

Hauling leg

Lazy deckie

Quarter rope

3161

ES | fijera

Redes de arrastre gemelas con puertas.

DA | hanefod

Betegner en konstruktion af stropper, der forbinder stroppernes adskilte fastgørelsespunkter til en enkelt wire eller tov. Stjertene på en trawl og bagstropperne på en trawlskovl kan således betegnes som hanefodskonstruktioner.

DE | Hahnepot

Bindeglied zwischen Scherbrett und Ochsenauge.

GR | παραδέτης· μαραφούντι

Δίδυμες τράτες με πόρτες, τράτες διπλού εξαρτισμού.

EN | crowfoot

Otter twin trawls; double rig trawls.

FR | patte d'oie de fune; patte d'oie

Dans le chalutage au gréement double, désigne l'ensemble de deux ou trois câbles reliant chaque fune aux panneaux (chalut simple) ou aux panneaux et au patin médian (chaluts jumeaux). Plus généralement, ensemble de deux câbles ou davantage, reliés en un seul point.

IT | forca

Sdoppiamento del cavo prima dei divergenti per la pesca con due reti.

NL | spruit; touw

Touw met twee of meer poten waaraan iets bevestigd kan worden.
Gebruikt bij de tweelingtrawl.

PT | pé-de-galinha dos cabos reais; pé-de-galinha

No arrasto com armamento duplo designa o conjunto dos dois ou três cabos que ligam cada cabo real às portas de arrasto (armamento simples) ou às portas de arrasto e ao patim mediano (armamento duplo geminado).
Mais genericamente designa o conjunto de dois ou três cabos que se juntam num só ponto.

3162

ES | vientre; plan bajo

DA | underpanelet i kroppen

Kroppen er den kegleformede del af en trawl mellem vingerne og posen. Består ofte af flere sektioner med forskelligt materiale, maskevidde og skæring langs sømmen.
På dansk skelnes ikke mellem over- og underside i kroppen. Se 3250.

DE | Bauchstück; Belly

Besteht bei größeren Schleppnetzen aus mehreren kegelstumpfförmigen Ringen mit abgestufter Maschenweite und Netzfadendurchmesser.

GR | κοιλιά

Περιοχή του κατώτερου φύλλου δικτυώματος, μεταξύ κατώτερων φτερών και τμήματος προέκτασης.

EN | belly

Section of lower panel between lower wings and extension piece of a trawl.

FR | ventre

Partie de la face inférieure d'un chalut, située entre les ailes inférieures et l'amorce, constituée d'une ou de plusieurs pièces de largeurs décroissantes.

IT | letto; ventre; tassello *

Parte inferiore della rete da traino.
* Rete a strascico mediterranea.

NL | buik

Gedeelte van de onderzijde van een trawlnet, begrensd door de middeling van de onderpees en de ondernaden of zijnaden bij respectievelijk twee of vier netperken.

PT | barriga inferior

Secção da face inferior das redes de arrasto situada entre as asas inferiores e a boca do saco.

3161

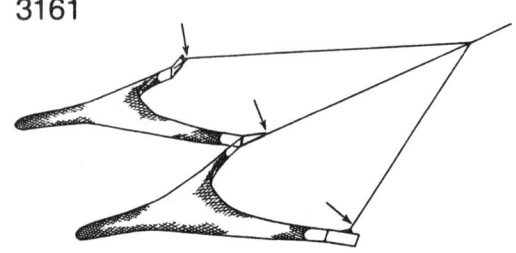

3163

ES gancho de remolque

Gancho especialmente diseñado para encapillar el cabo o cable de remolque.

DA slæbekrog

Krog specielt udformet til fastgørelse af en slæbetrosse.

DE Schlepphaken

Haken zum Befestigen einer Schlepptrosse.

GR γάντζος ρυμούλκησης

Γάντζος ειδικά κατασκευασμένος για την προσάρτηση ενός σχοινιού ρυμούλκησης (ρυμούλκιο).

EN towing hook

A hook specially designed for the attachment of a towrope.

FR croc de remorque

Croc spécialement conçu pour la fixation d'un câble de remorque.

IT gancio di rimorchio

Gancio progettato appositamente per incocciarvi un cavo di rimorchio.

NL sleephaak

Haak die speciaal is ontworpen voor het bevestigen van en slepen met een sleeptros.

PT gancho de engate; gato de engate

Gato especialmente concebido para a fixação de um cabo de reboque.

3164

ES arte de arrastre

Conjunto formado por los cables, puertas, malletas, vientos y el arte propiamente dicho.

DA trawlgrej; trawlsystem

Samlebetegnelse for alle komponenterne i en trawl.

DE Schleppnetzgeschirr

Sammelbegriff für alle Komponenten eines Schleppnetzes.

GR εργαλεία τράτας· σύνεργα τράτας· εφόδια τράτας

EN trawl gear

Assembly of warps, otter boards, bridles, legs and net which comprise a complete fishing unit.

FR train de pêche au chalut

Désigne l'ensemble de l'engin de chalutage, y compris les funes, panneaux, bras, entremises et le chalut lui-même.

IT attrezzatura per rete da traino

NL vistuig gebruikt door de trawlers

PT trem de pesca de arrasto; arte de pesca de arrasto

Designa o conjunto da rede de arrasto e compreende os cabos reais, malhetas, tirantes, portas de arrasto e a rede propriamente dita.

3162

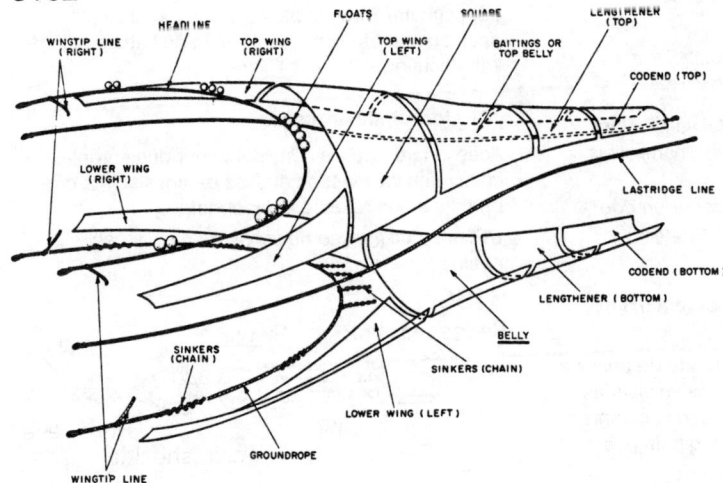

WINGTIP LINE (RIGHT) — HEADLINE — TOP WING (RIGHT) — FLOATS — SQUARE — TOP WING (LEFT) — BAITINGS OR TOP BELLY — LENGTHENER (TOP) — CODEND (TOP) — LASTRIDGE LINE — LOWER WING (RIGHT) — CODEND (BOTTOM) — LENGTHENER (BOTTOM) — BELLY — SINKERS (CHAIN) — SINKERS (CHAIN) — LOWER WING (LEFT) — GROUNDROPE — WINGTIP LINE (LEFT)

3165

ES captura accesoria

DA bifangst

I forbindelse med fiskerireguleringer: den del af fangsten, der ikke er målet for fiskeriet eller udgør den største andel. Den uønskede del af fangsten.
Ofte er bifangsten defineret af en bestemt fiskeriregulering, der lægger en begrænsning på, hvor stor en andel af hele fangsten bifangsten må udgøre. Der er her tale om fiskerier, hvor hovedparten kræver anvendelse af en maskevidde, der er mindre end almindeligt anvendt til konsumfiskeri. I visse af disse fiskerier er bifangsten imidlertid en væsentlig indtægtskilde, fordi den består af værdifulde fisk eller skaldyr, og den tilladte bifangst vil blive ilandbragt og solgt.

DE Beifang

Sammelbezeichnung für Fische und Meerestiere, die mitgefangen, aber nicht zu Speisezwecken verwendet werden.
Man unterscheidet echten und speziellen Beifang; echter Beifang gerät unbeabsichtigt ins Schleppnetz (Anteil am Gesamtfang meist gering). Die Nachfrage nach Fischmehl hat zur Ausübung des speziellen Beifangs geführt (Anteil liegt weit höher). Zum Schutz wurden bereits Maßnahmen zur Einschränkung des speziellen Beifangs eingeführt.

GR βοηθητικό εργαλείο για τη σύλληψη

EN by-catch

That part of the catch which is not the targeted species but which may be retained for sale.

FR prises accessoires

Partie des captures d'un engin de pêche, ne constituant pas la ou les espèces principales recherchées ou visées par cet engin, mais faisant l'objet d'une pêche accidentelle et inévitable dans l'état actuel de la technologie des pêches.

IT cattura accessoria

NL bijvangst

Gedeelte van de vangst, dat in aantal veel kleiner is dan de vissoort waarop de visserij voornamelijk is gericht.
Meestal vangt men verschillende vissoorten door elkaar; de bijvangst kan toch een belangrijke inkomstenbron zijn.

PT captura incidental; captura acidental; «by--catch»

Parte das capturas obtidas com uma arte de pesca e que não constituem a ou as espécies alvo visadas, mas antes são o resultado de uma pesca acidental e inevitável no estado actual da tecnologia da pesca.

3166

ES giratorio; eslabón giratorio

Una de sus mitades es un grueso perno de ojo que pasa por un agujero practicado en la otra mitad, de forma particular y remachado por dentro, pero dejándolo con juego para que pueda girar libremente.

DA drejeled

Kraftig svirvel, der, anvendt som samleled, tillader at delene frit kan dreje i forhold til hinanden.

DE Wirbelschäkel; Kettenwirbel

Schäkel mit einem Wirbel am oberen Teil des Bügels.

GR στριφτάρι

Δύο κρίκοι ενωμένοι με ένα στροφέα (άξονα) έτσι ώστε να επιτρέπει στον έναν κρίκο να περιστρέφεται ανεξάρτητα από τον άλλο.

EN swivel

Two links joined end to end by a pivot, thus allowing one link to rotate without the other moving.

FR émerillon

Accessoire comportant un axe central et présentant en général un anneau à chaque extrémité. Utilisé habituellement pour relier deux cordages (ou deux lignes) dont l'un d'entre eux peut tourner librement, sans vriller.
Employé également à la liaison d'un cordage, à une ferrure ou à un engin.

IT tornichetto

Congegno formato da due parti unite, ognuna delle quali può ruotare indipendentemente dall'altra.

NL kettingwartel; wartel

Oog of schalm, zodanig verbonden aan een ander oog, schalm, haak of blok, dat het ten opzichte daarvan kan draaien, om de as in de ketting of in de kabelrichting.

PT tornel; destorcedor

Acessório de pesca constituído por duas argolas colocadas topo a topo através de um sistema que permite a sua rotação independente.
Utiliza-se na ligação de linhas de pesca, cabos e amarras.

3166

Recessed link   Swivel   Warp
Link   Warp shackle   Hard eye

3167

ES calón

Parte del «danleno» en forma de bumerán.

DA butterfly

Del af et danlenoarrangement. Hængslet metalplade, der forbinder danlenokuglens aksel med de bagved værende to wirer eller kæder til henholdsvis undertælle og bobbinsgear.

DE Joch

Teil des Vorgeschirrs eines Grundschleppnetzes. Es dient zur Stabilisierung der Netzform. Weiter wird das Joch verwendet, um ausreichenden Abstand des Jagers vom Grund zu gewährleisten, wodurch Netzschäden infolge Grundberührung herabgemindert werden.

GR πεταλούδα

Τμήμα του συστήματος dan leno. Ατσάλινο έλασμα σχήματος L μεταξύ του άξονα του dan leno και των σκελών και προσαρτημένο στα σκέλη με κλειδιά.
Τράτα

EN butterfly; spreader bar

Part of a dan leno assembly. L-shaped or triangular steel plate shackled between dan leno spindle and legs.
Trawl.

FR pistolet; guindineau

Pièce d'acier en forme de crosse (ou L aplati), placée entre la sphère de guindineau et les entremises du chalut.

IT mazzetta «butterfly»

Non è usata sulle reti italiane mediterranee.

NL danlenospreider; spreider; butterfly

Onderdeel van een danleno waaraan de voorlopers naar de nokken van het net worden verbonden, in de vorm van een zwaluwstaart, om de verticale opening te verkrijgen.

PT borboleta

Dispositivo de aço do conjunto do calão das redes de arrasto pelo fundo, com a forma de L aberto, situado entre os tirantes das malhetas e a esfera do calão.

3167
Dan leno assembly

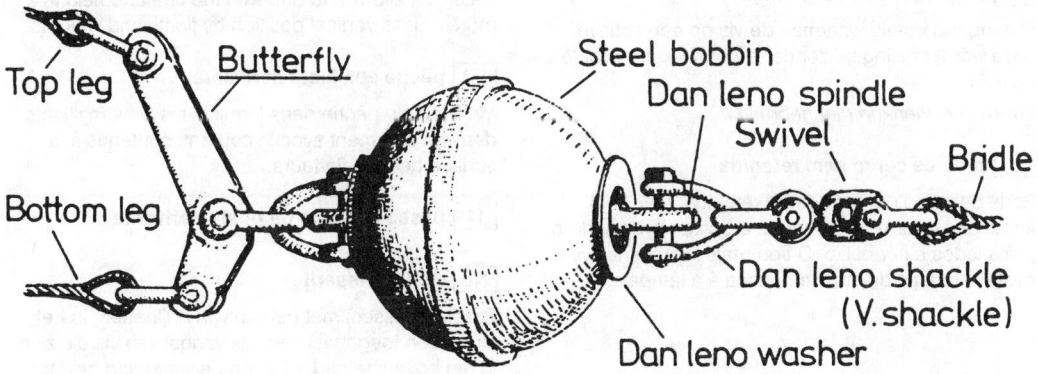

Top leg • Butterfly • Steel bobbin • Dan leno spindle • Swivel • Bridle • Bottom leg • Dan leno shackle (V. shackle) • Dan leno washer

3168

ES red de cerco sin jareta

DA not uden snurpeline

Redskabet bruges ikke i Danmark.

DE Umschließungsnetz ohne Purseleine

Eingesetzt zum Fang nahe der Oberfläche georteter Fischschwärme, wird es mit einer Schwimmleine an der Oberfläche gehalten. Handhabung von einem oder mehreren Booten aus.

In Ost- und Südostasien sind eine Reihe von Umschließungsnetzen ohne Purseleine in Gebrauch wie z. B. das Lampara-Netz, ähnlich den europäischen Waden ohne Netzsack.

GR κυκλωτικό δίχτυ χωρίς στίγγο

EN surrounding net without purse line

A surrounding net without a purse line but with a special shape which still allows it to encircle fish both from the sides and from underneath.

FR filet tournant sans coulisse

Filet tournant dépourvu de coulisse, mais dont la forme particulière permet l'encerclement du poisson à la fois par les côtés et par le bas.

Le type représentatif de cette catégorie d'engins est le lamparo.

IT rete da circuizione senza chiusura

Rete sprovvista di cavo di chiusura sulla corda dei piombi.

NL ringzegen zonder sluitlijn

Omringend vistuig waarmee de vis op een actieve wijze wordt omsingeld zonder de onderkant dicht te trekken.

Wordt in Nederland niet gebruikt.

PT rede de cerco sem retenida

Rede de cerco desprovida de retenida, mas cuja forma particular permite cercar o peixe envolvendo-o pelos lados e por baixo. O tipo mais representativo desta categoria de artes de pesca é a lâmpara.

3169

ES pesca con red de deriva

Las redes de deriva se calan en el mar, pendientes de sus relingas de corchos, y de boyas en sus extremos, y que se mantienen verticales por el peso de sus relingas de plomos. No se ponen en contacto con el fondo, es decir, no se fondean ni por anclas, lastres o muertos, por lo que quedan a merced de los movimientos del mar, olas, corrientes, mareas, etc., que las hacen derivar de un sitio para otro, capturando en su decurso a los peces que encuentran y que intentan pasar a través.

DA drivgarnsfiskeri

Passiv fiskerimetode. Garnene sættes i overfladen og driver passivt med strømmen. Fiskene sætter sig i nettets masker.

DE Treibnetzfischerei

Fangmethode der Hochseefischerei und Küstenfischerei. Sie ist im Gegensatz zur Schleppnetzfischerei eine passive Fangmethode, bei der der Fang durch Vermaschen des Fisches bei seinem Auftreffen auf die Netzwand erfolgt.

Mit Treibnetzfischerei werden Fische des Pelagials gefangen.

GR αλιεία με παρασυρόμενο δίχτυ

Αλιευτική μέθοδος κατά την οποία τα δίχτυα δεν σύρονται στο νερό αλλά αφήνονται να παρασύρονται ή να οδηγούνται από την παλίρροια και σε θέση λίγο-πολύ κατακόρυφη.

EN drift net fishing; drifting

A method of fishing with nets not drawn through the water but allowed to drift with the tide and held in a more or less vertical position by floats and weights.

FR pêche aux filets dérivants

Méthode de pêche dans laquelle les filets maillants dérivent librement avec le courant, soutenus à la surface par des flotteurs.

IT pesca con rete da posta derivante

NL drijfnetvisserij

Passieve visserij met behulp van drijfnetten; in het algemeen toegepast voor de vangst van vis die zich in het bovenste gedeelte van de waterlaag bevindt, zoals zalm, haring en makreel.

Kan ook op de bodem worden gebruikt.

PT pesca com redes de emalhar de deriva

3170

ES cacea; curricán

Aparejo de pesca consistente en una cuerda sujeta a un palo colocado en la popa o en un costado de la embarcación; lleva un anzuelo en el que en vez de cebo, se pone un trapo rojo o un manojo de plumas o de paja, que, flotando tras el barco, atrae a los peces a morderlo, especialmente a los atunes, bonitos y doradas, haciendo sonar un cascabel sujeto al extremo del palo, lo que sirve de aviso de que hay pesca.

DA dørgeline

En af flere liner med en eller flere kroge hver, der slæbes efter et fartøj i moderat fart.
Finder anvendelse ved fiskeri efter makrel, tun, sild og andre fisk, der jager i overfladen.

DE Schleppangel; Dorre; Schleppangelleine

Fanggerät der Hochsee- und Küstenfischerei zum Fang von Thun, Makrele und Schwertfisch. Vom Fangfahrzeug werden eine oder mehrere Angelleinen geschleppt. An den Enden befinden sich Haken mit natürlichen oder künstlichen Ködern. Durch Beiholerleinen können die unterschiedlich langen Angelleinen während des Schleppens zur Kontrolle oder Entnahme des Fangs an Bord genommen werden.

GR συρτή

Απλή ορμιά, εφοδιασμένη με δόλωμα φυσικό ή τεχνητό (δέλεαρ) που σύρεται κοντά στην επιφάνεια της θάλασσας ή σε κάποιο βάθος από ένα σκάφος. Γενικά πολλές ορμιές ρυμουλκούνται συγχρόνως με τη βοήθεια προωστών.

EN trolling line; troll line

One of the lines carried by the fish poles winged out on each side of a trolling boat. Usually made of hard-laid cotton with smaller linen leader lines followed by a piano wire leader to the spoon or bait.

FR ligne de traîne

Ligne simple, munie d'appât naturel ou artificiel (leurre) et traînée près de la surface ou à une certaine profondeur par un bateau. En général, plusieurs lignes sont remorquées simultanément, à l'aide de tangons.

IT lenza trainata; lenza al traino; traina; lenza trainata da natante

Lenza trainata a velocità da natanti.

NL vislijnen van de troller; sleeplijnen

Door een troller gesleept stel lijnen, bevestigd aan lange staken, die zodanig zijn opgetuigd dat de lijnen van elkaar vrijlopen; elke lijn verdeelt zich achter het schip in twee lijnen, elk aan het einde voorzien van een haak.
Deze visserijmethode komt in Nederland niet voor.

PT corrico; corripo

Arte de pesca rebocada por uma embarcação e constituída por uma linha simples que na extremidade livre é dotada de um troço de arame na ponta do qual se fixa um anzol simples ou duplo. Este anzol não é geralmente iscado com isco natural mas sim com uma amostra (pedaço de pano, penas, plumas, etc.).
As embarcações de pesca ao corrico utilizam várias destas linhas simultaneamente.

3170

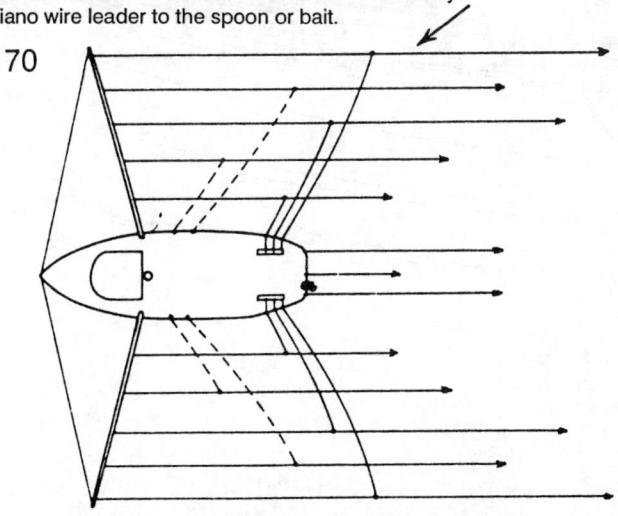

3171

ES | palo bípode; palo de popa

Estructura a popa del puente que sirve para sujetar los aparejos y para subir el saco por la rampa.

DA | tværbom; agtergalge

Kraftig opstående jernkonstruktion agter på en hæktrawler. Montering for trawlblokkene og løfteblokkene for posen m.m.

DE | Heckportal; Heckgalgenportal

Tragendes Konstruktionsteil für die Kurrleinenblöcke.
In der pelagischen Fischerei tragendes Teil für die Ausleger mit den dazugehörigen Kurrleinenblöcken.

GR | πρυμναίο καπόνι

Μηχανότρατα πρυμναίας σύρσης.

EN | stern gantry

Goal post structure mounted above the stern of a stern trawler to carry towing blocks, codend lifting blocks, outhauls etc.
Stern trawler.

FR | portique; portique arrière

Structure placée transversalement à l'arrière du chalutier et servant de support aux potences; utilisée également pour la manœuvre du chalut et pour l'embarquement du poisson.
Chalutier à pêche arrière.

IT | arcone

NL | hekportaalmast; portaal

Constructie in de vorm van een portaal, bevestigd op het hek van een hektrawler, ter geleiding van vislijnen, visborden, netsondekabel, voorlopers en het net, tijdens het uitzetten en binnenhalen.
Bij Nederlandse hektrawlers bevinden zich hierin ook de schijven ter geleiding van het net tijdens het halen, het zogenaamde „jojoën".

PT | pórtico de popa

Estrutura metálica colocada transversalmente à popa dos arrastões e que serve de suporte às patescas de arrasto; é ainda utilizado para a manobra da rede de arrasto e para o embarque das capturas.
Arrastões pela popa.

3171

3172

ES varilla graduada; calibrador de malla; galga

Instrumento utilizado para determinar la abertura de las mallas.

DA maskemåler

Instrument til bestemmelse af et nets maskestørrelse.

DE Maschenmeßgerät

Zur Bestimmung der Maschenöffnung verwendetes Instrument.

GR μετρητής ματιού

Όργανο που χρησιμοποιείται για τον προσδιορισμό του ανοίγματος ενός ματιού.

EN mesh gauge

Instrument used for determining the opening of mesh.

FR jauge de maille

Instrument utilisé pour déterminer l'ouverture de la maille.

IT misuratore di maglia

Strumento impiegato per determinare l'apertura delle maglie.

NL maaswijdtemeter; schiel

Instrument waarmee de maaswijdte kan worden gemeten.

- Er bestaan verschillende systemen, maar ze zijn alle gebaseerd op het strekken van de garens in de N-richting met een bepaalde voorgeschreven kracht.
- De wigvormige maaswijdtemeter wordt „schiel" genoemd.

PT bitola

Instrumento utilizado para determinar a malhagem.

3173

ES longitud de la malla

a) Para la red anudada, distancia existente entre los centros de dos nudos opuestos en una misma malla completamente tensa en el sentido N.
b) Para la red sin nudos, distancia existente entre los centros de dos cruces opuestos en una misma malla completamente tensa en el sentido que da a la dimensión su valor máximo.

DA strakt maskestørrelse; helmaske

Afstanden mellem centrum af to modstående knuder i en strakt maske.

På dansk anvendes maskevidde og maskestørrelse ofte synonymt, og man skelner mellem:
a) maskestørrelse eller halvmaske: afstanden mellem centrum af to knuder i siden af en maske, se 3310
b) strakt maskestørrelse eller helmaske: afstanden mellem centrum af to modstående knuder i en maske (denne)
c) indvendig maskevidde, se 3142.
Se tillige 3008, 3013 og 3315.

DE Maschenlänge

a) Bei geknoteten Netztuchen, Abstand zwischen den Mitten zweier einander gegenüberliegender Knoten einer in N-Richtung gestreckten Masche;
b) bei knotenlosen Netztuchen, Abstand zwischen den Mitten zweier einander gegenüberliegender Verbindungsstellen einer Masche, die im Sinne der N-Richtung gestreckt ist.

GR μήκος ματιού

α) Για δικτύωμα με κόμπους: η απόσταση μεταξύ των κέντρων δύο απέναντι κόμπων ενός ματιού, όταν αυτό εκτείνεται πλήρως κατά τη N-διεύθυνση.
β) για δικτύωμα χωρίς κόμπους: η απόσταση μεταξύ των κέντρων δύο απέναντι ενώσεων ενός ματιού, όταν αυτό εκτείνεται πλήρως κατά μήκος του μεγαλύτερου άξονά του.

EN length of mesh

(a) For knotted netting, the distance between the centres of two opposite knots in the same mesh when fully extended in the N-direction;
(b) for knotless netting, the distance between the centres of two opposite joints in the same mesh when fully extended along its longest possible axis.

3173

FR **longueur de maille**

a) Pour la nappe nouée, distance existant entre les centres de deux nœuds opposés dans une même maille complètement tendue dans le sens N.
b) Pour la nappe sans nœuds, distance existant entre les centres de deux croisements opposés dans une même maille complètement tendue dans le sens qui donne à la dimension sa valeur maximale.

IT **lunghezza di maglia**

Per la rete annodata: distanza esistente tra i centri di 2 nodi opposti di una stessa maglia completamente tesa nella direzione N. Per la rete senza nodi: distanza esistente tra i centri di 2 incroci opposti di una stessa maglia completamente tesa nella direzione che fornisce alla dimensione il valore maggiore.

NL **maaslengte**

De afstand tussen twee tegenover elkaar liggende knopen of verbindingen van dezelfde maas, gemeten van hart tot hart terwijl het netwerk in de lengterichting is gestrekt.

PT **malhagem**

a) Para as redes com nós, é a distância entre os meios de dois nós opostos e consecutivos de uma malha completamente estirada segundo a direcção normal.
b) Para as redes sem nós, é a distância entre os meios de dois entrelaçamentos opostos e consecutivos de uma malha completamente estirada segundo a direcção que permite o seu máximo valor.
Normalmente expressa-se em milímetros.

3174

ES **vigilancia de pesca**

DA **fiskeriinspektion; fiskerikontrol**

DE **Fischereischutz**

Überwachung und Unterstützung der Fischerei.
Die rechtliche Grundlage bildet das Gesetz über die Aufgaben des Bundes auf dem Gebiet der Seeschiffahrt.

GR **προστασία της αλιείας**

Διαδικασίες που αποσκοπούν στον έλεγχο των δραστηριοτήτων των αλιευτικών σκαφών και των εργαλείων που χρησιμοποιούνται από αυτά.
Στην Ελλάδα ασκείται από το Λιμενικό Σώμα.

EN **fisheries protection**

Surveillance and support operations for fishing vessels and gear.

FR **surveillance des pêches**

Opérations visant à contrôler l'activité des navires de pêche et les engins qu'ils utilisent.

IT **vigilanza sulla pesca**

NL **visserij-inspectie**

Toezicht op het visserijbeleid door hiertoe bevoegde organen.

PT **vigilância de pesca; fiscalização de pesca**

3173

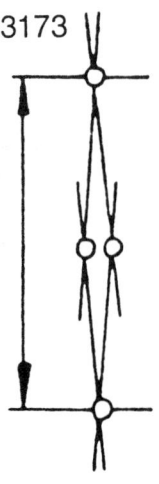

3175

ES curricanero

Buque pesquero para la pesca pelágica «al curricán» o «a la cacea», que remolca cierto número de líneas de pesca, dotadas con cebos adecuados. Para la faena de halar las líneas de pesca se emplean carreteles accionados hidráulicamente o eléctricamente.

DA fartøj, der driver dørgefiskeri

Fartøjer specielt indrettet til dette fiskeri vil ofte være forsynet med to meget lange tynde bomme, der kan svinges ud til siden for at sprede dørgelinerne.

DE Rollangelfischer; Schleppleinenfischer

Fahrzeug mit bestückten Leinen, die eine gewisse Fläche hinter dem Boot bestreichen. Zur Bestreichung einer noch größeren Fläche werden Angelleinen an bis zu 20 Meter breiten Bäumen ausgesetzt. Schleppgeschwindigkeit ca. 5 Knoten.
Überwiegend zum Fang des weißen Thunfischs eingesetzt.

GR αλιευτικό σκάφος με συρτή· αλιευτικό σκάφος με συρτές

Γενική ονομασία για βενζινάκατο, καΐκι ή βάρκα με κουπιά που αλιεύει με συρτή(-ές).

EN troller; trolling boat

A vessel used for the catching of pelagic fish by towing a number of lines fitted with lures.
The lines are attached to trolling booms which are raised and lowered by topping lifts and fore and aft stays. Hydraulic or electrically powered reels (or gurdies) are frequently used to haul in the lines.

FR bateau de pêche à la traîne

Bateau utilisant des lignes de traîne.

IT peschereccio con lenze trainate

Peschereccio utilizzato per la cattura di pesce pelagico trainando lenze con ami escati. Le lenze sono collegate a canne che vengono sollevate e abbassate azionando gli amantagli e i tiranti a prora e a poppa.

NL troller

Type vissersvaartuig voor de sleeplijnvisserij.

PT embarcação de pesca ao corrico; embarcação de corrico

Barco de pesca que utiliza linhas de corrico.

3175

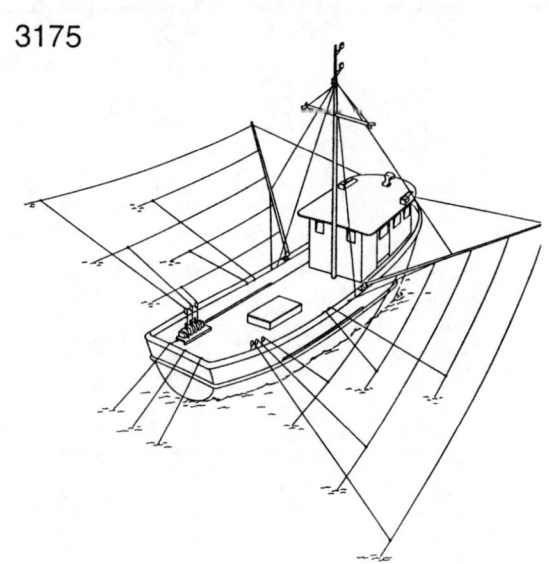

3176

ES buque guardapesca; guardapesca

Buque usado para la protección de los caladeros y la vigilancia de las faenas de pesca. Estos buques pueden ser unidades navales con armamento ligero. En algunos casos pueden operar en colaboración con helicópteros o aviones navales y contar con la asistencia de sus propias embarcaciones auxiliares.

DA fiskeriinspektionsskib

Fartøj — normalt let bevæbnet — tilhørende den nationale flåde, helikopterførende og medførende hurtiggående motorbåde. Udfører ligeledes redningstjeneste.

DE Fischereischutzboot; FSB

Schiff zur Überwachung und Unterstützung der Fischerei.

GR σκάφος για την προστασία της αλιείας

Σκάφος προορισμένο για την προστασία των αλιευτικών δραστηριοτήτων στην ανοιχτή θάλασσα. Μπορεί να προσφέρει κατά περίπτωση, υλική βοήθεια στους ψαράδες περιλαμβανομένης και της ιατρικής περίθαλψης.

EN fishery protection vessel; fisheries protection vessel

A vessel used for the protection of the fishery grounds and surveillance of fishing activities.

These vessels may be lightly-armed naval vessels, may be equipped to operate sea-planes or helicopters and the larger vessels normally possess small fast launches.

FR navire de surveillance des pêches; navire de surveillance et de protection des pêches; navire garde-pêche

Navire chargé de contrôler l'exploitation des zones de pêche et de surveiller les bateaux qui opèrent dans les eaux territoriales et dans la ZEE (zone économique exclusive).

Leur taille varie selon la zone où ils opèrent: taille moyenne dans la zone hauturière ou pour la ZEE, taille plus petite dans la zone côtière.

IT nave guardapesca

Nave utilizzata per la protezione delle zone di pesca e per la sorveglianza delle attività di pesca.

NL visserij-patrouillevaartuig; visserij-inspectievaartuig; visserijwachtschip

Vaartuig dat toeziet op de visserij ter voorkoming van verboden visserij-operaties: vissoort, tuigen, maaswijdte, quota, vermogen, enzovoort.

Het schip kan licht bewapend zijn en geschikt voor watervliegtuigen of helikopters. De grotere schepen hebben meestal snelle bijboten.

PT navio de fiscalização de pesca; lancha de fiscalização de pesca

Navio militar especialmente concebido para fiscalizar as actividades pesqueiras e patrulhar as zonas protegidas. Pode estar equipado para colaborar com meios aéreos e, normalmente, possui embarcações rápidas para exercer o direito de visita nos navios de pesca fiscalizados.

3177

ES inspector; inspector de pesca

DA fiskeribetjent; fiskeriinspektør

DE Fischereiinspektor

GR αξιωματικοί Λιμενικού Σώματος

Προσωπικό που ανήκει στο Υπουργείο Εμπορικής Ναυτιλίας (Λιμενικό Σώμα), επιφορτισμένο και με την προστασία της παράκτιας αλιείας.

EN fishing guard; fishing patrol; protection officer

A person undertaking fishery protection duties.

FR garde-pêche

Personnel dépendant de la marine marchande, chargé de la surveillance des pêches côtières; ce personnel arme, à l'occasion, des vedettes garde-pêche.

Pluriel: gardes-pêche

IT guardapesca; guardiapesca

Agente adetto alla sorveglianza della pesca.

NL visserij-inspecteur

Persoon belast met het toezicht op naleving van nationale en internationale visserijvoorschriften.

PT inspector de pesca

Agente encarregado das actividades de fiscalização da pesca.

Em Portugal é pessoal de Marinha de Guerra que desempenha a missão.

3178

ES desembarque

DA landing

Udtryk for at bringe fartøjet til land for at losse (eller lande) fangsten.

DE Anlandung

Die Ladung eines Schiffs, insbesondere den Fang eines Fischereifahrzeugs, an Land geben.

GR εκφόρτωση

EN landing

The action of bringing a fishing vessel to a port to unload its catch.

Landing may be used to refer to the catch itself.

FR mise à terre

– Opération de déchargement de la capture au port;
– quantité de poissons ramenée à terre par un navire de pêche.

IT sbarco

Far discendere la cattura da un peschereccio.

NL aanvoer

Hoevelheid vis, schelp- en schaaldieren die een visserijvaartuig aan wal brengt.

PT desembarque

Das capturas.

3179

ES pesca costera

DA kystfiskeri

Udtryk for et simpelt fiskeri, f.eks. fra kysten, med åbne både, af kort varighed, med simple redskaber eller lignende.

DE Küstenfischerei

Fischerei innerhalb des Küstengewässers.

GR παράκτια αλιεία

EN inshore fishing; inshore fishery; coast fishery

One of the fisheries along the seashore carried on by small open boats and other craft usually within territorial waters.

They include the catching of shrimp, lobsters, crabs, prawns; the breeding and fattening of oysters; the gathering of cockles and mussels, and so on.

FR pêche côtière

Pêche pratiquée par des petits bateaux, d'une longueur inférieure à 24 mètres, s'absentant du port pour une durée comprise entre 24 et 96 heures.

IT pesca costiera

Pesca esercitata con barche che si trattengono in mare per pochi giorni, non lontane dalla costa.

NL kustvisserij

Visserij uitgeoefend op visgronden, die zich op geringe afstand van de kustlijn bevinden.

- De reizen duren meestal relatief kort, enkele dagen, of minder dan een etmaal.
- In vele gevallen is het vistuig vanaf de kust te bedienen.

PT pesca costeira

3180

ES caladero; pesquería; paraje de pesca; puesto de pesca, región de pesca

DA fiskeplads; fiskebanke

DE Fangplatz; Fanggrund; Fischgrund

GR αλιευτικό πεδίο

Θαλάσσια περιοχή όπου η αλιεία είναι κανονική ή ασκείται με δεδομένη συχνότητα.

EN fishing ground

Area of the sea in which fishing is normally or frequently carried on.

FR lieu de pêche; zone de pêche; région de pêche; fond de pêche; pêcherie *

Endroit en mer où la pêche est régulièrement ou fréquemment pratiquée.

* Rarement utilisé dans ce sens.

IT zona di pesca; fondale di pesca

NL visgrond; visserijgebied

Gebied op zee waar de visserij wordt uitgeoefend.

Men spreekt meestal ook van visgronden waar het de pelagische visserij betreft.

PT caladouro; banco de pesca; pesqueiro; fundo de pesca

Local onde a pesca é normal e frequentemente exercida.

3181

ES pesca en pareja

Conjunto de dos embarcaciones iguales con las que se remolca un arte de arrastre.

DA partrawling; tvillingslæb

Fiskerimetode, hvor to fartøjer slæber på samme trawl. Fartøjernes indbyrdes afstand spiler trawlen åben horisontalt, hvorfor der ikke anvendes skovle.

Partrawling er den foretrukne metode ved flydetrawlsfiskeri i overfladen, hvor støjen fra de to fartøjer kan skræmme fiskene ind mod midten i trawlens bane.

DE Gespannfischerei

Fischfangmethode der Schleppnetzfischerei, bei der ein Zweischiffschleppnetz von zwei Fangfahrzeugen an Kurrleinen geschleppt wird.

GR καλάρισμα τράτας με ζευγαρωτά σκάφη

Μέθοδος σύμφωνα με την οποία η τράτα σύρεται από δύο σκάφη. Η απόσταση μεταξύ των δύο σκαφών εξασφαλίζει το οριζόντιο άνοιγμα του διχτυού.

EN pair trawling; bull trawling

Method in which the trawl is towed by two boats of similar power. The separation of the boats controls the opening of the net.

FR chalutage en bœufs; chalutage à deux bateaux

Méthode selon laquelle le chalut est traîné par deux bateaux. L'écartement des deux bateaux assure l'ouverture horizontale du filet.

IT pesca con rete da traino a coppia

NL spanvisserij; spannetvisserij

Vismethode waarbij een net wordt gesleept door twee schepen, die op een afstand van elkaar op evenwijdige koers varen, en zodoende voor een grote horizontale opening zorgen.
Zie wonderkuil.

PT pesca de parelha

Pesca com rede de arrasto rebocada por duas embarcações com potências aproximadamente iguais.

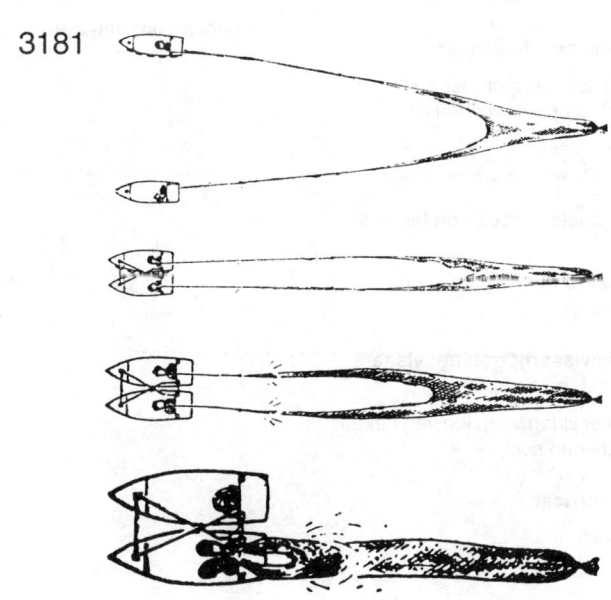

3181

3182

ES pesca a la cacea

DA dørgefiskeri

Krogfiskeri, hvor en line med en eller flere kroge med kunstig eller naturlig agn slæbes efter fartøjet.

DE Schleppleinenfischerei; Schleppangelfischerei; Rollangelfischerei; Laufangelfischerei

Vom Fangfahrzeug werden eine oder mehrere Angelleinen geschleppt. An den Enden befinden sich Haken mit natürlichen oder künstlichen Ködern. Durch Beiholerleinen können die unterschiedlich langen Angelleinen während des Schleppens zur Kontrolle oder Entnahme des Fangs an Bord genommen werden.

GR αλιεία με συρτή

Μέθοδος αλιείας η οποία συνίσταται βασικά στη σύρση (στο νερό) ενός δολώματος ή λαμπερού αντικειμένου όπου είναι προσδεδεμένο ένα αγκίστρι. Χρησιμοποιείται για τη σύλληψη θηρευτών που διατρέφονται με αφρόψαρα και σύρεται από καΐκια, βενζινακάτους και βάρκες με κουπιά.

EN troll lining; trolling; whiffing; railing

A method of fishing which consists essentially of dragging through the water a bait or bright object to which a hook is attached. It is employed for the capture of predaceous surface-feeding fishes.

FR pêche à la traîne; pêche à la caille *

Méthode de pêche consistant à remorquer en surface des lignes simples garnies d'appâts naturels ou artificiels.

* Terme local.

IT pesca con lenza trainata; pesca con lenza al traino; pesca alla traina

Si effettua rimorchiando lenze inescate dalla poppa di un'imbarcazione in movimento.

NL sleeplijnvisserij; lijnvisserij; trolling; visserij met sleeplijn; sleephengelen

Het slepen van één of meer lijnen voorzien van haken met aas of kunstaas achter een schip.

PT pesca ao corrico; corricar

Pesca com linhas de corrico.

3183

ES pesca de palangre

DA langlinefiskeri

Passiv fiskerimetode med langliner.

DE Langleinenfischerei; Großangelfischerei

Fischen mit gewichtbeschwerten am Grund ausgelegten langen Leinen, die in Abständen mit Vorfächern bestückt sind, und an deren Enden der beköderte Angelhaken befestigt ist.

GR αλιεία με παραγάδια

Μέθοδος αλιείας που περιλαμβάνει τη χρήση ορμιάς μεγάλου μήκους που φέρει πολυάριθμα αγκίστρια.

EN long lining

Fishing with long lines.

FR pêche à la palangre

Méthode de pêche comprenant l'emploi de lignes très longues (palangres) munies de nombreux hameçons.

IT pesca con palangresi; pesca col palangaro

NL visserij met de beug; beugvisserij

Passieve vismethode waarbij wordt gevist met de beug.

PT pesca com palangre

3184

Aparejo formado por un largo cordel con ramales y
un anzuelo en el extremo de cada uno de éstos. El
cordel principal se llama madre y cada ramal,
brazolada.

DA langline

Fiskeredskab bestående af en lang hovedline, hvortil
der med en vis afstand er fæstnet tavser, korte
sideliner med kroge og agn. Langliner kan være flere
km lange og lægges ud på bunden eller forsynes
med flåd og forbliver så ved overfladen eller i
midtvand. Krogstørrelse, afstand mellem tavser, agn
og positionering afpasses efter det bestemte fiskeri.

DE Langleine

Fanggerät der Hochsee-, Küsten- und
Binnenfischerei (Aalschnur). Sie wird zum Fang von
großen, einzeln lebenden und schnell
schwimmenden Fischen (Thun, Hai, Lachs, Kabeljau,
Aal) oder auch auf Fangplätzen, die mit anderen
Fanggeräten nicht befischt werden können,
eingesetzt (unreiner Meeresgrund). Die Konstruktion
der Langleine ist dem Fangobjekt und dem
Fangfahrzeug angepaßt. Die Langleine besteht aus
der Hauptleine (aus Sektionen zusammengesetzt),
deren Tiefe durch die Länge des Bojenreeps
bestimmt wird. Die Länge der Hauptleine erreicht bei
Thunangelleinen 160 km. Von der Hauptleine führen
die Mundschnüre zu den Haken, die mit natürlichen
Ködern besteckt werden.

GR παραγάδι

Αριθμός συνδεδεμένων ορμιών, οι οποίες είτε είναι
τοποθετημένες στον πυθμένα είτε παρασύρονται. Η
κάθε μία φέρει μεγάλο αριθμό δολωμένων
αγκιστριών.

EN long line; line

A number of connected lines, either set at the bottom
or drifting, each bearing a large number of baited
hooks.

FR palangre; corde

Une palangre comprend une ligne principale sur
laquelle sont fixés de nombreux hameçons par
l'intermédiaire d'avançons de longueur et
d'écartement variables selon l'espèce recherchée et
le type de palangres.

*On distingue les palangres de fond, mouillées au
fond, et les palangres dérivantes, supportées par des
flotteurs en surface.*

IT palangaro; palangrese; lenzara; parangale;
palamito

Attrezzo da pesca formato da un insieme di ami
collegati ad intervalli regolari ad un unico sostegno
(trave) mediante spezzoni di filo detti braccioli.

NL beug; beuglijn

Vistuig bestaande uit een lange zware lijn met op
bepaalde afstanden dunnere dwarslijnen, sneuen,
aan het uiteinde waarvan zich een haak bevindt.

PT aparelho de anzol; palangre; espinel

Arte de pesca constituída por uma linha de grande
comprimento (madre) à qual se ligam numerosas
linhas de pequeno comprimento (estralhos) na
extremidade livre das quais se empata um anzol.

3185

ES | filar; arriar

Aflojar, soltar poco a poco aunque comúnmente se entiende por soltar todo o de una vez.

DA | fire; give los; slække

DE | fieren

Eine Leine oder Kette mittels Winde oder Handkraft ablaufen lassen, nachlassen oder gleiten lassen.

GR | μαϊνάρω

Το λασκάρισμα σχοινιού ή παλαμαριού (καλώδιου).

EN | veer (verb); veer out (verb); pay out (verb); lower (verb)

To let out one end of a rope or cable in a controlled manner when the other end is fixed.

Used particularly for trawl wires or seine ropes.

FR | filer

Laisser aller une chaîne ou un câble dont l'une des extrémités est fixée.

S'emploie en particulier pour les funes en chalutage.

IT | filare

Lasciare scorrere un cavo o una catena a cui è applicata una resistenza.

NL | uitzetten; uitvieren; uitwinden

Laten uitlopen of schieten van het vistuig.

PT | arriar

Acto de deixar sair de uma embarcação para a água uma amarra ou um cabo fixos a bordo por uma das extremidades.

Os termos largar e calar aplicam-se no caso das artes de pesca.

3186

ES | halar; tirar; virar

Dar vueltas al cabrestante para levar las anclas o subir o bajar cosas a bordo.

DA | hale; hive; tage ind

DE | hieven; einholen *; einhieven *

Eine Leine oder Kette mittels Winde einholen oder steifholen; eine Last anheben.

* Netz.

GR | βιράρω

Η έλξη σχοινιού ή παλαμαριού με μηχανικό μέσο (π.χ. βαρούλκο).

EN | heave (verb)

To pull on a rope or cable with mechanical aid, as distinguished from hauling by hand.

FR | virer

Exercer un effort sur un cordage ou sur une chaîne par enroulement sur une bobine ou sur une poupée de treuil.

S'emploie en particulier pour les funes en chalutage.

IT | virare

Fare forza su di un cavo per tesarlo o per sollevare un peso. Lo sforzo può essere fornito da uomini o da un mezzo meccanico, e ridotto impiegando paranchi. La via del cavo può essere verticale o portata orizzontalmente per mezzo di bozzelli o pastecche.

NL | halen; hieuwen

Binnenhalen, scheephalen van het vistuig en de vangst.

PT | virar; alar

Acto de recolher para bordo uma amarra, um cabo ou uma arte de pesca.

3187

ES maquinilla de arrastre; chigre de pesca

Máquina de vapor o eléctrica, que, situada generalmente en cubierta, recibe el nombre de chigre cuando está destinada al servicio de carga y descarga, y de molinete si se emplea en levar anclas y maquinilla en pesca.

DA trawlspil

Spil til trawlwiren. Mekanisk, hydraulisk eller elektrisk drevet.

DE Kurrleinenwinde; Schleppnetzwinde; Netzwinde

Winde zum Einholen und Aussetzen eines Schleppnetzes, besonders zum Speichern der Kurrleine auf Trawlern.

GR βαρούλκο αλιείας· βαρούλκο τράτας· βίντσι τράτας

Βαρούλκο το οποίο έχει ένα ή περισσότερα τύμπανα που λειτουργούν ανεξάρτητα ή ταυτόχρονα για να κρατήσουν, βιράρουν ή μαϊνάρουν τα σχοινιά της τράτας. Μπορεί να έχει συσκευές ασφάλειας συνεχούς ελέγχου και οδηγούς σχοινιών για την προστασία σχοινιών, διχτυού και εργαλείων. Επίσης, μπορούν να προσαρμοστούν σ' αυτό βοηθητικά τύμπανα και κεφαλάρια.

EN trawl winch

A winch having one or more drums which may operate independently or simultaneously to hold, haul in or pay out the trawl ropes. It may have monitoring safety devices and spooling gear to protect ropes, net and gear. Auxiliary drums and warping ends may be fitted.

FR treuil de pêche; treuil de chalut

Treuil à un ou à plusieurs tambours qui peuvent fonctionner séparément ou ensemble pour maintenir, virer ou filer les funes. Il peut être équipé de dispositifs de sécurité de contrôle et de guide-funes, pour assurer le bon fonctionnement du train de pêche. Il peut protéger les funes, le chalut et le réducteur. Il peut avoir aussi des tambours auxiliaires et des poupées.

IT verricello per rete da traino; salpareti

NL trawllier

Lier voor het inhieuwen en vieren van de vislijnen aan boord van een trawler.

PT guincho de arrasto

Aparelho de força destinado à manobra dos tambores dos cabos reais.

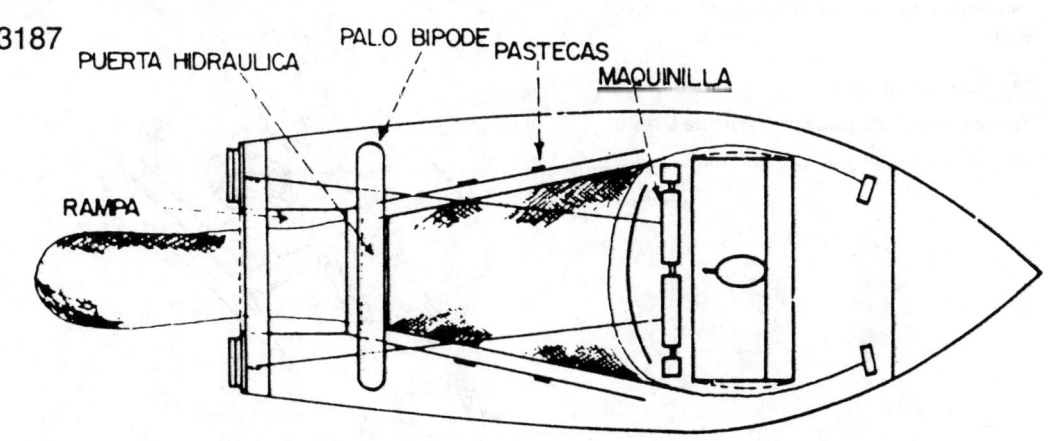

3187

PUERTA HIDRAULICA — PALO BIPODE — PASTECAS — MAQUINILLA

RAMPA

3188

ES tambor para estiba de la red; estibador de la red; tambor de red

DA **nettromle**

Tromle, der drives af et kraftigt spil, og hvorpå trawlen rulles op.
Et fartøj kan have flere tromler til forskellige trawl. Anbringes normalt i styrbord side (sidetrawlere) eller på agterdækket (hækken).

DE **Netztrommel**

Zumeist hydraulisch angetriebene Speichertrommel zur Aufnahme der Jager und des gesamten Netzes.

GR **τύμπανο περιέλιξης διχτυού**

Τύμπανο με μηχανική κίνηση, που χρησιμοποιείται στο βιράρισμα ή μάζεμα διχτυών μεγάλων διαστάσεων.

EN **net drum; net roller; transporter**

Wide, powered spool (usually hydraulic) on which a trawl or salmon purse seine net is wound when hauling.
May be powerful enough to haul a trawl aboard unaided.

FR **enrouleur de filet; tambour de chalut**

Tambour, entraîné mécaniquement, servant à virer ou à emmagasiner un filet de grande dimension (chalut, filet maillant, senne).

IT **tamburo avvolgirete; tamburo salparete**

NL **nettentrommel**

Mechanisch aangedreven trommel waarop het net kan worden gewonden na loskoppelen van het voertuig.

PT **tambor da rede**

Tambor onde é enrolada uma rede de arrasto.

3189

ES **maquinilla de enrollar la red**

DA **nettromlespil**

Spil til oprulning og opbevaring af et net.

DE **Netzwinde; Netzaufwickelwinde**

Winde mit netzaufwickelnder Trommel zum mechanisierten Aussetzen und Einholen von Fangnetzen.
Die Netzwinde wird vorzugsweise zur Handhabung von Schleppnetzen auf Trawlern verwendet.

GR **βαρούλκο διχτυού**

EN **net winch; net-winding winch**

Mechanically driven winch which hauls and stores a net.
Trawl mainly.

FR **treuil à filet; vire-filet**

Appareil servant au halage des filets.

IT **verricello salpareti**

NL **netlier; visnetlier**

Hijswerktuig met horizontale, draaibare spil, voorzien van een cilindrische trommel waarop touw, staaldraad of tros kan worden gewonden, en dat dient voor het ophalen van visnetten.

PT **guincho da rede**

Aparelho de força destinado à manobra do tambor da rede.

3188

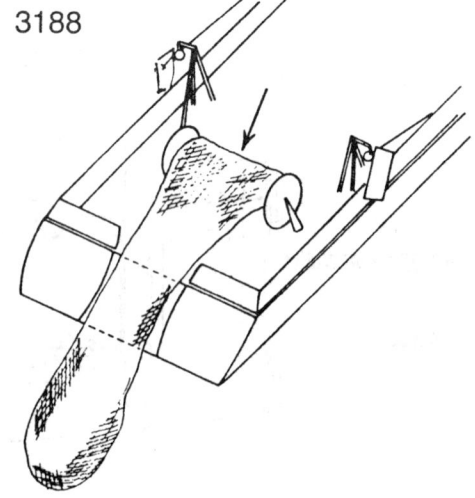

3190

ES canal

Ranura de un motón o cuadernal en la parte exterior de las quijadas para que no se corra la gaza.

DA skølp

Rille til fastgørelse af et tov udvendigt på en blok.

DE Kerbe; Blockkerbe

Einkerbung in der Backe eines Tau- oder Hangerblocks.

GR αυλάκωση· γλυφή· εγκοπή· εντομή

Αυλάκωση των πλαϊνών τοιχωμάτων μιας ξύλινης τροχαλίας για την υποδοχή του ιμάντα. Μια τροχαλία μπορεί να έχει μία ή δύο αυλακώσεις.

EN score

The groove made in the cheeks of a wooden block for receiving the strap.
A block may be single- or double-scored.

FR engoujure

Une ou deux entailles dans les joues d'une poulie en bois, destinées à recevoir une ou deux estropes.

IT cavatoia

Alloggio ricavato nei bozzelli, per farvi passare un cavo.

NL neut

Gleuf in de wang van een stropblok waarin een touw- of staaldraadstrop ligt.

PT goivado

Cavado na parte exterior das faces de um moitão por onde passa a alça de suspensão do moitão.

3191

ES carretel dividido

DA delt spiltromle

Spil, hvor akslen er forsynet med en ekstra flange, midterskjold, på midten.

DE geteilte Trommel

Trommel mit Zwischenschild.

GR σχιστό τύμπανο· τύμπανο με ενδιάμεση φλάντζα

EN split drum

Drum with a supplementary flange between the two ends.

FR tambour cloisonné

Tambour comportant un flasque supplémentaire entre les deux extrémités. Le flasque supplémentaire peut comporter une fente ou non.

IT tamburo con flangia intermedia

Tamburo dotato di tre flange delle quali due lo delimitano alle estremità mentre l'altra ne suddivide la capacità.

NL trommel met scheiding

Trommel die door een tussenschild is opgedeeld in twee gedeelten, waarop verschillende kabels kunnen worden gewikkeld.

PT tambor duplo

3190

Goivado

3192

ES pasteca; pasteca de retorno

Especie de motón herrado que tiene abierta una de sus caras laterales por un punto superior al lugar que ocupa el círculo de la roldana, a fin de que pueda meterse por seno el cabo que ha de laborear por ella, y también para sacarlo sin que haga falta pasarlo por el chicote.

DA kasteblok

Blok med en enkelt skive, hvis blokhus i den ene side er indrettet til at lukke op ved drejning om et hængsel, således at man kan lægge et tov eller en trosse ind i blokken.

DE Fußblock; Klappblock; aufklappbare Umlenkrolle

Einscheibiger Stahlblock, bei dem eine Backe aufgeklappt und das Tauwerk in den Tauraum eingelegt werden kann. Wird vorwiegend zur Änderung der Zugrichtung von laufendem Gut verwendet.

GR λυκίσκος· ματσαπλί σχιστό· σπαστή τροχαλία

Τροχαλία μακριά και λεπτή, της οποίας η θήκη είναι ανοιχτή στο ένα από τα δύο μάγουλα με τέτοιο τρόπο ώστε να μπορούμε να τοποθετήσουμε μέσα σ' αυτή το σχοινί που θέλουμε να βιράρουμε.

EN snatch block

A single block so fitted that the bight of a rope may be passed through it, without the delay of reeving or unreeving. The iron strap is hinged on one side and the shell is divided to allow the rope to be shipped into the sheave.

FR poulie coupée; galoche

Poulie longue et plate, dont la caisse est ouverte sur l'une de ses joues, de façon que l'on puisse y introduire librement le cordage que l'on veut virer.

IT bozzello apribile

Grosso bozzello in legno o ferro con gancio girevole in testa, una cavatoia con puleggia, e con sportello apribile e chiudibile, che consente di immettere nella cavatoia un cavo anche se il capo non è libero.

NL voetblok; kinnebaksblok

Lang blok, meestal met een wartelhaak en een dwarse gleuf in een der haken, zodat de loper zonder meer op de schijf kan worden gelegd.

PT patesca; patesga

Poleame de laborar semelhante ao moitão, tendo uma abertura lateral fechada com dobradiça e chaveta, para gornir o cabo pelo seio.

É especialmente empregada para dar retorno a cabos.

3192

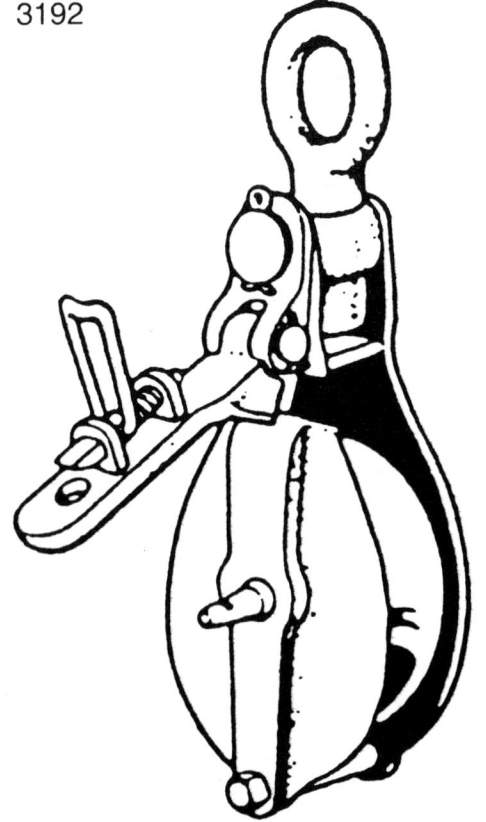

3193

ES arte de pesca

DA fiskeredskab

DE Fischfanggerät; Fanggerät

Fangausrüstung des Fischers. Dazu gehören Angeln, Fangkammern und die verschiedenen Arten von Netzen.

GR αλιευτικά εργαλεία

Είδη αλιευτικού εξοπλισμού που βρίσκονται στο νερό κατά τη διαδικασία της αλιείας.

EN fishing gear

Apparatus for catching fish which is lowered into the sea in the course of fish-catching operations, and at other times is carried on board. It includes nets, warps, hooks, lines, otter boards etc.

FR engin de pêche; gréement de pêche; train de pêche; apparaux de pêche

Ensemble des éléments du dispositif de capture du poisson, mis à l'eau durant l'opération de pêche.

IT attrezzo da pesca

NL vistuig

Samenstel van kabels, net, gewichten en eventueel borden of palen waarmee passief of actief vis wordt gevist.

PT arte de pesca

Conjunto de elementos do dispositivo de captura de peixe que é colocado na água durante a operação de pesca.

3194

ES ojo

Del motón.

DA hundsvot

Metalbøjle eller øje til fastgørelse af den faste del af en blok, blokhuset.

DE Hundsfott; Hundsvott

Unterbügel oder unteres Auge eines hölzernen Blocks.

GR δακτύλιος τροχαλίας

Δακτύλιος τοποθετημένος στη θήκη μιας τροχαλίας.

EN block eye

A rope grommet or metal eye at the bottom of a block for securing the standing end of a fall.

In metal blocks the becket and shell are cast in one piece.

FR ringot

Anneau fixé à la caisse d'une poulie.

IT occhiello

Del bozzello.

NL hondsvot

Oog aan een blok waaraan het vaste part van de loper wordt gezet.

PT alça do moitão

Aro metálico ou anel de cabo de fibra na base de um moitão.

3194

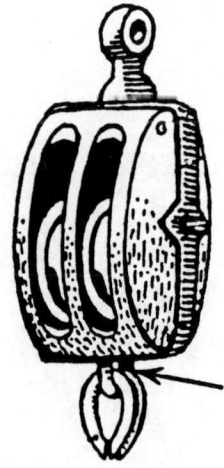

3195

chigre de remolque

DA slæbespil

Et spil med to eller flere tromler til at hale slæbetove eller -wirer.

DE Schleppwinde

Zwei- bzw. Mehrtrommelwinde für Kurrleinen, Stander und Gien.

GR βαρούλκο ρυμούλκησης

Βαρούλκο με ένα ή περισσότερα τύμπανα αποθήκευσης σχοινιού για το μαϊνάρισμα, βιράρισμα ή τη συγκράτηση σχοινιών ρυμούλκησης.

EN towing winch; towing machine; towing engine

A winch with one or more rope storage drums for paying out, hauling in or making fast tow ropes.

FR treuil de remorque

Treuil muni d'un ou de plusieurs tambour(s) de stockage servant à filer, à virer ou à maintenir des câbles de remorque.

IT verricello di rimorchio

NL sleeplier

PT molinete de reboque

Aparelho de força munido com um ou mais tambores e destinado a armazenar, largar e virar cabos de reboque.

3196

ES pescado que no alcanza la talla mínima reglamentada

DA undermålsfisk

Fisk, hvis længde ikke har det foreskrevne mindstemål.

DE untermaßiger Fisch

Fisch, der das vorgeschriebene Mindestmaß nicht erreicht.

GR ψάρι με μέγεθος κατώτερο του ζητούμενου

EN undersized fish

Fish not of legally marketable size - the minimum size may vary depending on species.

FR poisson n'atteignant pas la taille minimale

La taille minimale de chaque espèce est fixée par la réglementation.

IT pesce sotto misura

Rispetto alle taglie minime fissate dalla legislazione nazionale.

NL ondermaatse vis

Vis die kleiner is dan een voorgeschreven minimale maat.

PT pescado que não apresenta o tamanho mínimo legal

3197

ES abrazadera; mordaza; sujetacables

Cualquier pieza que rodea algo para ceñirlo o sujetarlo.

DA kabelklemme; wireklemme

U-formet bolt med specielt udformet skive. Anvendes til at lave øjer på wirer eller samle to wirer.

DE Kabelklemme; Klemmbügel

Befestigungsschelle.

GR σφιγκτήρας· μπουλντόγκ

Γόμφος σχήματος U που χρησιμοποιείται για τη σύνδεση δύο συρματόσχοινων.

EN bulldog grip

U-bolt with specially shaped sheave used to clamp together two wire ropes.

FR serre-câble; étrier

Ferrure munie de boulons servant à maintenir ensemble deux cordages.

IT morsetto

Piccolo congegno, generalmente a vite e a forma di U, che serve a unire due cavi d'acciaio.

NL trekbeugel; klamp

Inrichting die om een kabel kan worden geklemd zodat op die kabel een trekkracht kan worden uitgeoefend zonder gebruik te maken van een oog of splits.

PT cerra-cabos; grampo; braçadeira; alfinete

Ferragem munida de sistemas de «prisão» e que se destina a prender em conjunto dois cabos.

3197

3198

ES total autorizado de capturas; TAC

DA samlede tilladte fangstmængder; TAC

Angivelse af den mængde af en bestemt fiskebestand, der må tages ved fiskeri inden for en tidsperiode (normalt et år). Fastlægges af myndighederne, evt. på grundlag af videnskabelig rådgivning. Såfremt flere nationer kan gøre krav på bestanden, fordeles TAC'en evt. som nationale kvoter.

DE zulässige Gesamtfangmenge; TAC

Gemeinschaftsquote für den Fischfang.

GR συνολική ποσότητα επιτρεπόμενης σύλληψης· TAC

Για ορισμένα αλιευτικά αποθέματα που υπάρχουν μέσα στην καθορισμένη από την Κοινότητα αλιευτική ζώνη.

EN total allowable catches; TAC

Catch limit for a particular fishery over a given period — may be a recommendation by a scientific body or a legal requirement by a fishery management authority.

For certain fish stocks existing within the European Community fisheries zone.

FR total admissible des captures; TAC

IT totale delle catture ammesse; TAC

NL totaal toegestane vangsten; TAC

PT total admissível de capturas; TAC

3199

ES zapata

Parte inferior del patín o de una puerta de arrastre que va en contacto con el fondo.

DA sko

Flad stålplade svejset på undersiden af bommen på en bomtrawl.

DE **Kurrschuh; Schlitten**

GR πέδιλο· πέλμα

Πλατιά ατσάλινη βάση, συγκολλημένη στο κάτω μπροστινό μέρος μιας δοκότρατας.

EN **sole plate; shoe**

Flat steel plate welded to the bottom of the otter board or trawl head of a beam trawl.

FR **patin**

Semelle incurvée fixée sous l'étrier d'un chalut à perche.

IT **scarpa**

NL **slee**

Vlakke onderzijde aan het korijzer van de boomkor.

PT **sapata**

Parte inferior do patim das redes de arrasto de vara ou das portas de arrasto pelo fundo.

3200

ES **cuota de captura; contingente de captura**

DA **fangstkvote**

Fangstbegrænsning, som bliver pålagt et bestemt land, en fartøjsgruppe, et fartøj, eller en gruppe af fiskere, som en bevaringsforanstaltning. Angives normalt for en bestemt fiskeart og bestand i vægtenheder.

DE **Fangquote**

GR **ποσόστωση**

EN **catch quota**

Catch limit, usually specified by weight, for a particular country, fleet, boat or group of fishermen, set for the purposes of fishery management.

FR **quota de capture**

Limitation du total des captures effectuées par la flottille d'un pays dans une région et pour une espèce donnée, dans un but d'aménagement des pêches.

IT **contingente di cattura**

NL **vangstquotum**

Voorgeschreven hoeveelheid te vangen vis van een bepaalde soort.
Wordt jaarlijks bepaald aan de hand van schattingen van de aanwezige biomassa van die soort.

PT **quota de pesca; quota de captura**

3199

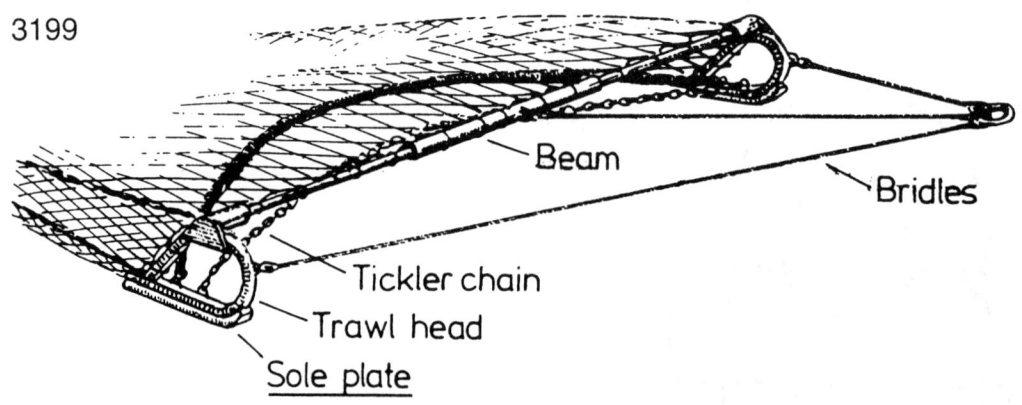

Beam

Bridles

Tickler chain

Trawl head

Sole plate

3201

ES largar; calar

Disponer en el agua en forma adecuada un arte de pesca.
Red.

DA sætte ud

Udtryk for at bringe redskabet i vandet ved fiskeriets begyndelse.

DE aussetzen

Fischfangtechnologische Bezeichnung der Arbeitsphase des Fangzyklus, die mit dem Inwasserlassen des Fanggeräts beginnt und bei Erreichen der Fangstellung beendet ist.

GR καλάρω· ρίχνω· πετώ

Ρίχνω τα δίχτυα στη θάλασσα για αλιεία.
Το δίχτυ· την ορμιά, μπετονιά.

EN set out (verb); shoot (verb)

To lower into the water.
The net or the line.

FR filer; mouiller

Mettre à l'eau un filet, un cordage ou une ligne.

IT calare

La rete o gli attrezzi p.e. nasse, palangari.

NL uitzetten; schieten

Het in het water brengen van een vistuig.

PT largar; calar

Colocar na água de forma adequada uma arte de pesca.

3202

ES chigre de amantillo en vacío

DA hangerspil

Særligt spil til wirerne på en bomtrawler.

DE Hangerwinde

Winde zum Verstellen des unbelasteten Baumes.

GR βαρούλκο ανύψωσης

EN topping winch

A winch with a rope storage capacity, used for topping, lowering and supporting under load and no load.

FR treuil d'apiquage à vide

IT verricello d'amantiglio a vuoto

NL topperlier

Gedeelte van een lier waarop de kabels worden gewonden voor de beweging van de gieken.
Geldt voor boomkorvaartuigen, die meestal een samengestelde lier hebben met verschillende trommels. Tijdens het vissen worden de gieken in horizontale positie gebracht.

PT molinete da amarra

3203

ES patín

Cada una de las piezas metálicas colocadas en los extremos de la barra rígida.

Red de arrastre de vara.

DA slæde

Den solidt udformede bøjle, som sidder for hver ende af bommen på en bomtrawl. Foruden påsvejsningen af selve bommen sidder der på slæden også fastgørelser til hanefodskædetrækket, skrabekæderne og nettet.

DE Klaue; Bügel; Kufe

Der herzförmige Kopfbeschlag an jedem Baumende eines Schleppnetzbaumes.

Baumkurre.

GR έλκηθρο

Μεταλλικό κομμάτι εφαρμοσμένο σε κάθε άκρο μιας δοκότρατας είναι εφοδιασμένο με ένα πέδιλο καμπυλωτού σχήματος για να κινείται πάνω στον πυθμένα.

EN trawl head; beam head; sledge

A strong heart-shaped iron frame fitted at each end of a trawl beam. The after side is straight and slopes upward of each head to stake the ropes or wires by which the trawl is towed. The sides of the net are seized or lashed at a point close to the ground.

Beam trawl.

FR étrier

Pièce métallique fixée à chaque extrémité d'un chalut à perche munie d'un patin pour passer sur le fond.

IT slitta

Sfogliara.

NL slof

Samenstel van verticale en horizontale platen, bevestigd aan beide einden van de korboom.

Wordt gebruikt ter geleiding van de boom over de bodem.

PT patim

Cada um dos aros de ferro com forma aproximadamente trapezoidal colocados nas extremidades de vara das redes de arrasto de vara e cuja parte inferior (sapata) contacta com o fundo.

3203

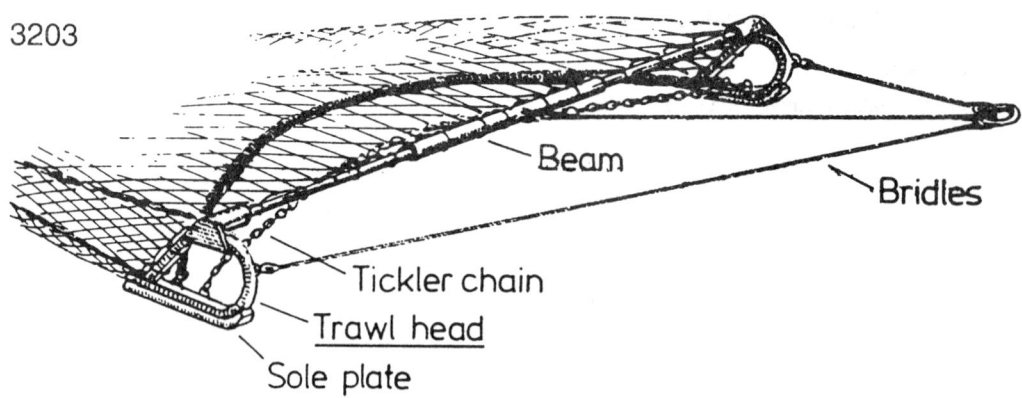

Beam

Bridles

Tickler chain

Trawl head

Sole plate

3204

ES perro

Robusta mordaza con la que se fijan ambos cables de arrastre en la pesca lateral con red, para que el arrastre se efectúe desde un punto lo más cerca posible de la popa.

DA kasteblok

En speciel blok placeret agter på en sidetrawler, som de to trawlwirer kan samles i under fiskeri. Benyttes for at lægge trækket fra trawlen langt agter. I Danmark bygges den ofte sammen med trawlblokken i den agterste galge. Denne specielle blok har da to skiver: en almindelig, der fører den agterste wire, og en, der kan åbnes til den forreste wire.

DE Kurrleinensliphaken; Hakentauroller

Kräftiger, an einer Kette beweglich befestigter Haken, der durch einen mit Splint gesicherten Ring geschlossen wird. Gebräuchlich auf Seitenfängern.
Der Sliphaken hat seinen Festpunkt an der Reling hinter dem achteren Galgen.

GR τροχαλία ρυμούλκησης· μακαράς ρυμούλκησης

Μικρή σιδερένια τροχαλία που χρησιμοποιείται στις μηχανότρατες και είναι προσδεδεμένη στην κορυφή της κουπαστής πίσω από τα πρυμναία καπόνια στην πλευρά ρυμούλκησης. Κρατά τα ρυμούλκα κοντά το ένα με το άλλο. Δίνει τη δυνατότητα στο πλήρωμα να καταλάβει την τάση, έτσι ώστε αν γίνει έντονη η ταχύτητα, να μπορεί να μειωθεί για να μην προκληθεί ζημιά στο εργαλείο.

EN towing block

A special iron frame usually fastened to the rail near the stern of a side trawler into which both warps are hauled at the start of a tow.

FR chien

Mâchoire d'acier placée à l'arrière et en abord d'un chalutier à pêche latérale. Elle réunit les deux funes et les maintient éloignées de l'hélice.
N'est pas utilisé sur les chalutiers à pêche arrière.

IT gancio per pesca laterale

Gancio a scocco o bozzello con cui si fissano i due cavi di traino nella pesca laterale.

NL slipblok

Kantelbare schijf aan de verschansing achter de achtergalg, waarin de vislijnen tijdens het vissen rusten, zodat de trawler bestuurbaar blijft.
Voordat het halen van het net begint, laat men het slipblok kantelen waardoor de vislijnen eruit slippen.

PT patesca de arrasto

Patesca de aço colocada à popa e a seguir ao aro de arrasto da popa de um arrastão lateral. Tem por função reunir os dois cabos reais na popa da embarcação mantendo-os afastados do hélice.

3204

3205

ES redero

Especialista en montaje y fabricación de redes.

DA vodbinder

En person, der fremstiller eller reparerer fiskeredskaber af net.
Her om denne aktivitet på land. Se 3208.

DE Netzmacher

GR κατασκευαστής διχτυών

EN net maker; netter

Person who makes or repairs nets by hand, plaiting and knotting twine using netting needle or shuttle.

FR monteur de filet; fabricant de filet; filetier

Ouvrier spécialisé dans la confection ou le montage des filets.

IT retiere; retante

NL nettenmaker

Persoon die met een boetnaald en garen netten breit en repareert.
Persoon, die met boetnaald en garen stukken netwerk aan elkaar verbindt, zodat een vistuig ontstaat.

PT redeiro; mestre de redes

Operário especializado na confecção e montagem de redes de pesca.

3206

ES calón de puerta; pony

DA »pony board«

Lille skovl eller skærebræt, der sættes ved undervingen på en trawl.
Bruges ikke i Danmark.

DE Ponyscherbrett; Ponybrett; Pony

Kleines, dem Rollengeschirr vorgeschaltetes Scherbrett.

GR πόρτα dan leno

Μικρή πόρτα που χρησιμοποιείται στη θέση ενός dan leno.

EN pony board; dan leno board

Small otter board used in place of a dan leno.

FR poney

Petit panneau renforcé utilisé à la place du guindineau sur les chaluts à fonds durs pour la grande pêche.

IT divergente «pony»

Non esiste in Italia.

NL ponybord

Klein, zwaar visbord, gebruikt in plaats van een danleno om kabels van een voortuig meer verticale opening te geven.
Niet in Nederland gebruikt.

PT porta de calão

Porta de arrasto nas redes de arrasto sem malhetas.

3207

ES armazón

Del arte.

DA rigning

Den måde, hvorpå de enkelte komponenter i et fiskeredskab er samlet eller indstillet, tilrigget.

Redskabernes rigning kan ofte ændres betydeligt for at opnå specielle egenskaber: f.eks. længde af stjerter, mængde af vægt og opdrift.

DE Fischereigeschirr

Pauschale Bezeichnung für Zusammenstellung bzw. variable Änderungen am Geschirr, z. B. Änderung der Jager- und Standerlängen und/oder der Beflottung.

GR αρματωσιά· εξαρτία

Τρόπος με τον οποίο είναι συναρμολογημένα τα εξαρτήματα ενός αλιευτικού εργαλείου.

EN rig

The way the components of fishing gear are assembled.

FR gréement de l'engin

Façon d'assembler les éléments constitutifs d'un engin de pêche.

IT attrezzatura di un attrezzo da pesca

Insieme delle parti accessorie che rendono possibile la pesca con un determinato attrezzo.

NL optuiging; voortuig

Samenstel van kabels waarmee het vistuig gesleept wordt (kettingen, lijnen en eventueel visborden).

Hieronder kunnen ook bijkomende hulpstukken zoals drijvers, scheerborden en ballast vallen.

PT armamento

Constituição, arranjo e disposição de todos os elementos constituintes de uma arte de pesca.

3208

ES redero

Marinero cualificado para la reparación de redes a bordo de los buques.

DA vodbinder

Person, der fremstiller eller reparerer fiskeredskaber af net.

Her om en person, der har dette hverv om bord på et fartøj, se 3205.

DE Netzmacher; Netzflicker

Mitglied der Besatzung eines Fischereifahrzeugs. Aufgrund seiner vertieften Kenntnisse und Fertigkeiten bei der Reparatur der Fanggeräte (bzw. des Schleppnetzes) nimmt er die besondere Stellung des Netzmachers ein.

GR

EN net mender

A person who repairs nets on board.

FR ramendeur

Spécialiste chargé de la réparation des filets à bord des navires de pêche.

IT retante; rammendatore

NL nettenboeter

Persoon die netten samenstelt en repareert (aan boord van vissersschepen).

Tegenwoordig worden reparaties steeds meer op de wal uitgevoerd.

PT mestre de pesca

Pescador que, a bordo de um navio de pesca, é especializado e encarregado de, entre outras tarefas, proceder e orientar o trabalho de remendar as artes de pesca avariadas.

3209

ES salabardo; salabre

Manga de red montada en un aro con mango. Se emplea en extraer la pesca de las redes grandes o directamente del agua.

Con múltiples variantes y nombres se halla muy extendido.

DA ketsjer; kes

Mindre net monteret på en jernring; det bruges til at tage fangsten om bord fra en not, samt fra fartøjets last til kajen.

Overførslen fra noten til lasten sker på moderne fartøjer med en fiskepumpe.

DE Kescher

Gerät zur Entnahme des Fisches nach dem Anbordnehmen eines Teils der Ringwade aus dem mit einem dickeren Netzfaden und kleinerer Maschenweite gefertigten Bunt.

GR απόχη· κόφα

Δίχτυ χρησιμοποιούμενο για τη μεταφορά του αλιεύματος ενός γρίπου βαθιών νερών, αφού αυτός έχει τοποθετηθεί παραπλεύρως του σκάφους. Ο χειρισμός του γίνεται ή εξ ολοκλήρου με το χέρι ή με το χέρι και με μηχανικό μέσο.

EN dip net; scoop; brailer; spoon net; brail net; hand brailer; landing net

A net used for transferring the catch of a deep-sea seine after it has been brought alongside.

It is operated either entirely by hand or partly by hand and partly by power.

FR épuisette; haveneau; salabarde; salabre *

Filet en forme de poche, monté sur une armature circulaire fixée à l'extrémité d'un manche.
Épuisette de grande taille, manœuvrée manuellement ou mécaniquement, servant au transfert de la capture du filet au bateau sur un senneur.

* Terme régional.

IT coppo; guadino; voliga

Particolare tipo di rete da raccolta di piccole dimensioni. La rete è sostenuta da una intelaiatura fissa che viene manovrata per mezzo di un manico.

NL schepnet

Zakvormig net aan een ronde beugel met een steel, waarmee vis uit een ben of kaar, of een ander net wordt geschept, dan wel gebruikt voor het vissen als zodanig.

PT chalavar; enchalavar

Arte de pesca com a forma de saco, montada numa armação circular fixada na extremidade de um cabo de madeira ou metal e manobrada à mão.

Em Portugal usa-se o mesmo termo para designar um apetrecho também com a forma de saco e montado numa armação circular mas sem cabo e manobrado por meio de cabos e de um pau de carga e utilizado a bordo para desenvazar o peixe capturado por uma rede de cerco.

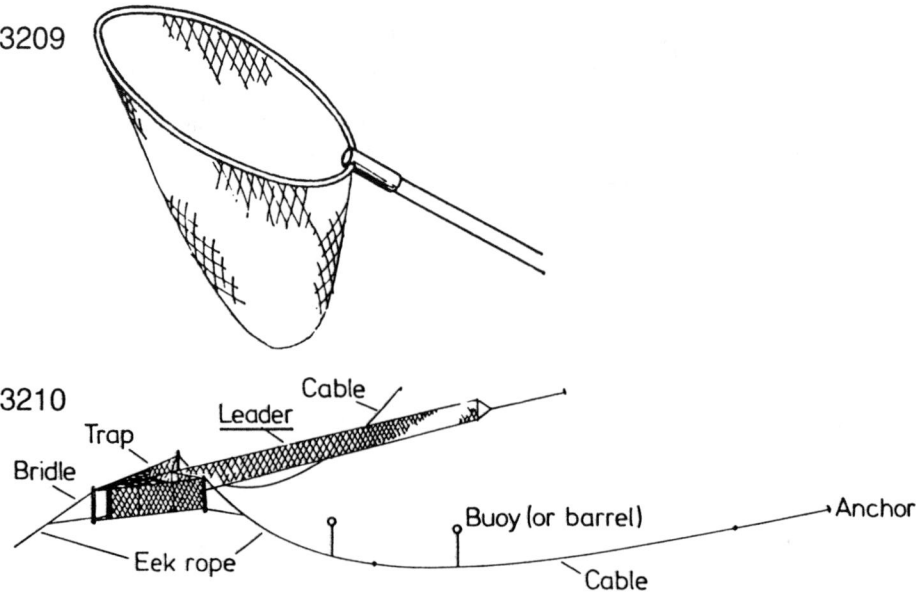

3209

3210

Cable

Leader

Trap

Bridle

Buoy (or barrel)

Anchor

Eek rope

Cable

3210

ES barrera guía

DA rad; ledegarn

Netvæg, der spændes ud i vandet for at lede fiskene ind mod indgangen til et fiskeredskab: f.eks. bundgarn, ruse eller lignende. Støttes ofte af pæle.

DE Leitwehr

Stehendes Gerät für die stille Fischerei mit Netzwerk, an dem der Fisch einem Fangbau zugeleitet wird.

GR δίχτυ οδηγός

Τοίχος αποτελούμενος από δικτύωμα υποστηριζόμενος συνήθως από πασσάλους, που οδηγεί το αλίευμα προς την είσοδο ενός περίφρακτου με δίχτυ χώρου ή σε μια παγίδα.

EN leader; leader net

Wall of netting, often supported by stakes, which guides fish to the entrance of a net, enclosure or trap.

FR guideau; filet barrière

Barrière en filet utilisée dans les filets pièges du type «verveux»; sert à rabattre et à guider les poissons vers le corps du filet et sa chambre de capture.

IT braccio d'incanalamento

Nel caso della tonnara si chiama «pedale».

NL schutnet; schutwant; geleidenet

Netwerk, al dan niet bevestigd aan palen, ter geleiding van de vis naar de ingang van een vangnet, kom of visval.

Wordt gebruikt in passieve vistuigen.

PT andiche; endiche

Panos de rede armados de diferentes formas, utilizados nas barreiras, barragens, estacadas, armações, nassas, etc., e que se destinam a concentrar e a conduzir o peixe para o corpo da arte de pesca e em direcção à câmara de captura.

Também se designam em português pelo mesmo termo as aberturas das armadilhas.

3211

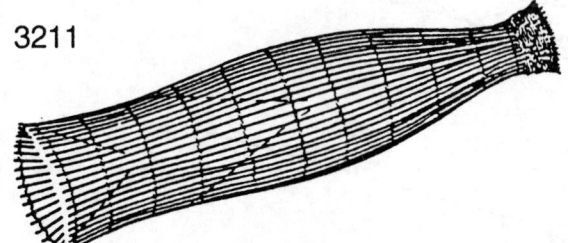

3211

ES nasa

DA tejne

Kasse- eller tøndeformet redskab af tremmeværk (træ eller plastik), net eller fletværk (vidjer) med 1-2 tragtformede indgange. Passivt fiskeredskab, der virker som skjul (uden agn) eller lokker (forsynet med agn).

DE Fangkorb

Aus Weidenruten geflochtener, trichterförmiger, mit einer Kehle versehener Korb (Aalfischerei). Kasten- oder walzenförmiger, mit Netzwerk bespannter Korb mit nach innen gerichteten, sich verjüngenden Eingangsöffnungen bzw. glockenförmiger Korb mit oben liegender Eingangsöffnung (Hummer-, Langustenfang).

GR κοφινέλο

Κατασκευή σχήματος βαρελιού φτιαγμένη από λυγαριά, με ένα ή δύο κωνικά ανοίγματα.

EN basket trap

Barrel-shaped structure made out of willow, with one or two funnel-shaped openings.

FR nasse; bourrache *

Piège destiné à la capture des crustacés, poissons ou mollusques; en forme de cage ou de panier et comportant une ou plusieurs ouvertures. Muni ou non d'appât, il est mouillé, en général sur le fond, isolément ou en filière, relié par un orin à une bouée en surface.

* Terme local.

IT nassa in vimini

NL korf; kubbe; kobbe; mand

Tonvormig vistuig, gevlochten uit wilgetenen, met 1 of 2 trechtervormige openingen.

PT nassa; covo

Arte de pesca tipo armadilha fixa que se utiliza para capturar peixes, moluscos ou crustáceos, com a forma de caixa, cesto ou pote; pode ser construída com diversos materiais (madeira, varas de metal, rede de pesca, rede de metal, rede de plástico) e possui uma ou mais aberturas ou entradas (boca e endiche). São colocados no fundo com ou sem isco, isoladas ou em teias, e ligadas a um ou mais cabos de alagem referenciados à superfície por meio de bóias.

3212

ES red de enmalle de superficie

Beta, emballo.

DA drivgarn

Garn, der ikke sættes på bunden. Omfatter drivgarn og opankrede garn i midtvand eller ved overfladen. *Se 3113.*

DE Schwebenetz

Frei im Wasser treibendes, an einer Schwimmleine im Pelagial gehaltenes Setznetz, an einem Ende mit dem Fangfahrzeug verbunden.

GR επιπλέον απλάδι

EN floating gill net

A set gill net anchored to the seabed or to a boat so that the netting is in midwater or near the surface.

FR filet maillant flottant

Filet maillant dont la ralingue supérieure apparaît à la surface.

IT rete da posta galleggiante; rete da imbrocco galleggiante; rete da parata galleggiante

NL drijfvleet

Kieuwnet dat als een gordijn in het water hangt, en in verticale positie wordt gehouden door drijflichamen aan het wateroppervlak en zinkers op de bodem.

PT rede de emalhar de superfície

3213

ES red de arrastre con puertas

DA trawl med skovle; skovltrawl

Enbåds trawl med to store skovle, som spiler trawlen åben horisontalt.

DE Scherbrettnetz

Trichterförmiger Netzsack mit weiter Öffnung nach vorn und verjüngend. Die Öffnung des geschleppten Netzes wird durch die aus Zug und Scherwinkel resultierende Kraft bewirkt. Die Scherbretter schließen entweder unmittelbar an den Netzflügeln an, oder sind über den Knüppel und Jager mit dem Netz verbunden.

GR τράτα με πόρτες

Μεγάλο, κωνικό δίχτυ εφοδιασμένο με δύο πόρτες, οι οποίες κρατούν το στόμιο του διχτυού ανοιχτό οριζόντια.

EN otter trawl

A large, conical net supplied with two otter boards which keep the mouth of the net open horizontally. *Single boat operation only.*

FR chalut à panneaux

Chalut dont l'ouverture horizontale est assurée par l'écartement de panneaux divergents.

IT rete da traino a divergenti; rete da traino a porte; rete da traino a un solo natante

Rete la cui apertura orizzontale è assicurata da divergenti o porte o tavoloni.

NL ottertrawl; bordentrawl; bordnet; visbordnet; planknet

Kor of trawl opengehouden door visborden.

PT rede de arrasto com portas

Rede de arrasto cuja abertura horizontal é conseguida por meio de portas de arrasto.

3213

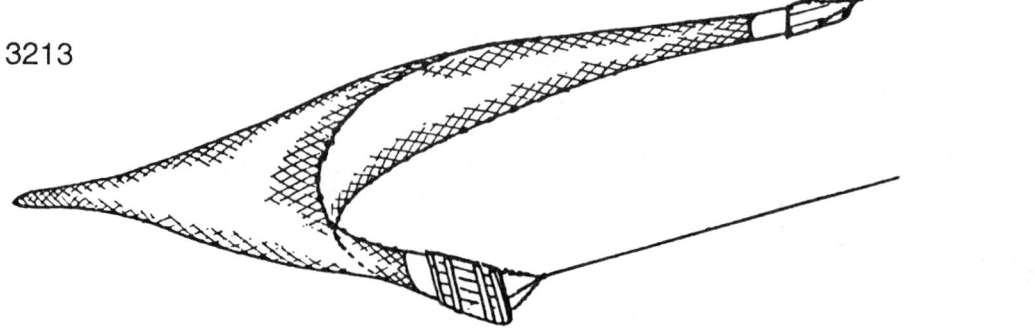

3214

ES patilla del anzuelo

DA plade

Den øverste del af en fiskekrog, der har et øje eller er banket ud til en plade, hvor forfanget fastgøres.

DE Kopf

Abgeplatteter Teil des Hakenschenkels, an welchem das Vorfach angeknotet wird. Bei modernen Angelhaken ist der Kopf als Öse ausgebildet.

GR κεφαλή αγκιστριού χωρίς μάτι

EN plate of a hook

The flattened part of the shark of a hook.

FR palette d'un hameçon

Partie aplatie à l'extrémité de la hampe d'un hameçon, servant à maintenir le nœud de fixation de l'avançon.

IT paletta di un amo

NL palet; bledje

Het platte, verbrede deel van de steel van een haak, dat dienst doet om het afglijden van de bevestigde lijn te voorkomen.

PT pata de um anzol

Extremidade achatada da haste do anzol à qual se fixa o estralho.
Anzol de pata.

3214

3215

ES lance; echada; calada

DA fangst

Her som udtryk for fangsten i et slæb (trawl), sæt (garn og snurrevod) eller kast (not).

DE Hol

Die mit einem Zuge gefangene Fischmenge.

GR αλίευμα· ψαριά· διχτυά

Αλίευμα που συλλέγεται κατά τη διάρκεια μιας αλιευτικής προσπάθειας.

EN haul; catch

The fish caught by hauling a net once.
In fishery management, catch can also mean the fish retained for sale over a given period by a defined fleet of vessels — e.g. total allowable catch.

FR prise

Capture réalisée au cours d'une seule opération de pêche.

IT cattura $(^1)(^3)$; pescata $(^1)(^2)(^3)$; saccata $(^2)$; retata $(^2)(^3)$

Quantità di pesce che risulta catturata al ritiro della rete dall'acqua.
$(^1)$ Palangaro; $(^2)$ rete da traino; $(^3)$ rete da posta.

NL vangst; trek

Hoeveelheid gevangen vis die in één visserijhandeling of -operatie wordt binnengehaald.

PT lanço; lance

Operação de pesca realizada com uma rede de pesca (arrasto, cerco, envolvente-arrastante) por cada vez que é largada.

3216

ES brazolada

DA forfang; tavs

Kort line (eller tynd wire) i den ene ende forsynet med en krog og i den anden fastgjort til en hovedline; line- og krogfiskeri.

DE Mundschnur

Die in der Langleinenfischerei an die Hauptleine angesteckte kurze Leine, an die sich das Vorfach anschließt.
Diese Kombination wird für den Fang starker Fische, z. B. Thun und Lachs, eingesetzt.

GR παράμαλλο

Λεπτή, γερή ορμιά που φέρει το ή τα αγκίστρια και είναι προσδεδεμένη στη μάνα.
Συρτή· παραγάδι.

EN branch line; snood; snell; ganging; gangion

A thin strong line by which a hook is attached to the main or back line of a troll or long line.

FR avançon; ligne secondaire; empile

Partie de la ligne constituée d'un fil plus fin, reliée à la ligne principale et qui porte l'hameçon.

IT bracciolo

Monofilo, filo o cavetto che collega l'amo alla madre nel palangaro.

NL sneu

Aan de beuglijn op bepaalde afstanden bevestigde dunne dwarslijn, aan het uiteinde waarvan zich een haak bevindt.

PT estralho; baixada; estrobo

Parte do palangre constituída por um fio geralmente de diâmetro reduzido, que se liga à madre do aparelho e cuja parte terminal livre leva um ou mais anzóis empatados.

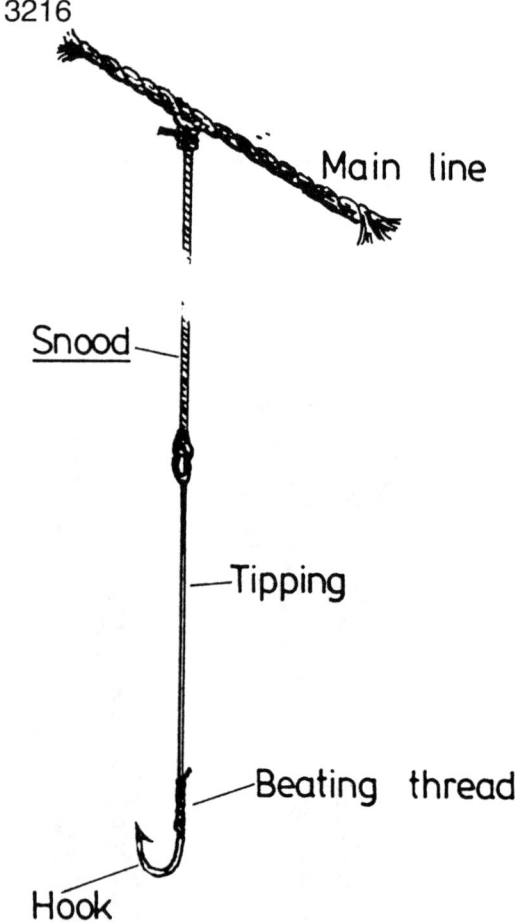

3216

Main line

Snood

Tipping

Beating thread

Hook

3217

ES | anilla del anzuelo

DA | øje

Den øverste del af en krog, der er bukket om til et øje til fastgørelse af forfanget.

DE | Auge; Hakenauge

Teil eines Hakens zur Befestigung des Blinkers bzw. des Vorfachs.

GR | μάτι αγκιστριού· κρίκος αγκιστριού

Οπή που βρίσκεται στη μία άκρη του αγκιστριού και στην οποία προσδένεται το παράμαλλο.

EN | eye

Head of the hook which serves for fastening the line and is shaped as an eye.

FR | anneau de l'hameçon

Œil placé à la tête de l'hameçon et sur lequel se fixe l'avançon.

IT | occhiello; anello

Parte di un amo per fissarlo alla lenza.

NL | oog

Boveneinde van een haak, ter bevestiging van de lijn.

PT | argola de um anzol; olhal de anzol

Olhal da extremidade da haste do anzol à qual se fixa o estralho.
Anzol de argola.

3218

ES | caña del anzuelo

DA | stang

Den rette del af en fiskekrog.

DE | Schaft

Angelhaken.

GR | λαβή αγκιστριού

Τμήμα του αγκιστριού που βρίσκεται απέναντι από τη μύτη, συνήθως ευθύγραμμου ή καμπύλου σχήματος. Το παράμαλλο προσδένεται στην άκρη της λαβής του αγκιστριού.

EN | shank

Part of a fish hook.
Can be of varying length and form. Its cross section can be round (regular) or flattened (forged).

FR | hampe

Partie de l'hameçon opposée à la pointe, en général de forme droite ou légèrement incurvée. L'avançon se fixe à l'extrémité de la hampe.

IT | gambo di un amo

NL | steel

Deel van een vishaak tussen oog en bocht.

PT | haste de um anzol

Parte do anzol oposta à ponta, geralmente com forma recta ou ligeiramente curva, a cuja extremidade se liga o estralho.

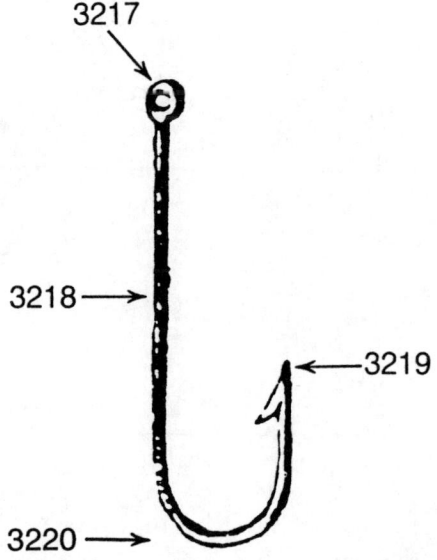

3217

3218 →

←—3219

3220 →

3219

ES punta del anzuelo

DA modhage; barbe; od

Den spidsede del af en krog.

DE Hakenspitze

Angelhaken.

GR μύτη αγκιστριού· κεντρίδα αγκιστριού

Αιχμηρή άκρη αγκιστριού εφοδιασμένη ή όχι με αρπάδι που δεν επιτρέπει στο ψάρι να ξεαγκιστρωθεί.

EN point; point of a hook

Part of a fish hook.
May be either straight or even reversed and curved.

FR pointe de l'hameçon

Extrémité pointue de l'hameçon, munie ou non d'un ardillon empêchant le poisson de se décrocher.

IT punta di un amo

NL punt

Scherp, puntig einde van de bocht, dat bij het vasttikken na aanbeet het zetten van de haak in de vissebek bevordert.

PT ponta de um anzol

Extremidade ponteaguda do anzol, provida ou não de barbela que impede que o peixe desferre do anzol.

3220

ES seno del anzuelo

DA bue

Den krumme del af en krog.

DE Hakenkrümmung

GR αγκώνας αγκιστριού

EN bend

Part of a fish hook.

FR courbure de l'hameçon

Partie cintrée de l'hameçon, entre la pointe et la hampe.

IT collo di un amo

NL bocht

Deel van een haak.

PT curvatura de um anzol

Parte central curva do anzol, situada entre as respectivas haste e ponta.

3221

ES **potera**

Aparejo para la pesca de potas, grandes calamares y a veces pulpos.

En Cataluña, se da este nombre a una calamarera.

DA **pirk til blæksprutter**

Særlig krog med talrige opadbøjede barber uden modhager.

Anvendes alene eller mange på samme line. Sænkes dybt ned og hales så til overfladen i én bevægelse.

DE **Tintenfischhaken; Reißangel**

In Form eines Bündels an einem Stabende dicht beieinander angeordnete Haken.

In Ostasien und im Mittelmeerraum sind Reißangeln mit zweifachem Hakenring in Gebrauch. Anwendung mit handbetriebener Leinenwinde. Leinenführung über eine Auslegerrolle.

GR **καλαμαριέρα**

Μολυβένιος σκελετός τυλιγμένος με άσπρο χασέ, με ένα στεφάνι από μυτερές βελόνες στο κάτω μέρος. Μπορεί να είναι, επίσης, μολυβένιος κώνος βαμένος άσπρος με δύο στεφάνια από μυτερές βελόνες, το κεφάλι των οποίων είναι χωμένο μέσα στον κώνο, η δε μύτη τους γυρισμένη προς τα πάνω.

EN **squid jig; ripper; pirn**

Name given by Newfoundland bankers to a small piece of lead about 3 in. long and of oval shape which has a number of radially arranged fishhooks curving upwards. It is kept constantly moving from a boat, and is used for catching squid for baiting the cod lines.

FR **turlutte**

Plomb ou corps en fuseau, portant de nombreux hameçons sans ardillon à sa partie inférieure. Fixé sur une ligne verticale animée d'un mouvement saccadé de bas en haut. Utilisé principalement pour la pêche des encornets.

IT **amo da calamaro; ancoretta; polpara**

NL **inktvisdreg**

PT **toneira**

Amostra de chumbo, ferro ou plástico dotada de um ou mais anéis com numerosos dentes ponteagudos sem barbela e que é utilizada na pesca de lulas e afins.

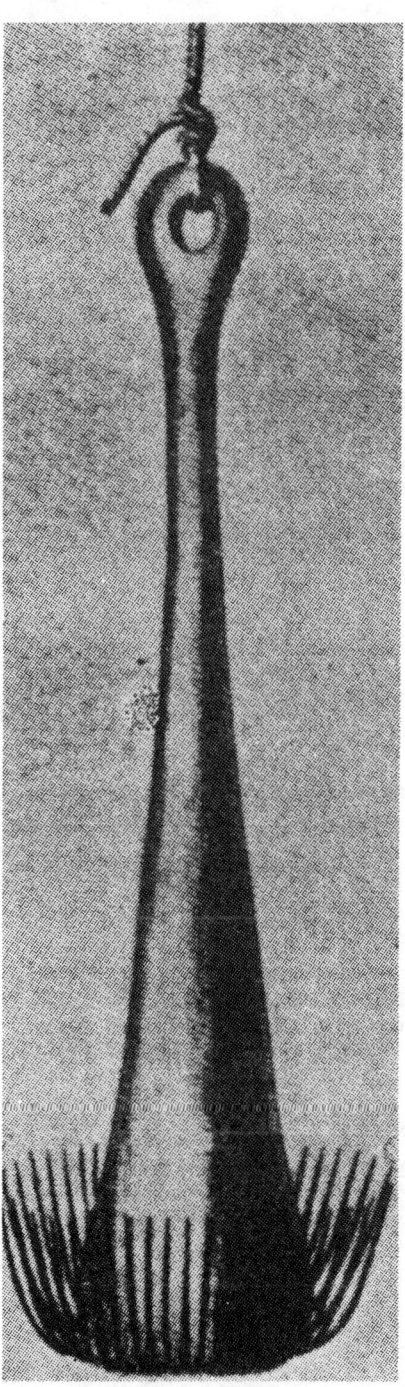

3222

ES curricán

DA devonspinner; kunstfisk

To eller flere kroge fastgjort til et metallegeme af form som en fisk. Vinger på »fisken« får den til at spinne. Til dørgefiskeri.

DE Devonspinner

Länglich runder Körper mit zwei dem Vorfach zugekehrten Propellerflügeln, die auf Zug durch das Wasser den Spinner mit dem am Ende dreiteiligen Haken in Drehung versetzen.

GR ψαράκι συρτής

Συνίσταται από δύο ή τρία αγκίστρια ενσωματωμένα ή στερεωμένα σε ένα μικρό ψεύτικο ψάρι (δέλεαρ)· χρησιμοποιείται στην αλιεία με συρτή.

EN Devonspinner; Devon

Consists of two or more hooks embedded in or fixed to a small artificial fish (lure), used for trolling.

FR devon

Consiste en deux hameçons ou plus incorporés ou fixés à un petit poisson artificiel (leurre), utilisé pour la pêche à la traîne.

IT devon

Esca artificiale metallica, di gomma o altra materia.
Il termine anglosassone è entrato nell'uso internazionale.

NL devon

Metalen, houten of kunststof kunstaas, in de vorm van een visje dat voorzien is van één, twee of drie dreggen, en al of niet verstelbare schoepen, vooral gebruikt bij het vissen op (zee)forel, zalm en zeebaars.

PT amostra

Isco artificial constituído por um objecto (metálico, plástico, madeira, pano, penas, etc.) simulando uma presa, provido de um ou mais anzóis e utilizado na pesca ao corrico.

3223

ES cucharilla

DA skeblink

Fiskekrog med metalplade af skinnende metal. Spinner rundt, når der dørges, og lokker ved at udsende lysreflekser, der minder om småfisk.

DE Löffeltoder; Spoon; Blinker

GR κουτάλι· κουταλάκι

Ψεύτικο δόλωμα (δέλεαρ) με ένα ή περισσότερα αγκίστρια σχήματος κουταλιού που χρησιμοποιείται στην αλιεία με συρτή.

EN spoon; spoon lure; spoon bait

An artificial bait (lure) with one or more hooks shaped like a spoon used for trolling.

FR cuiller

Leurre en forme de cuiller, de métal brillant ou de couleur vive avec un ou plusieurs hameçons, utilisé pour la pêche à la traîne.

IT cucchiaino

Esca artificiale in metallo lucido o colorato, ruotante od ondeggiante durante la trazione e armata di uno o più ami, usata per la pesca alla traina dello sgombro, della spigola e di altre specie.

NL lepel

Type kunstaas bestaande uit een hol, bol, gewelfd, gegolfd of plat metalen blad, voorzien van één of meer haken, dat door zijn bewegingen en lichtreflexen de roofvissen aantrekt.

PT colher

Amostra com a forma de colher, de metal brilhante ou de cor viva.
Pesca ao corrico.
Em Portugal também se designa pelo mesmo termo uma arte de pesca manobrada de e por uma embarcação e que é constituída por uma bolsa de rede com forma triangular, em que dois dos lados estão montados em duas varas (ou canas) e o terceiro lado é livre e se situa entre as extremidades distais das varas (ou canas).

3222

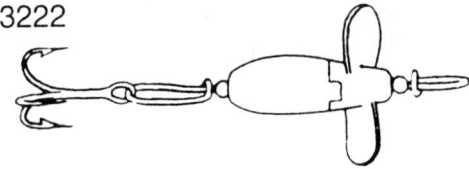

3224

ES | anzuelo doble sin agalla

DA | dobbelt tunkrog uden modhager

DE | doppelter Thunhaken ohne Widerhaken

GR | διπλό αγκίστρι χωρίς αρπάδι, για τόνο
Αλιεία με συρτή.

EN | barbless double tunny hook
Trolling.

FR | hameçon double à thon sans ardillons
Pêche à la traîne.

IT | doppio amo da tonno senza ardiglioni;
doppio amo da tonno senza barba

NL | dubbele tonijnhaak zonder weerhaken

PT | anzol duplo sem barbela para o atum
Pesca ao corrico.

3225

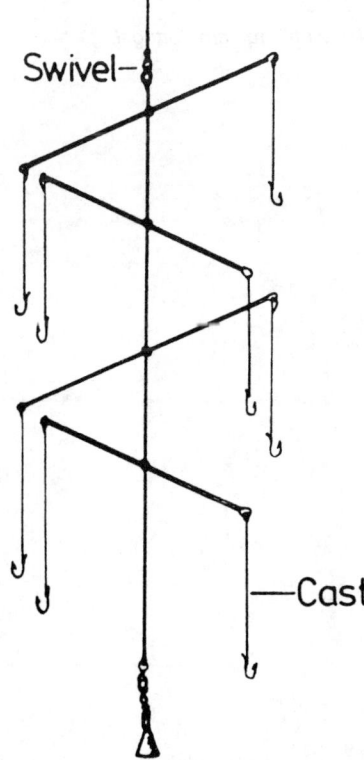

Swivel—

—Cast

3225

ES | balancín

Fuerte cordel del cual pende un plomo con o sin gaza atravesado transversal o verticalmente por una vara o varilla flexible de madera o metal, a menudo reforzada con pies de gallo, en cuyos extremos se suspenden las punteras o las tanzas.

DA | paternoster

Fiskeline, der ender i en spreder, hvorfra to forfang kan hænge ned.

DE | Brandungsangel

Angel, die mit Gewichten beschwert ist, damit sie sich nicht verwickeln kann. Wird im Küstenbereich eingesetzt.

GR | paternoster

Ελαφριά συρμάτινη ράβδος προσαρτημένη στην καθετή, στις άκρες της οποίας υπάρχουν δολωμένα αγκίστρια.
Δεν χρησιμοποιείται στην Ελλάδα.

EN | paternoster; dandy

Light wire rod attached to handline on which to 'outrig' a baited hook.

FR | pater-noster

Ligne pourvue, au-dessus du plomb, de plusieurs avançons et hameçons, fixés à un support léger perpendiculaire à la ligne (clipot).
Ligne à main.

IT | paternoster

Lenza a mano con più ami.
Non è conosciuto in Italia.

NL | paternoster

Snoer of onderlijn van een werphengel, voorzien van korte dwarslijnen met daaraan een haak, die op de grond ligt, of zwevend wordt gehouden d.m.v. dobbers; dient voor de vangst van kabeljauw, schar, bot, schol, makreel, geep, aal en tong.

PT | varestilha; guarda-chuva; barqueira

Linha de mão armada, acima do lastro, com numerosos estralhos e anzóis presos a um ou mais suportes finos fixados perpendicularmente à linha.
Aparelho de anzol.

3226

ES | **palangre sin fin**

DA | **endeløs line**

En lang løkke med tavser, der sættes i fartøjets ene side og hales i den anden. Anvendes i forbindelse med en dørgemaskine.

DE | **Paternosterangel**

Endlose Leine, die, in Abständen beschwert, mit Maulleinen und Haken bestückt ist. Eine auf dem Schandeck montierte Winde holt die Leine ein und läßt sie über eine am getoppten Baum befindliche Leitrolle aufgrund der Schwerkraft wieder zu Wasser.

GR | **ατέρμων συρτή· κυκλική συρτή**

Ορμιά που σχηματίζει δακτύλιο και η οποία κινείται αργά. Είναι εφοδιασμένη με βάρη, παράμαλλα και δολωμένα αγκίστρια. Εισέρχεται στο νερό σε συγκεκριμένο βάθος, επιστρέφει στην επιφάνεια και οδηγείται πάνω από το σκάφος, έτσι ώστε το αλίευμα να μπορεί να καμακωθεί και τα αγκίστρια να ξαναδολωθούν.

Δεν χρησιμοποιείται στην Ελλάδα.

EN | **endless trolling line; roundhauler**

Slowly moved loop of line fitted with weights, snoods and baited hooks, that enters the water down to a specific depth, returns to the surface and is led over the vessel so that the catch can be gaffed and the hooks rebaited.

FR | **ligne sans fin**

Technique de pêche au maquereau dans laquelle la ligne principale a ses deux extrémités nouées l'une à l'autre et passe d'un bord à l'autre au-dessus du bateau.

IT | **palangaro senza fine; palangaro ad anello**

Non usato in Italia.

NL | **doorhaler; continue lijn**

Vistuig waarbij een lange, van haken voorziene lijn-zonder-eind wordt gevoerd door een samenstel van twee buizen, waarvan er een op het schip bevestigd is, en de andere zich onder water bevindt.

Vooral gebruikt voor de vangst van makreel; niet gebruikt in Nederland.

PT | **palangre sem fim**

Palangre cujas extremidades se encontram ligadas uma à outra e que tem um funcionamento em contínuo.

Não é utilizado em Portugal.

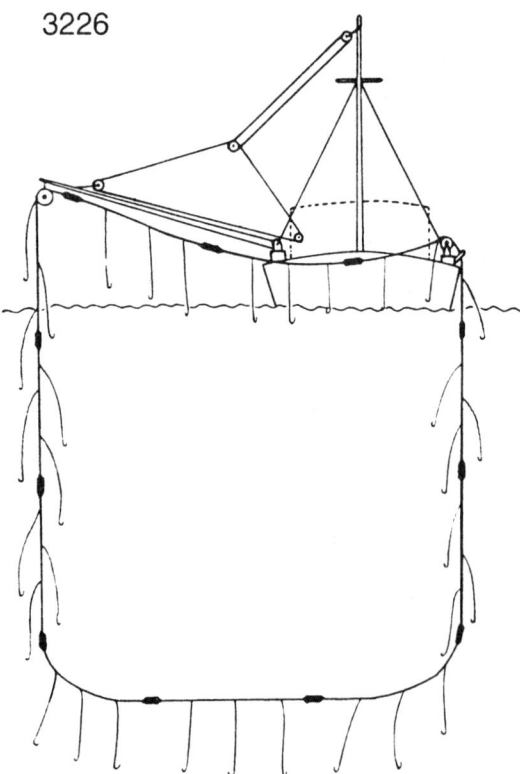

3226

3227

ES fisga; fítora; tridente

Instrumento de pesca derivado de los primitivos arpones. Consiste en un largo mango, a veces de muchos metros de longitud, que se maneja desde embarcaciones menores y que va terminado en dientes, como tenedores, que pueden estar dotados de puntas de flecha o no, según la especie de que se trata.

DA lyster; ålejern

Parallelle klinger, savtakkede eller med barber. Ved hjælp af et langt skaft stødes lysteren f.eks. mod bunden, hvorved eventuelle fisk klemmes fast mellem klingerne.
Anvendes i Danmark til isfiskeri efter ål.

DE Speer

Vier parallel stehende, gezähnte flache Klingen, die sich auf Druck scherenartig öffnen und den Fisch einklemmen.
In Nordeuropa wird das Gerät auf zugefrorenen Gewässern (Eisloch) in der Blindfischerei auf Aale eingesetzt. Durch blindes Zustoßen werden die inaktiven Aale erfaßt und auf den Schlamm gezogen.

GR τρίαινα

Καμάκι με 2 έως 5 ακιδωτές προεξοχές που χρησιμοποιείται για τη σύλληψη ορισμένων ειδών ψαριών, όπως οι παλαμίδες.

EN fish spear

A spear with 3 or 5 barbed prongs used for the capture of some species of fishes, such as bonitos.

FR foëne

Sorte de harpon à plusieurs pointes, utilisé à la main, pour la pêche à pied ou en embarcation en eau peu profonde.

IT fiocina

NL speer

PT fisga

Espécie de arpão com vários dentes com ou sem barbela, operado à mão, a pé ou de uma embarcação, em águas pouco profundas.

3228

ES anillo de jareta; llave

Anillo situado en la parte inferior de una red de cerco, sujeto a la religa de plomo por un chicote o pie de gallo, por el que pasa la jareta.

DA notring; snurpering

Kraftig karabinhage, der ved en hanefod er fastgjort til en nots undertælle. Gennem sådanne notringe langs hele undertællen løber snurpewiren.

DE Wadenring; Schnürleinenring

GR κρίκος στίγγου· χαλκάς στίγγου

Κρίκος ενωμένος με σχοινί (είδος ληγαδούρας) με το κάτω γραντί σε ένα γρι-γρι και από τον οποίο περνάει ο στίγγος. Ανάλογα με τον τύπο του γρίπου, οι κρίκοι του στίγγου μπορεί να είναι ανοιχτού ή κλειστού τύπου.

EN pursing ring; purse ring

Strong metal clip ring with spring loaded gate through which the pursing wire passes.

FR anneau de coulisse

Anneau relié par une pantoire (sorte de patte d'oie) à la ralingue inférieure d'une senne coulissante et dans lequel passe la coulisse servant au boursage. Selon la senne, les anneaux de coulisse peuvent être du type fermé ou du type ouvrant.

IT anello del cavo di chiusura; anello del cianciolo

NL ring

Wordt in Nederland niet gebruikt.

PT argola

Argola ligada à tralha inferior (dos chumbos) das redes de cerco com retenida por uma espécie de pé--de-galinha —a aranha— e através da qual passa a retenida utilizada para fechar a rede por baixo. As argolas podem ser de dois tipos: fechadas ou de abrir.

3227

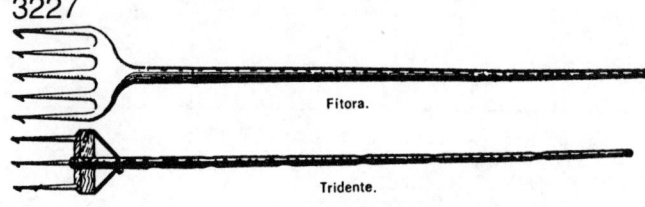

Fítora.

Tridente.

3229

ES | nasa para gambas

DA | rejeruse; rusekurv

DE | Garnelenreuse

Reuse aus Netzwerk mit ca. 2 m langen Flügeln. Der sich nach hinten verjüngende Netzsack ist etwa 2 m lang und 0,5 m hoch. Die Reusen werden paarweise gesetzt. Die beiden äußeren Flügel eines Paares erhalten gewöhnlich noch vorgesetzte Netzstücke bis zu 12 m Länge.

GR | κιούρτος για γαρίδες

EN | prawn trap; prawn creel

FR | nasse à crevettes

IT | nassa da gamberetti

NL | garnalenkorf

Korf van latten, twijgen, rotan of plastic, voor de vangst van garnalen.

PT | nassa para camarões

Arte de pesca tipo armadilha fixa que se utiliza para capturar crustáceos, com a forma de caixa; pode ser construída com diversos materiais (madeira, varas de metal, rede de pesca, rede de metal, rede de plástico) e possui uma ou mais aberturas ou entradas (boca e endiche). São colocadas no fundo com isco, isoladas ou em teias, e ligadas a um ou mais cabos de alagem referenciados à superfície por meio de bóias.

É normalmente forrada com rede de malha muito reduzida.

3230

ES | nasa para anguilas

DA | åleruse; rusekurv

DE | Aalreuse

Für den Aalfang verwendete Reuse aus Weidenruten oder Garn. Ihre Form ist konisch, und sie besitzt zwei Kehlen. Sie ist etwa 0,90 m lang und hat vorn einen Durchmesser von 0,20 m. Hinten wird sie mit einem Holzpflock verschlossen. Die beköderten Reusen werden, untereinander mit einer mit Steinen beschwerten Reusenleine verbunden, in einer Reihe ausgesetzt.

GR | παγίδα για χέλια

EN | eel basket; eel trap; eel pot

FR | nasse à anguilles

IT | nassa per anguille

NL | kubbe; palingkorf; aalkorf; kobbe

Van dicht gevlochten twijgen vervaardigde korf voor de vangst van aal, paling.

PT | nassa para enguias; galricho; cofre

Arte de pesca tipo armadilha fixa que se utiliza para capturar peixes, com a forma de caixa ou cesto; pode ser construída com diversos materiais (madeira, varas de metal, rede de pesca, rede de metal, rede de plástico) e possui uma ou mais aberturas ou entradas (boca e endiche). São colocadas no fundo com isco, isoladas ou em teias, e ligadas a um ou mais cabos de alagem referenciados à superfície por meio de bóias.

3230

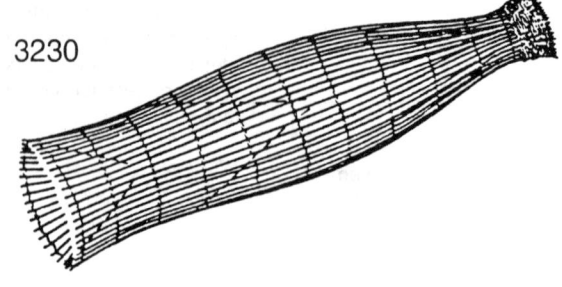

3231

ES doris

Embarcación de fondo plano, proa y popa lanzada que usan en los bancos de Terranova algunos pescadores de bacalao con liña. Con una eslora entre 4 y 6 metros es llevada a bordo de los buques bacaladeros en número de hasta veinte.

DA dory; jolle

Mindre, åben, fladbundet jolle. Et antal af disse medføres på større fartøjer, og udsættes på fangstpladsen. Herfra pilkes eller sættes langliner.

DE Dory-Boot; Dory

Kleines flachbordiges, mit Hilfssegel ausgerüstetes Boot zum Fischfang mit Angel, von Mutterschiffen mitgeführt und am Fangplatz mit einem Fischer ausgesetzt.

GR δορίς

Μικρή βάρκα με ισοπεδωμένο πυθμένα αμερικάνικης προέλευσης, που χρησιμοποιείται από τους ψαράδες κυρίως για το μαϊνάρισμα των σχοινιών της τράτας.

EN dory; dory boat

A small flat-bottomed open rowboat of American origin, chiefly used by fishermen for setting their troll lines.

FR doris

Embarcation légère, à fond plat, utilisée en particulier pour la pêche à la morue dans la région de Terre-Neuve. Employé à l'origine comme embarcation annexe opérant à partir d'un navire mère; le doris existe aussi en version motorisée pour la pêche côtière.

IT dory

Leggero battello piatto dei pescatori di merluzzo di Terranova; i dory vengono portati sul luogo di pesca da un grosso peschereccio, dal quale, una volta in mare, si allontanano a raggiera per effettuare la pesca con lenze da fondo.

NL dory; dorie

Roeiboot, meestal in grotere aantallen meegevoerd door „bankers" naar de Newfoundlandse banken en aldaar gebruikt voor de hoekwantvisserij.

PT dóri

Antiga embarcação de madeira, fundo chato, bancos desmontáveis e reduzidas dimensões, utilizada particularmente na pesca do bacalhau na Terra Nova. Originariamente empregues como embarcações operando a partir de navios mães.

3232

ES cordón metálico

DA dugt; kordel

Et antal dugter spindes til en wire.

DE Drahtseillitze; Stahllitze

Drahtseile, die dem Seewasser ausgesetzt sind, werden verzinkt. Eine der Seilkonstruktion entsprechende Anzahl von Einzeldrähten wird zu einer Litze verdreht.

GR συρμάτινο έμβολο· συρμάτινο έμπουλο· συρμάτινος κλώνος

EN wire strand

One of the component parts of a rope, made of wire.

FR toron métallique

IT trefolo

NL kardeel; streng

PT cabo de aço

3233

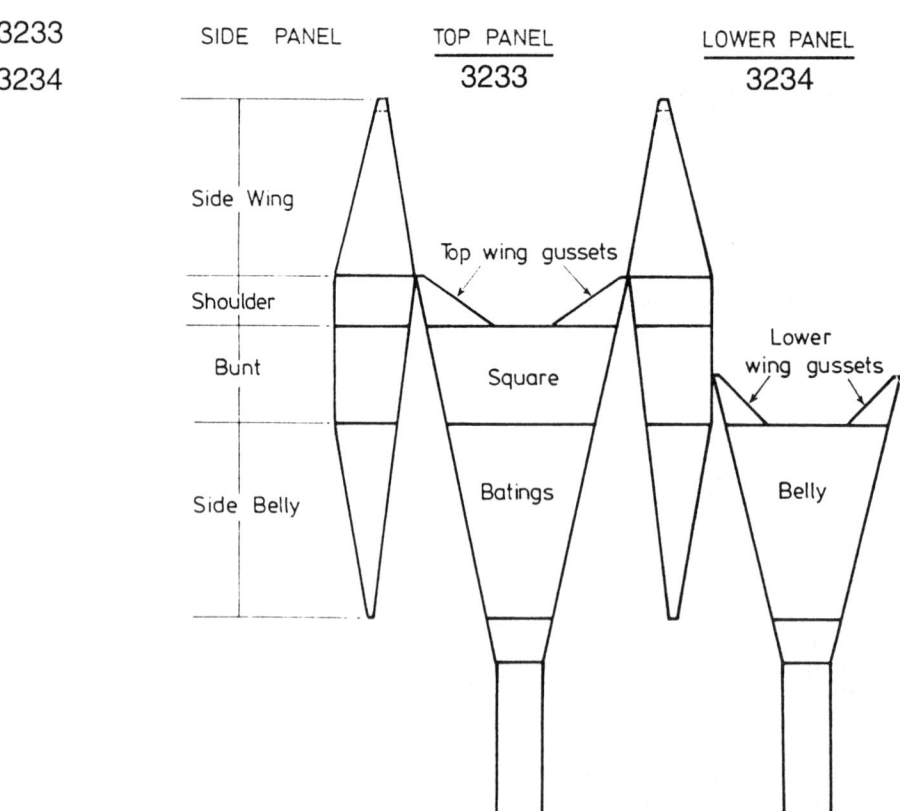

3233
3234

SIDE PANEL   TOP PANEL   LOWER PANEL
3233   3234

Side Wing
Top wing gussets
Shoulder
Lower wing gussets
Bunt
Square
Side Belly
Batings
Belly

3234

ES parte inferior de la red

DA underpanel

Alle netsektionerne i en trawls underside.

DE untere Netzhälfte; Unterblatt; Unternetz

Besteht aus den gleichen Teilen wie das Obernetz. Der vordere Teil ist das Bauchstück. Ihm folgen das Hundertmaschenstück, Tunnel und Steert.

Zum Schutz des Steertes gegen das Schleifen über Grund werden an seiner Unterseite bis zu vier Scheuerlappen befestigt. Sie können aus alten Steertstücken oder Kuhhäuten bestehen.

GR κατώτερο φύλλο δικτυώματος

Συμπεριλαμβάνει όλες τις περιοχές διχτυού του κατώτερου τμήματος της τράτας, δηλαδή κάτω φτερά, κοιλιά, κατώτερο τμήμα προέκτασης (κόψες).

EN lower panel

Comprises all the net sections of the lower (underside) part of the trawl net, i.e. lower wings, belly, lower extension piece.

FR face inférieure

Partie inférieure d'un chalut, constituée de plusieurs pièces de filet.

IT parte inferiore; tassello *

Della rete.
* *Rete a strascico mediterranea.*

NL ondernet; onderzijde; onderkant; onderboel

Onderste paneel van een trawlnet.

PT face inferior

Parte inferior de uma rede de arrasto, constituída por vários panos de rede.
Rede de arrasto.

3235

ES red Larsen

DA Larsen-trawl

Flydetrawl, hvor alle fire paneler er ens.
Ældre betegnelse. Opkaldt efter en vodbinder i Skagen, Danmark.

DE Larsen-Schleppnetz

Dänische Entwicklung (1948) eines pelagischen Netzes aus vier kongruenten Netzblättern. Handhabung mit zwei Booten.

GR τράτα Larsen

Κωνικός σάκος δικτυώματος, κατασκευασμένος από τέσσερα ίσα τμήματα.
Πελαγική τράτα.

EN Larsen trawl

Conical bag of netting, made up of four equal parts.
Floating trawl.

FR chalut Larsen

Type de chalut pélagique, constitué de quatre faces identiques.

IT rete larsen; rete volante; rete da traino pelagica

NL Larsentrawl

Pelagische trawl met vier gelijke bladen: onder-, boven- en twee zijkanten, en met vierkante netmond.
Genoemd naar de uitvinder R. Larsen; 1948; kenmerkend voor de ontwikkeling van de trawl en voor de toepassing van het echolood.

PT rede de arrasto Larsen

Rede de arrasto de superfície com forma de saco piramidal, constituída por quatro faces iguais.
Rede de arrasto pelágico.

3236

ES monofilamento; hilo simple

Filástica constituida por un solo hilo simple.

DA monofilament

En line bestående af et enkelt filament.

DE Monofilgarn; Monofil

Filamentgarn, bestehend aus einem einzelnen Filament, ohne oder mit Drehung hergestellt.
Monofilgarn mit einem Durchmesser von mehr als etwa 0,1 mm wird nur „Monofil" oder auch „Draht" genannt.

GR νήμα συνεχούς ίνας

EN monofilament; monofilament yarn

A twine composed of a single yarn.

FR monofil; monofilament

Filament en textile synthétique obtenu par extrusion et d'un diamètre supérieur à 0,1 mm (en dessous, on parle de filament continu).

IT monofilamento

NL monofilament; monofil vezel

Enkelvoudige continu doorlopende vezel.

PT monofilamento; fio singelo

3237

ES lupa de pesca

DA fiskelup

Oscilloskop, der viser ekkoet fra en udsendt lydimpuls som vandrette udslag fra en lodret lysstribe. Som tilbehør til et ekkolod gav det en god diskrimination af ekkointensiteten ved forskellige fiskestimer. Er nu teknisk passé, erstattet bl.a. af farveekkolod.

DE Fischfinder; Fischlupe

Elektro-akustisches Gerät, das speziell für Vertikal- und Horizontalortung eingesetzt wird. Das über den Bodenschwinger ausgesendete und zum Empfänger zurückkommende elektro-akustische Signal wird auf der Fischlupe (Braunsches Rohr) als vertikaler Strich dargestellt. Die Echodarstellungen über Grund oder im Pelagial stehender Fische werden durch horizontale Auslenkungen des Kathodenstrahls sichtbar. Die Echodarstellung ist je nach Fischart verschieden.

GR ιχθυοανιχνευτής

Λυχνία καθοδικών ακτίνων· μπορεί κανείς να δει στην οθόνη τις ανακλάσεις του πάγκου των ψαριών και του πυθμένα, με τη μορφή οριζόντιας προβολής μιας κάθετης φωτεινής δέσμης.

EN fish-finder; fish loop

Cathode-ray tube which shows an echo of the fish shoal and of the bottom as horizontal deflections from a vertical stripe of light.
Equipment seldom used now.

FR loupe de pêche

Tube à rayons cathodiques; on voit sur l'écran les échos du banc de poissons et du sol sous forme d'excroissances horizontales d'un rai lumineux vertical.
Équipement techniquement dépassé.

IT ittioscopio

NL visloep

Een op het echolood aangesloten kathodestraalbuis, waarop een gedeelte van de waterkolom tussen oscillator en zeebodem vergroot wordt weergegeven; op vissersvaartuigen gebruikt om meer informatie te verkrijgen over eventuele visconcentraties.

PT «fish-finder»

Equipamento acústico já ultrapassado e actualmente não utilizado. Foi substituído por equipamento mais moderno e evoluído fornecendo informações muito mais precisas e completas e genericamente designado por «sondas».

3238

ES escandallada horizontal

Detección de los bancos de pescado por medio de sonar.

DA horisontal ekkopejling

Ekkoopsamling fra fiskestimer og bund fra lydimpulser, der udsendes vandret. Fortæller om afstand og retning til disse. Impulserne gives af en drejelig lydgiver. Dette princip anvendes i sonar og adskiller sig fra ekkolod, der kun sender lyden lodret ned.

DE Horizontallotung

Aussendung eines elektro-akustischen Signals unterhalb 30 kHz. Der Lotschwinger (ausfahrbares Gerät) ist auf einer kipp- und schwenkbaren Basis montiert und ermöglicht Lotungen nach vorn (Kippwinkel ca. 10 Grad), seitlich und nach unten. Die Horizontallotung gibt Aufschluß über das Schwarmverhalten.

GR οριζόντιος ακουστικός εντοπισμός

EN horizontal echo ranging

Detection of underwater objects by sonar echoes transmitted at an angle to the vertical, compared to echo sounders which transmit only vertically downwards.

FR détection acoustique horizontale

Équipement servant à la localisation des bancs de poissons ou à la reconnaissance des modifications du fond en avant ou autour du bateau.

IT scandaglio orizzontale; ecoscandaglio orizzontale; scandaglio acustico orizzontale

NL horizontale echoloding

Door middel van een sonar horizontaal waarnemen onder water, bij de visserij toegepast voor het observeren en lokaliseren van visscholen.

PT detecção acústica horizontal

Sonar, asdic.

3239

ES intensidad de pesca

DA fiskeriintensitet

Fiskeriindsats (effort) pr. arealenhed.
Se fiskeriindsats 3148.

DE Intensität der Fischerei

Fischereiaufwand pro Gebietseinheit.

GR αλιευτική ένταση

EN fishing intensity

Fishing effort per unit area.
Fishing effort.

FR intensité de pêche

Mesure de la puissance de pêche déployée par unité de surface de la pêcherie à un temps donné.

IT intensità di pesca

NL visserij-intensiteit; bevissing; bevissingsintensiteit

De grootte, het motorvermogen, het aantal schepen, en de doelmatigheid van de vistuigen.

PT intensidade de pesca

3240

ES | anzuelo sin agalla

DA | tværpind

Art fiskekrog uden bue, som sluges af fisken.

DE | Knebel

Vorläufer des modernen Angelhakens. Gerader Haken aus Stahl, mit Hakenauge auf halber Länge, beidseitig mit Widerhaken, wird zum Aalfang benutzt.

GR | ίσιο αγκίστρι

Λεπτό και ευθύ στέλεχος που χρησιμοποιείται στη θέση ενός αγκιστριού και το οποίο εύκολα καταπίνεται από ένα ψάρι αλλά δύσκολα αποβάλλεται.

EN | gorge

A line fishing device used instead of a fish hook that consists of an object easy to swallow, but difficult to eject.

FR | aiguille; hameçon droit

Employée à la place d'un hameçon, sorte de tige fixée à une ligne, facile à avaler, mais difficile à rejeter par le poisson.

IT | amo diritto

NL | knevelhaak

In het aas verborgen haak, oorspronkelijk in de vorm van een stokje met twee punten, later van metaal met weerhaken, die gemakkelijk wordt ingeslikt, maar niet meer kan worden uitgespuugd.
Voorloper van de vishaak.

PT | anzol direito; agulha

Utensílio de pesca com a forma de estilete utilizado em vez de um anzol, fácil de engolir mas difícil de expelir pelo peixe.

3241

ES | nasa para fondo

DA | tejne; rusekurv

Fiskeredskab af træ eller fletværk til fangst af ål, hummer eller krabber.

DE | Bodenreuse

Reuse aus Holz oder Weidengeflecht für den Krabben-, Hummer- oder Aalfang.

GR | κιούρτοι βυθού· κοφινέλα βυθού

EN | ground basket; bottom set pot

Pot or basket made of wood or osier and used to catch crabs, lobsters or eels.

FR | casier calé au fond

IT | nassa di fondo

NL | grondkubbe

Cilindrisch, rechthoekig of kegelvormig vistuig makkelijk door vis en schaaldieren te betreden, maar moeilijk te verlaten.

PT | nassa; covo

Arte de pesca tipo armadilha fixa que se utiliza para capturar peixes, moluscos ou crustáceos, com a forma de caixa, cesto ou pote; pode ser construída com diversos materiais (madeira, varas de metal, rede de pesca, rede de metal, rede de plástico) e possui uma ou mais aberturas ou entradas (boca e endiche). São colocadas no fundo com ou sem isco, isoladas ou em teias, e ligadas a um ou mais cabos de alagem referenciados à superfície por meio de bóias.

3240

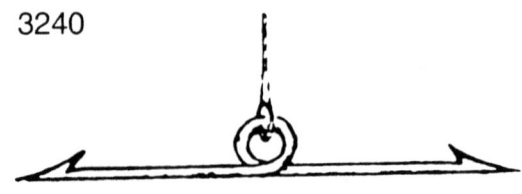

3242

ES salabardo de mango doble

DA sakse-ketsjer

Et redskab af net udspændt mellem to stokke. Til fiskeri på lavt vand.

DE Scherenhamen; Handhamen; Scherhamen

Netz, das zwischen zwei scherenartig verbundenen Stangen gespannt ist.

GR απόχη δύο πασσάλων

Είδος απόχης που στηρίζεται σε δύο πασσάλους.

EN skimming net

Net mounted on a light frame of wood or other material and operated by hand in shallow water or from a boat.

FR filet à l'étalage de surface

IT nichessa

Tipo di coppo sostenuto da due pali.

NL scheersaaiingskor; gebbe; aalgebbe; steekhaam; stokwade; gigonet

Net, bevestigd aan twee kruisende palen, dat de visser voor zich uitduwt.

PT chalrão

Arte de pesca manobrada a pé em águas pouco profundas e constituída por um saco de rede montado em duas canas (ou varas) de comprimento variável e amarradas em cruz perto das extremidades próximas do operador. A rede é montada nas canas (ou varas) desde o ponto onde estas se cruzam até às extremidades distais e o bordo livre do saco é lastrado de modo a manter-se permanentemente cobro o fundo durante o arrasto.
Rede de arrasto pelo fundo.
No caso de apresentar grandes dimensões e ser manobrada de e por uma embarcação designa-se por colher.

3243

ES arpón lanzado a mano

DA håndharpun

DE Handharpune

Wurfspeer, dessen Widerhakenspitze sich nach dem Auftreffen vom Harpunenschaft löst. Die an die Harpunenspitze geknotete Harpunenleine ist mit dem Fangboot verbunden.

GR καμάκι χειρός

EN hand-harpoon

FR harpon à main

Sorte de lance projetée à la main, dont la pointe comporte des barbes ou une partie basculante servant à retenir l'animal frappé.

IT arpione scagliato a mano

NL handharpoen

Vistuig voor de (wal)visvangst, met de hand geworpen, pijlvormig met weerhaken, op een schacht gestoken, waarbij een lijn aan de pijlpunt met weerhaak is bevestigd en de schacht na het treffen loslaat; de schacht blijft dan aan de oppervlakte drijven en leidt de visser naar de prooi; soms blijft de visser verbonden met schacht en prooi d.m.v. de lijn.

PT arpão manual; arpão de arremeço

3242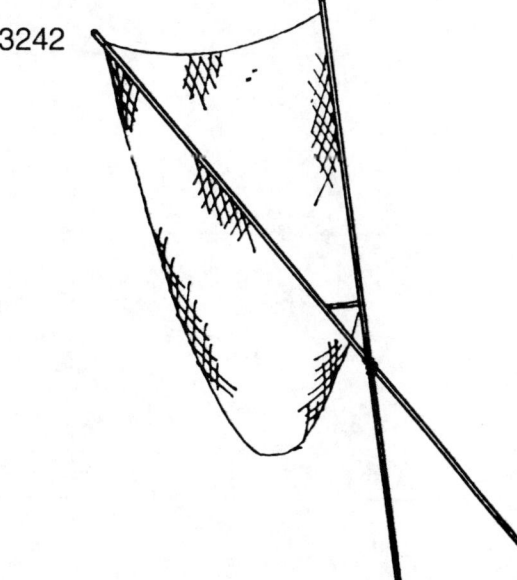

3244

ES arpón lanzado con fusil

DA harpungevær

Våben til fiskefangst. Drives af trykluft, fjederkraft eller elastisk kraft fra et gummibånd. Benyttes fortrinsvis af sportsdykkere.

DE Gewehrharpune

Diverse Ausführungen für Sporttaucher. Bei dem französischen Modell bleibt die Harpunenspitze nach dem Schuß über eine Ablaufleine mit dem Gewehr verbunden. Das Gewehr arbeitet mit Federkraft. Die Modelle anderer Länder benutzen als Treibmittel Druckluft (USA) und CO_2 (Italien). Einfachere Unterwassergewehre werden mit gespannter elastischer Kordel oder Gummiband betrieben (Japan, Hawaii, China).

GR ψαροντούφεκο

EN rifle-harpoon

FR harpon lancé au fusil

IT arpione scagliato col fucile

NL geweerharpoen

Harpoen die uit een geweer geschoten wordt, gebruikt bij de walvisvangst of in de sportvisserij.

PT arpão mecanizado lançado por espingarda

3245

ES arpón lanzado con cañón

DA harpunkanon

Våben til hvalfangst.

DE Kanonenharpune

Harpunenkanone in Gebrauch seit 1731. Mitte des letzten Jahrhunderts konstruierte Svend Foyn eine Walharpune mit Explosionsgeschoß, um den Wal schneller zu töten. Moderne Harpunenkanonen sind um 360 Grad schwenkbar. Ein größerer Erfolg in der Waltötung wird mit der elektrifizierten Harpune erreicht. Anstelle der Granate dringt die Harpune mit einem stromführenden Kabel in den Walkörper ein.

GR πυροβόλο με καμάκι

EN gun-harpoon

FR harpon lancé au canon

IT arpione scagliato col cannone

Arpione che viene lanciato con uno speciale cannoncino, detto lancia arpioni.
Usato nei casi di prede molto grosse.

NL kanonharpoen

Harpoen door een kanon afgeschoten.

PT arpão mecanizado lançado por canhão

3246

| ES | anzuelo curvo |

| DA | buet krog |

Fiskekrog med næsten cirkulær udformning.

| DE | **Bogenhaken** |

Allgemein in Abwandlungen übliche Hakenform.

| GR | κυρτωμένο αγκίστρι |

| EN | **curved hook** |

| FR | **hameçon recourbé** |

Hameçon fortement courbé, de forme presque circulaire, utilisé surtout pour la pêche du thon à la palangre.

| IT | amo ricurvo; amo storto |

| NL | **ronde haak** |

Soort vishaak.

| PT | anzol curvo |

3247

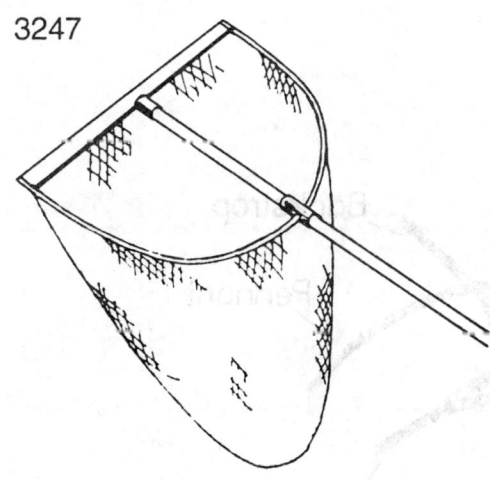

3247

| ES | **rastrillo para gambas; angaza para gambas** |

Pequeña draga, manejada a mano por medio de una pértiga empleada para la pesca de gambas.

| DA | **glib** |

Poseformet fiskenet på en lang stang, der med hånden skubbes hen over havbunden. Til reje- og ålefiskeri på lavt vand.

| DE | **Schiebehamen; Stoßwade; Scheerenwade; Schiebenetz; Bügelhamen** |

Handgerät zum Garnelenfang im seichten Wasser. Gerades Holz mit darüber im Halbkreis gespanntem Bügel, mit daran anschließendem Netzsack. Am geraden Holz schließt im rechten Winkel der Stiel an, der mit dem Netzbügel verbunden ist.

| GR | **απόχη· γαριδολόγος** |

Μικρό δίχτυ προσαρτημένο σε μία μακριά λαβή, το οποίο σπρώχνεται στο νερό μπροστά από τον ψαρά. Χρησιμοποιείται στην αλιεία γαρίδας.

| EN | **push net** |

A small net attached to a long handle and pushed by hand through the water in front of the fisherman.
Used for shrimps.

| FR | **pousseux; trouble** |

Filet de pêche en forme de poche, dont l'ouverture est montée sur une armature en bois munie d'un manche, poussé à la main par un pêcheur à pied; utilisé surtout pour les crevettes.

| IT | **guatta** |

Piccola rete a bocca fissa spinta con un palo.

| NL | **schuifhaam; steekhaam; gebbe** |

Trechtervormig net, waarvan de opening is bevestigd aan een raamwerk, en dat met de hand wordt bediend.
Gebruikt voor de garnalenvangst.

| PT | **camaroeiro; ganha-pão; ganapão** |

Arte de pesca com a forma de pequeno saco, montada numa armação circular fixada na extremidade de um cabo de madeira ou metal e manobrada à mão e a pé.
Utilizada principalmente na pesca de camarões.

3248

ES red de enmalle de monofilamento

Red formada por paños construidos con monofilamentos de poliamida.

DA monofil garn

Garn, hvor nettet er fremstillet af monofilamenter.

DE Kiemennetz aus Monogarn

Eingesetzt in der sogenannten stillen Fischerei als gesetztes Netz.

GR μονονηματικό απλάδι

EN monofilament gill net

Gill net made of monofilament netting.

FR filet maillant en monofilament

IT rete da posta in monofilo; rete da imbrocco in monofilamento; barracuda

Rete da posta armata con pezze costruite in monofilamento di poliammide.

NL monofilament kieuwnet

Kieuwnet, gemaakt van monofilament garens.
- Dit type garen vermindert de zichtbaarheid onder water, het is namelijk transparant.
- Monofilament = niet gevlochten.

PT rede de emalhar de monofilamento

3249

ES ocho de tope

Grillete que impide que siga corriendo la malleta una vez deslizada.

DA stopper

Samleled mellem indhalertampen og stjerterne. Udformet med skuldre, der hindrer passage gennem brillen.

DE Stopper; Stoppereisen

Das Jagerende trägt den Stopper (oder das Stoppereisen), der durch das Ochsenauge geführt wird. Das Stoppereisen kann durch einen Ring und das Ochsenauge durch einen Schäkel ersetzt werden.

GR εμπλοκέας

Σφυρήλατος διπλός κρίκος με εξοχές, ο οποίος συνδέει το ανεξάρτητο σύρμα με το χοντρό συρματόσχοινο.

EN stopper

Forged double eyed link with shoulders connecting pennant to bridle.
Trawl.

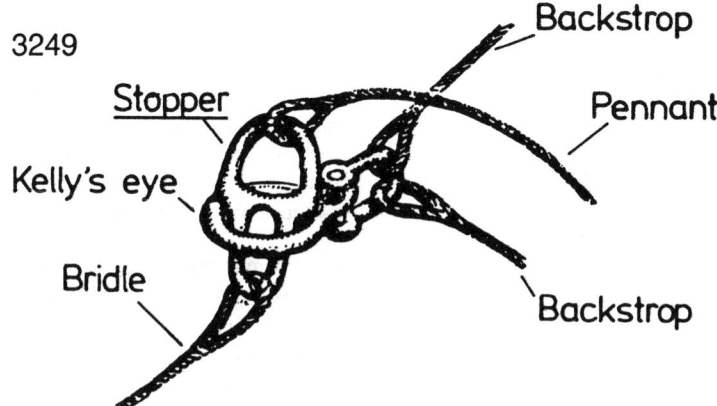

3249

Stopper

Kelly's eye

Bridle

Backstrop

Pennant

Backstrop

3249

FR stoppeur

Pièce forgée comportant deux anneaux et reliant le bras au rapporteur; des épaulements bloquent le stoppeur en traction sur le huit.

IT scontro

NL stopper

Gesmede schalm met schouders en twee ogen, die blijft steken in een ander onderdeel, de bril genoemd.
Wordt toegepast in de trawlvisserij om een verbinding te vormen tussen de bordstroppen en de breidels.

PT elo em oito; oito; «stopper»

Peça de ferro com dois orifícios, de igual ou diferente diâmetro, que faz a ligação entre o brinco da malheta e a malheta.
Rede de arrasto.

3250

ES espalda; plan alto

Parte superior de una red de arrastre entre la visera y el copo.

DA overpanelet i kroppen

De sektioner i trawlen, der udgør oversiden af kroppen, altså mellem taget og posen.
Se 3162.

DE Oberbelly

Teil des Obernetzes zwischen dem Dachstück und dem Zwischenstück (Hundertmaschenstück), an welches sich der Tunnel anschließt.

GR γούλες

Το ανώτερο τμήμα του διχτυού της τράτας μεταξύ τετράγωνου και σάκου.

EN batings; baitings

Tapered section of top panel between square and extension piece.

FR petit dos; bas de dos

Pièce de la face supérieure du chalut, entre le grand dos et l'amorce.

IT seconda parte del cielo

Di una rete a strascico.

NL rug

Bovenste deel van een trawl tussen de kap en het achtereind.

PT barrigas superiores

Parte superior das redes de arrasto entre o quadrado e a boca do saco.
Rede do arrasto.

TOP PANEL LOWER PANEL

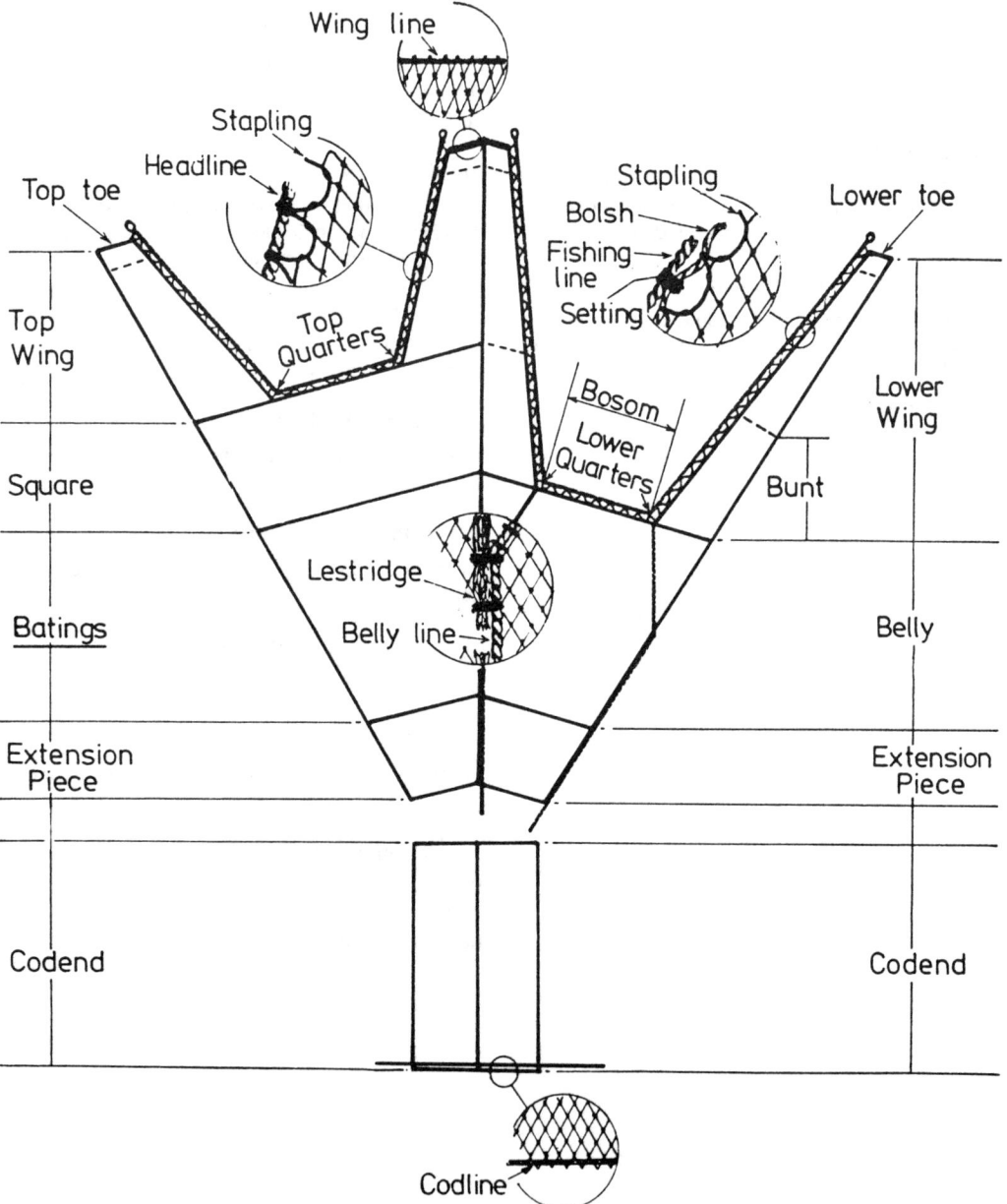

3251

ES nasa metálica

DA ståltrådsruse; ståltrådstejne

DE Drahtreuse

GR συρμάτινο καλάθι

EN wire basket

Trap made of wire or wire mesh.

FR casier en grillage; casier métallique

IT nassa metallica

NL metalen korf

Van ijzergaas vervaardigd, meestal cilindrisch of kegelvormig vistuig, voorzien van één of meer enkels.

PT nassa metálica; nassa forrada com rede metálica; murejona

Armadilha construída exclusivamente com materiais metálicos.

3252

ES piedra-bola; palangre de superficie

Palangre fijo al fondo y que, alternando flotadores y lastres, mantiene los anzuelos más o menos separados del fondo.
Se emplea para la captura de merluza y besugo en aguas comunitarias.

DA flydeliner

Langliner, der flåddes op til at stå midt i vandet eller ved overfladen.

DE Treibangel

GR επιπλέον παραγάδι βυθού

Τύπος παραγαδιού βυθού, εφοδιασμένο με πλωτήρες κατά διαστήματα, το οποίο εξυπηρετεί στη σύλληψη ψαριών σε συγκεκριμένη απόσταση από τον πυθμένα.

EN floated line; floating line; floating bottom longline

A type of bottom line equipped with floats and used to catch fish at a short distance from the bottom.

FR palangre de fond flottante

Type de palangre de fond, munie de flotteurs de place en place, servant à la capture des poissons à une certaine distance du fond.

IT palangaro galleggiante; palangaro fisso a mezz'acqua

Palangaro sostenuto da galleggianti a mezz'acqua o in superficie.

NL drijflijn

Gevlochten of gekloste zijde- of nylonlijn, die dusdanig is geprepareerd dat de lijn op het water blijft drijven.
Wordt hoofdzakelijk gebruikt bij het driftend vissen op snoek.

PT palangre fundeado de meia-água; palangre fundeado de superfície

Tipo de palangre de fundo, dotado de flutuadores espaçados, utilizado para a captura de peixes a uma certa distância do fundo.

3252

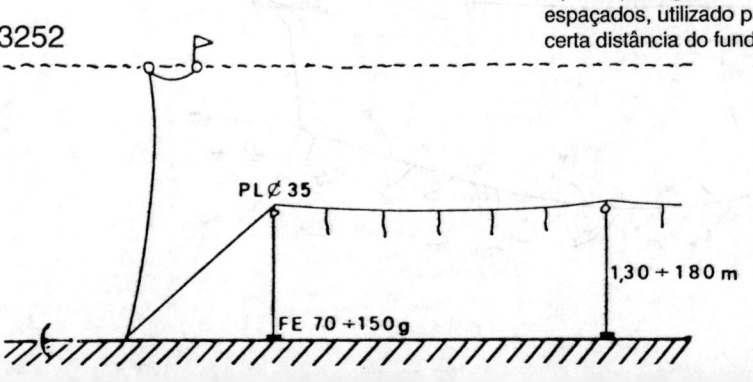

PL ∅ 35

1,30 ÷ 180 m

FE 70 ÷150 g

3253

ES cerco a la pareja; chinchorro a la pareja

DA vod trukket af to fartøjer

Vod, som bruges til en art snurrevodsfiskeri, der drives af to fartøjer. Metoden minder om partrawling, men der udlægges vodtove i stedet for, at der slæbes i wirer.

DE Zweischiffwadennetz

Zweischiffschleppnetz in pelagischer Netzkonstruktion oder als Grundschleppnetz.

GR γρίπος που τον χειρίζονται δύο σκάφη

EN pair seine

Similar to pair trawl but using long seine ropes instead of wires, and often a seine net. The boats heave in the ropes and come closer together near the end of the tow.

Methods vary within UK and are still being developed.

FR senne manœuvrée par deux bateaux; senne-bœuf

Senne de surface ou de fond mise en œuvre par deux bateaux de taille comparable; dans le cas d'une senne de fond, son utilisation par deux bateaux présente des points communs avec celle d'un chalut-bœuf de fond.

Le terme «senne-bœuf» est peu courant pour une senne.

IT sciabica a due natanti

NL spanzegen

Zegen gebruikt door twee schepen die in span vissen.

PT rede envolvente-arrastante de parelha

Rede envolvente-arrastante de alar para bordo manobrada por duas embarcações.

3254

ES palangre de fondo

DA langline til bundfiskeri

Samlebetegnelse for et redskab, der består af en hovedline med tavser og kroge. Forankres til bunden og afmærkes med bøjer. Krogene forsynes med madding.

DE Grundleine

Am Grund ausgelegte Langleine, bestückt mit bis zu 600 beköderten Haken.

GR παραγάδι βυθού

EN bottom set long line

Long line set near the seabed.

FR palangre de fond

Palangre ancrée au fond.

IT palangaro fisso; palangrese di fondo

Palangaro ancorato al fondo marino.

NL grondbeug

Beug waarmee op de bodem wordt gevist, en die ter plaatse met ankers is vastgelegd.

PT palangre de fundo; palangre fundeado de fundo

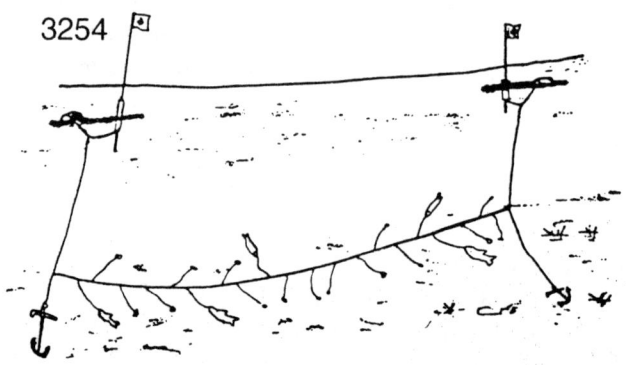

3254

3255

ES | palangre de deriva; palangre japonés

DA | flydeline

Langline, der ikke forankres. Med passende afstand er linen forsynet med flåd eller bøjer, der holder linen oppe i vandet eller ved overfladen. Driver med strømmen.

DE | treibende Langleine

An Schwimmern im Pelagial gehaltene, in Sets zusammengestellte Leine von je 150 bis 400 m Länge. Eingesetzt z. B. im Thunfisch- und Lachsfang.

GR | παρασυρόμενο παραγάδι

EN | drift line

Longline set in midwater or near surface and allowed to drift with the current.

The heavy drifting longlines with some thousands of hooks are typical in high seas fisheries.

FR | palangre dérivante

Palangre soutenue par des flotteurs en surface et dérivant librement avec le courant.

IT | palangaro derivante

Palangaro lasciato all'azione dei venti e delle correnti.

NL | drijvende beug

Beug die d.m.v. drijvers aan de oppervlakte in de juiste positie wordt gehouden.

Naast de Japanse variant voor tonijn bestaan er ook systemen voor de vangst van snoek, paling, zalm en krokodillen.

PT | palangre derivante

3256

ES | cerco escocés

DA | flyshootervod

Redskab, der minder meget om en trawl; anvendes af flyshootere på samme måde som en snurrevod. Forskellen ligger i, at under indhalingen ligger fartøjet ikke for anker, men gør fart fremad. Herved bliver det befiskede areal meget større.

DE | schottisches Wadennetz

Fangtechnisch ähnlich der dänischen Snurrewade, jedoch ohne Verankerung der Wadenleine.

GR | σκωτσέζικος γρίπος

Γρίπος βυθού.

EN | Scottish seine

A type of seine net which is set from a free-floating marker buoy. When the vessel has reached the buoy again, it is lifted aboard, the two ends of the hauling lines are connected to the winch and dragging and hauling begins from the forward-moving vessel (fly dragging).

Bottom seine.

FR | senne écossaise

Type de senne de fond utilisée en Écosse, différant de la senne danoise par sa conception, plus proche de celle d'un chalut.

Senne de fond.

IT | sciabica scozzese

NL | Schotse zegen

Zegen gelijkend op de snurrevaad, waarbij vanaf de boot wordt gewerkt die zich langzaam voortbeweegt terwijl de lijnen worden ingehaald.

Zie snurrevaad: bij de snurrevaad blijft het schip aan het anker liggen.

PT | rede envolvente-arrastante escocesa

Rede de fundo.

3255

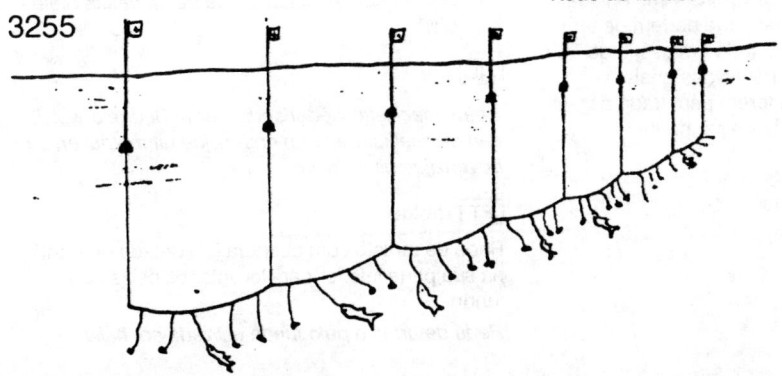

189

3257

ES almadraba

Arte fijo de trampa que se cala en lugares apropiados para interceptar el paso de los atunes (y otras especies).
Existen dos variantes: la denominada de «monte-leva», que sólo se cala cuando hay atunes a la vista, levantándose cuando desaparecen, y la llamada de «buche», que se instala al comienzo de la temporada de paso de los atunes, y no se levanta hasta que termina.

DA forankret tunfiskenet; tonnara

Stort bundgarnslignende redskab til fangst af tunfisk.

DE Tonnare

Verankertes, aus mehreren Kammern bestehendes Netz.

GR τονάρα· δίχτυ-παγίδα για τόνους

Ιχθυοπαγίδα μεγάλων διαστάσεων, ποντισμένη κοντά στην ακτή για τη σύλληψη τόνου.

EN madrague; pig catcher; tuna trap; trap net for tunas; tonnara; tunny-fishing net; tunny net

A trap used in the Mediterranean to catch tuna.

FR madrague; thonnaire

Filet-piège de grande dimension, mouillé près de la côte pour capturer le thon.

IT tonnara

Rete trappola che, situata in punti particolari sulle rotte dei tonni, ne permette la cattura.

NL tonnara; tonnaar

Type kom voor de vangst van tonijn in de Middellandse Zee, waarbij een vleugel loodrecht vanuit de kust wordt uitgezet, en leidt naar een langgerekte kamer met grote toegangsopening.

PT armação; almadrava; almadraba

Arte de pesca tipo armadilha fixa de grande extensão, constituída por redes verticais sustentadas por estacas ou cabos e âncoras, que partem de terra e entram pelo mar dentro, definindo uma série de canais, barreiras e câmaras através das quais os peixes são conduzidos, após terem penetrado na armadilha, até ao copo onde são apanhados.

3258

ES rápido

Arte de abertura horizontal fija con depresor para peces planos.

DA »rapido«

Fiskeredskab, der ligner en mellemting mellem en bomtrawl og en muslingeskraber. Forsynet med en depressorplade langs bommens overkant. Anvendes i Middelhavet til fangst af fladfisk.
Kendes ikke i Danmark.

DE Dredge

Hauptsächlich in Italien zum Seezungenfang eingesetzt.

GR τράτα με δόντια με σύστημα κατάδυσης

Τράτα σταθερού ανοίγματος με σύστημα κατάδυσης, εφοδιασμένη με δόντια στο κάτω μέρος της για να ανασηκώνει τα ψάρια από το βυθό.
Χρησιμοποιείται στην Ιταλία για την αλιεία της γλώσσας.

EN rake trawl with depressor

Type of beam trawl with diving board along its upper edge and a row of teeth along its lower edge to disturb flatfish.
— The faster the tow, the stronger would be the pressure of ascent, but the depth shearing effect of the depressors would also be increasing. Both forces could be held in balance so that the net would remain floating.
— Used in Italy for sole fishing.

FR rapido

Chalut à ouverture fixe, avec volet plongeur, muni de dents à sa partie inférieure pour soulever les poissons du fond.
Utilisé en Italie pour la pêche de la sole.

IT rapido

Rete da traino a bocca fissa armata con denti di ferro a guisa di rastrello e con una tavola che funziona da depressore.
Tale rete è usata principalmente per la pesca delle sogliole.

NL kor

Geen specifieke Nederlandse term. Bedoeld wordt een mosselkor met een onderzijde uit metaal en een bovenzijde uit netwerk.

PT rápido

Rede de arrasto com abertura fixa dotada de dentes na sua parte inferior para levantar os peixes do fundo.
Rede de arrasto pelo fundo utilizada em Itália.

3259

ES línea de mano

DA håndsnøre; håndline
Til pilkefiskeri.

DE Handangel; Handleine

GR καθετή
Ορμιά η οποία ψαρεύει πάνω από τον πυθμένα της θάλασσας και η οποία φέρει αγκίστρια και στην άκρη της βαρίδι.

EN handline
A hand-held line with weighted end and hooks used to fish above the sea bed.

FR ligne à main; palangrotte *
Ligne maintenue directement à la main, lestée à son extrémité et utilisée pour la pêche près du fond ou entre deux eaux.
* Terme local (Méditerranée).

IT lenza a mano
Lenza il cui filo è direttamente tenuto con la mano.

NL handlijn
Lijn die wordt gebruikt bij het vissen in zee vanaf pieren op het strand, en die, voorzien van haken en lood, met de hand (dus zonder hengel) te water wordt geworpen.
Soms voorzien van drijvers.

PT linha de mão
Aparelho de anzol manobrado à mão, lastrado e utilizado para a pesca no fundo ou a meia água.

3260

ES línea de caña mecanizada

DA pilkemaskine
Maskine, der automatiserer pilkefiskeriet og bevæger krogene i vandet, haler op, når fisken kroges, og i visse tilfælde også tager fisken af krogen, når den er halet om bord.

DE mechanisierte Angelrute
Ablösung der Handangel durch entlang der Reling festaufgestellte, elektrisch betriebene Maschinen. Durch kleine Auf- und Abwärtsbewegungen wird der Fisch angelockt. Nach dem Anbiß lösen der Zug und das Gewicht das Überschwingen des Fanges an Deck aus.
Mit modernen Glasfiberruten von 5 m Länge können bis zu 30 kg schwere Fische mit dieser Methode geangelt werden.

GR καλάμι με μηχανισμό

EN mechanized pole-line
The actions of moving the hook up and down while fishing, bringing the hooked fish on board, releasing it from the hook on deck and redeploying the hook in the water are done by a machine.
Used e.g. for tuna catching, with the pole movement being entirely automatic.

FR ligne avec canne mécanisée
Ligne employée au moyen d'une canne automatisée dont les mouvements reproduisent ceux du pêcheur.

IT lenza a canna meccanizzata

NL mechanisch aangedreven hengel
Machine die het systeem hengel + snoer + haak met aas aandrijft, en dezelfde handelingen als een mens produceert; de lijn wordt geworpen, de vis naar de haak gelokt met korte bewegingen en uiteindelijk aan de haak geslagen en binnenboord gebracht.
Onder andere op Japanse schepen gebruikt.

PT linha de vara mecanizada
Pesca do atum.

3258

DEPRESSORE

PARTICOLARE
BOCCA

ATTACCO BRAGA

RASTRELLO

SCIVOLO

3261

ES | línea de caña; sedal

DA | kastesnøre

DE | Angelleine

GR | καλάμι

EN | pole-line

A hook with or without natural or artificial bait is attached to a line on the end of a pole of similar length to the line. Pole-lines are usually operated from a boat.

FR | ligne avec canne

Type de ligne à main utilisée par l'intermédiaire d'une canne, munie ou non d'un moulinet.

IT | lenza a canna

Lenza il cui filo è sostenuto dalla canna da pesca con un mulinello su cui questo filo può avvolgersi o svolgersi.

NL | hengelsnoer

Draad waaraan de haken zijn bevestigd, en die aan de hengel vast zit.

PT | linha de vara

Pesca do atum.

3262

ES | línea fondeada

DA | forankret fiskeline

Langline, der ved dræg er sikret mod at drive.

DE | verankerte Angel

In der Langleinenfischerei Sicherung der ausgelegten Leine durch Anker in Verbindung mit einer Markierungsboje.

GR | αγκυροβολημένη ορμιά

EN | anchored line

A type of bottom line anchored to the sea bed at its end or at several points.

FR | ligne ancrée

Type de palangre ou ligne verticale, ancrée au fond par son extrémité ou de place en place.

IT | lenza ancorata

NL | verankerde vislijn

Beug die d.m.v. ankers op de bodem is vastgezet.

PT | linha fundeada

Aparelho de anzol.

3263

draga mecanizada; rastra mecanizada

DA **mekaniseret skraber**

Redskab til muslingefiskeri. En spulepumpe på selve skraberen spuler muslingerne fri af havbunden og op i posen. De overføres til fartøjet, enten når skraberen tømmes eller kontinuerligt ved at blive pumpet op eller ført på et transportbånd.

DE **motorisierter Bagger**

GR **δράγα με μηχανισμό**

Δράγα η οποία αποχωρίζει τους θαλάσσιους οργανισμούς από την άμμο και τη λάσπη, με τη βοήθεια μιας αντλίας.

EN **mechanized dredge**

Water jets dislodge molluscs from the seabed ahead of the dredge. The catch may be transferred to the boat by a conveyor belt device or by pump.

The Italian dredge operated in the area of Venice is an example of a heavy but good mechanized dredge operated by a special gallows from the bow of the vessel. When fishing, the boat is anchored with a long wire and, after setting the dredge, the vessel, together with the gear, is towed backward by a small winch. This is a technique very often used in dredging. Moreover, water jets from two tubes connected to the dredge, wash out deeper-sited shells, which are collected in the wire basket of the dredge. The catch can be brought on board the vessel when the dredge is hauled in, by a system which is often used with trucks which are emptied very quickly by lifting one end on the platform.

FR **drague mécanisée**

Drague comportant un système de pompage ou transfert mécanique permettant la récolte en continu de coquillages.

IT **draga meccanizzata**

Draga che separa gli organismi marini dalla sabbia e dal fango tramite l'azione di una pompa.

NL **motordreg**

Mechanisch bediende kor.

PT **draga mecanizada**

Draga equipada com um sistema de bombagem ou outro sistema mecânico que permita a recolha de bivalves em contínuo.

3264

ES **esparavel fijo**

DA **faststående kastenet**

Bruges ikke i Danmark.

DE **Standwurfnetz**

Tragendes Netzgerüst mit einverzinktem Stahlring von 2 m Durchmesser und größer. In der Ringmitte befindet sich eine Scheibe von ca. 100 mm Durchmesser mit einem Loch im Zentrum. Abgehend von dieser Scheibe sind etwa 10 Drähte mit dem Ring verspannt. Das über das Gerüst gespannte Netz hängt wie ein Schleier ca. 0,9 m lose nach unten. Der Simm ist mit Blei beschwert. In Abständen an den Simm geknüpfte Schnüre verlaufen innerhalb des Netzmantels nach oben, werden durch das Loch der Mittelscheibe geführt und mit der Schnürleine verbunden. Die sogenannte Hauptleine hält das Netz an einer Öse der Mittelscheibe am Ende des Baumes fest. Zum Wurf wird die Schnürleine soweit angezogen, daß die Bleikugeln dicht unter der Mittelplatte sitzen. Durch die Bleibeschwerung fällt der Netzmantel nach unten und schwingt aus. Gleichzeitig läßt man das Netzgestell fallen. Breitgefächert sinkt das Netz ins Wasser.

GR **υποστηριζόμενος πεζόβολος· υποστηριζόμενο δίχτυ πόντισης**

EN **supported cast net**

A type of cast net too heavy to use by hand and therefore operated from a support, which is usually installed on a boat.

FR **épervier fixe**

Sorte d'épervier manœuvré à partir d'un support, installé en général sur une embarcation.

IT **rezzaglio fisso**

Non è usato in Italia.

NL **groot werpnet**

Werpnet dat door zijn grootte niet meer door één man kan worden bediend.

PT **tarrafa fixa**

Espécie de tarrafa manobrada a partir de um suporte instalado normalmente numa embarcação.

3265

ES atarraya; esparavel a mano

Esparavel que se lanza y cobra a mano.

DA håndkastenet

DE Handwurfnetz

GR πεζόβολος χειρός· δίχτυ πόντισης χειρός

EN hand cast net

A cast net operated by hand.

FR épervier à main

Filet en forme de cône évasé à circonférence pouvant atteindre une vingtaine de mètres, lesté à sa périphérie et retenu par une ligne amarrée en son centre. L'épervier, lancé à la main du rivage ou d'une embarcation, capture les poissons en retombant et en se refermant sur eux.

IT rete da lancio a mano; giacchio lanciato a mano; rezzaglio *; sparviere *

* Termini dialettali.

NL handwerpnet; met de hand te werpen net

Er is een onderscheid tussen „hand" en „boat-cast net", laatstgenoemde is zó groot dat het werpen met de hand niet meer mogelijk is; er wordt dan op schepen gebruik gemaakt van bomen en gieken.

PT tarrafa de mão

Arte de pesca de arremeço manuseada por um só homem, em águas pouco profundas, a pé ou de uma pequena embarcação. É uma arte muito original, com forma cónica ou de funil e que se abre em círculo quando é arremeçada. A respectiva malhagem vai aumentando desde o vértice até à base, a qual é guarnecida com chumbos.

3266

ES reducción; menguado

DA indtagning

Udføres under håndbinding af et net, hvor der ønskes en reduktion af nettets bredde: den maske, der bindes, fastgøres til to masker i den foregående række.

DE Mindern; Schnitt; Minderung

GR σμίκρυνση· μάζεμα

Κατά το πλέξιμο, το αποτέλεσμα του δεσίματος με κόμπο ενός ματιού σε δύο γειτονικά μάτια στην προηγούμενη σειρά, για να μειωθεί το πλάτος του δικτυώματος.

EN bating; baiting; tapering

In hand braiding the result of knotting one mesh on to two adjacent meshes in the preceding row to reduce the width of the netting.

FR diminution

Réduction progressive de la largeur d'une pièce de filet, obtenue par laçage manuel ou coupe directe le long du ou des bord(s).

IT sematura

NL minderen

Het doen afnemen van het aantal mazen in de breedte door verschillende verbindingstechnieken met de voorlaatste rij mazen.

Bij machinaal gebreid netwerk kan men minderen door de mazen op verschillende manieren los te snijden.

PT amatar; reduzir

Processo utilizado na construção manual de panos de rede visando a redução do número de malhas em largura.

3267

ES lance; calada

Conjunto de trabajos para largar al agua el arte de pesca y volverlo a cobrar luego cuando se considera hecha la captura.
Tiempo de permanencia de la red en el caladero.

DA slæb *; sæt **; kast ***

En enkelt operation med et net fra udsætning til haling.
* Trawl; ** garn og snurrevod; *** not.

DE Hol

Aussetzen, Schleppen und Einholen des Netzes.

GR καλάδα· ψαριά

EN haul *; set **

A single fishing operation.
* Trawl; ** seine.

FR trait *; calée **; coup de senne ***

Opération de pêche effectuée avec l'engin de référence.
* Chalut; ** filet maillant; *** senne.

IT cala

Singola operazione di pesca.

NL trek

Het totaal van handelingen van het vissen met een sleepnet: uitzetten, vissen en halen.
De periode van het vissen.

PT lanço; lance

Operação de pesca realizada com uma rede de pesca (arrasto, cerco, envolvente-arrastante) por cada vez que é largada.

3268

ES parpalla

DA slidgarn

Tilbehør til visse trawl bestående af materiale (net, presenning, huder, se 3157), der fæstnes til oversiden, se 3011, eller undersiden, se 3012, af posen. Skal beskytte mod slid mod havbunden eller under halingen.

DE Scheuerschutz

GR στρώση

EN chafer; false belly; chafing gear

Replaceable material (hides, old netting etc.) fixed to underside of the codend for protection against chafing on the sea bed.

FR tablier

Pièce amovible en filet, cuir ou autre matériau, fixée au cul du chalut pour le protéger contre l'usure ou les déchirures au contact du fond.

IT foderone

NL sleeplap

Stuk zeildoek, netwerk of enig ander materiaal, bevestigd aan de onderzijde of ook bovenzijde van een trawl, om openscheuren door contact met obstakels op de bodem tegen te gaan.

PT forra

Material substituível (em couro, rede, etc.) fixado exteriormente ao saco das redes de arrasto e que se destina a proteger o saco.

3268

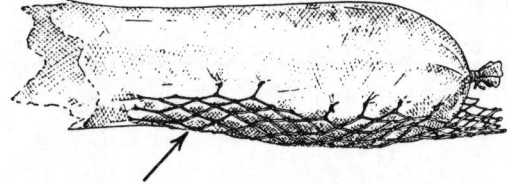

3269

ES anzuelo

Arponcillo de acero que está unido a un hilo y cebado para pescar.

DA krog

Fiskeredskab af bukket ståltråd, tilspidset og forsynet med modhager.

DE Haken; Fischhaken

GR αγκίστρι· αγκίστρι αλιείας

EN hook; fish hook

Bent, sharpened piece of steel wire usually with barb, for catching fish.

FR hameçon

Petite pièce d'acier en forme de crochet que l'on place au bout des lignes de pêche et qui reçoit l'appât.

IT amo

Piccolo attrezzo da pesca formato da un filo d'acciao piegato a uncino e terminante nell'estremità più corta a punta con ardiglione e nell'estremità più lunga con una paletta o un anello per fissarlo alla lenza.

NL haak; vishaak

Metalen voorwerp, al of niet voorzien van een of meer weerhaken, vervaardigd uit een bepaalde lengte, plat, ovaal, rond of vierkant staaldraad, waarvan de ene zijde in een bocht is gebogen en de andere zijde is voorzien van een oog of palet, ter bevestiging van de lijn.

PT anzol

Engenho de aço, normalmente com forma curva, com ou sem barbela, que se utiliza nas linhas de pesca e ao qual se prende o isco.

3270

ES puntera

DA forfang

Kort line eller ståltråd, som i den ene ende har en krog og i den anden er bundet til hovedlinen. Pilkefiskeri.

DE Vorfach; Mundschnur

Verbindung zwischen Schnur und Haken oder künstlichem Köder.

GR παράμαλο

Κοντό σχοινί από μεσινέζα ή σπάγγο που συνδέει το κεντρικό σχοινί («μάννα») ενός παραγαδιού με τα αγκίστρια.

EN cast

Terminal yarn, or strand, to which hooks of a handline are attached by short droppers.

FR bas-de-ligne

Partie terminale plus fine de la ligne à laquelle sont fixés les avançons.

IT terminale

Pezzo di filo o di cavetto di acciaio inossidabile, oppure di nylon, di diametro adeguato, sul quale viene fissato l'amo.

NL onderlijn

Lijn van bepaalde lengte, die de verbinding vormt tussen een hoofdlijn en de haak, en die is vervaardigd van ander materiaal —of die een andere dikte heeft— dan de hoofdlijn.

Het gebruik van een onderlijn geschiedt om vistechnische redenen: een bepaald systeem van vissen, of het vissen op bepaalde vissoorten.

PT estralho; baixada; estrobo

Parte dos aparelhos de anzol constituída por um fio de comprimento e diâmetro reduzidos que se liga à madre e em cuja extremidade terminal livre se empata um anzol.

3271

ES reparación de la red

Trabajo que consiste en alar las mallas rotas o reponer los daños más graves por trozos de paño.

DA bødning

Reparation af net.

DE Netzausbesserung

Manuelle Ausbesserung beschädigter oder zerrissener Maschen entsprechend der Anzahl und der Maschenweite.
Austausch von (vorgefertigten) Netzteilen.

GR επιδιόρθωση διχτυών αλιείας· μπάλωμα διχτυών

Χειρωνακτική επιδιόρθωση των φθορών ενός διχτυού με μπάλωμα ή μερική αντικατάσταση προκατασκευασμένων μερών αυτού.

EN net mending

FR ramendage; «bridage»

Réparation manuelle des filets après chaque usage. Il consiste à remplacer les mailles usées ou déchirées par un nombre correspondant de mailles nouvelles ou par l'insertion d'une pièce neuve.

IT riparazione delle reti

NL boeten van netten; repareren van netten

Herstellen van visnetten.

PT remenda; reparação de uma rede

Reparação manual de redes de pesca; consiste na reconstituição e/ou substituição das malhas partidas e/ou avariadas de um pano de rede por malhas novas.

3272

ES calón triangular

Triángulo de acero que sustituye al calón tradicional.

DA danlenotrekant

Kraftig metalplade, der kan sidde enten som forbindelse mellem mellemlinen og stjerterne eller understjerten og undertællen og bobbinsgearet.
Ved at trekanten adskiller disse, nedsættes risikoen for generende tørner.

DE Dan-Leno-Dreieck; Joch

Die Einheit von Jager, Kugel, Joch und Stander ist das Bindeglied zum Netz (Vornetz), d. h. zum Grundgeschirr und zur Headleine.

GR ματσέτα

EN dan leno triangle

A triangular piece of steel placed between the bridle and leg of a trawl net.

FR triangle

Pièce d'acier en forme de triangle placée entre le bras et les entremises d'un chalut.
Sorte de guindineau.

IT mazzetta triangolare

Rete a strascico.

NL danlenodriehoek

Driehoekige metalen plaat waaraan voorlopers en breidels kunnen worden bevestigd.
De breidels hebben dan meer verticale afstand.

PT triângulo do calão

Peça de aço com forma triangular colocada entre as malhetas e os tirantes das malhetas de uma rede de arrasto.

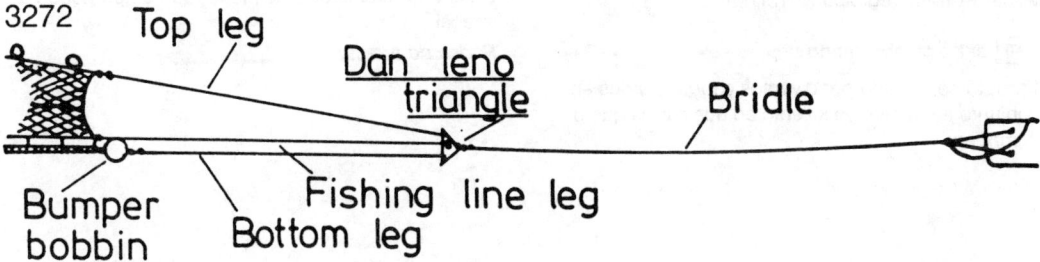

3272 Top leg — Dan leno triangle — Bridle — Bumper bobbin — Fishing line leg — Bottom leg

3273

ES calabrote

Producto compuesto por varias cuerdas cableadas conjuntamente en sentido inverso al de las cuerdas componentes.

DA kabelslået tov

Tovværk fremstillet ved flere på hinanden følgende samlinger af snoede liner. På hvert trin lægges slåningerne modsat den forrige.

DE Kabelschlagseil

Dreistufig verseiltes Seil. Die erste Stufe erfolgt durch das Verseilen der Kardeelgarne zum Kardeel. In der zweiten Stufe werden die Kardeele zur Trosse verseilt und in der dritten Verseilstufe erfolgt das Zusammenschlagen der Trossen zum Kabelschlagseil. Die Reihenfolge der Schlagrichtung (Drehrichtung) vom Kabelgarn bis zum Kabelschlagseil ist SZSZ.

GR δίπλοκο σχοινί· καρλίνο

Σχοινί αποτελούμενο από τρία μονόπλοκα, δηλαδή λεπτότερα σχοινιά, στριμμένα αριστερά.

EN cable laid rope

A product consisting of several ropes cabled together in the opposite direction to that of the constituent ropes.

FR grelin

Produit composé de plusieurs cordages câblés ensemble, en sens inverse de celui des cordages composants.

IT gherlino; corda torticcia

Prodotto composto di più corde commesse insieme, nel senso inverso a quello delle corde componenti.

NL geslagen touw

Touw samengesteld uit drie of vier strengen die in elkaar worden gedraaid (getwijnd).

PT cabo calabroteado

Produto constituído por diversos cabos torcidos em conjunto e em sentido inverso ao dos cabos que o compõem.

3274

ES trencilla; cabo de entrallar

DA bols

Anvendes, hvor maskerne i et net ikke sættes direkte på tællerne. Bolslinen fæstnes til hver enkelt maske og fastgøres med bændsler til tællen.
Anvendes bl.a. ved undertællen på en trawl for at forstærke og for at lette reparationer.

DE Bülschleine

Leine, die durch die fliegenden Maschen eines Netzes läuft und mit Bändseln am Grundtau befestigt wird.
In England ist dieses Verfahren nicht üblich, sondern es werden die Maschen an das Grundtau gemarlt.

GR ακρόσχοινο διχτυού

Σχοινί που διαπερνά τα (κυρίως) κάτω άκρα του πλέγματος ενός διχτυού και στερεώνεται με σήματα (κν. καμάρια) στο κυρίως σχοινί βυθού του διχτυού.

EN bolsh; bolch; bolch line

Rope attached along edge of lower wings and bosom netting for securing in bights to fishing line.

FR filière

Cordage servant au montage des ailes inférieures et du carré de ventre le long du bourrelet.

IT corda di sopporto

Corda, di materiale appropriato e di dimensioni convenienti, sulla quale la pezza di rete è fissata.

NL dunnepees

Touw dat langs het netwerk van de onderzijde vastgezet wordt, en dan later op de dikke, zware onderpees wordt bevestigd.

PT cabo de entralhe de asa inferior

Cabo de entralhe da asa inferior que é apontoado ao arraçal.
Redes de arrasto.

3275

ES intercalador

DA afstandsstykke

Rør af kraftigt gods, oftest metal, der indsættes mellem bobbinskuglerne på et bobbinsgear.

DE Klotje; Klote

Gußeisernes Bestückungselement am Rollengeschirr eines Grundschleppnetzes.

GR διαχωριστικό

EN spacer

Small diameter bobbin inserted between principal bobbins of a groundrope.

FR intermédiaire *; yoyo **

* Pièce cylindrique assez longue, généralement en métal, servant à espacer les sphères sur la ligne de bourrelet; elle peut être reliée au bourrelet par une chaîne (intermédiaire avec chaîne).
** Petite pièce métallique de faible diamètre, enfilée sur la ligne de sphères et reliée au bourrelet par une chaîne.

IT yoyo; distanziatore

NL klos; klosje

PT carrinho do arraçal

Pequenos roletes colocados no arraçal entre os roletes normais e aos quais se fixam as bichanas.

3276

ES viravira; parpallón; leva; trapa

Cabo que sirve para cerrar la red, uniendo ambas relingas y para izar éstas.

DA låringsline

Tov, der er fæstnet til undertællen af en trawl. Bruges til at hale ruppen op under indhalingen.

DE Knüppeltau

Auf Seitenfängern Hilfstau zum Beiholen des Vornetzes.

GR σχοινί ασφαλείας

Η μεσογειακή τράτα που χρησιμοποιείται στην Ελλάδα δεν έχει αυτό το σχοινί ασφαλείας.

EN quarter rope; leech line

Handling rope used in side trawling to bring the bosom section of the groundrope to the ship's side.

3274

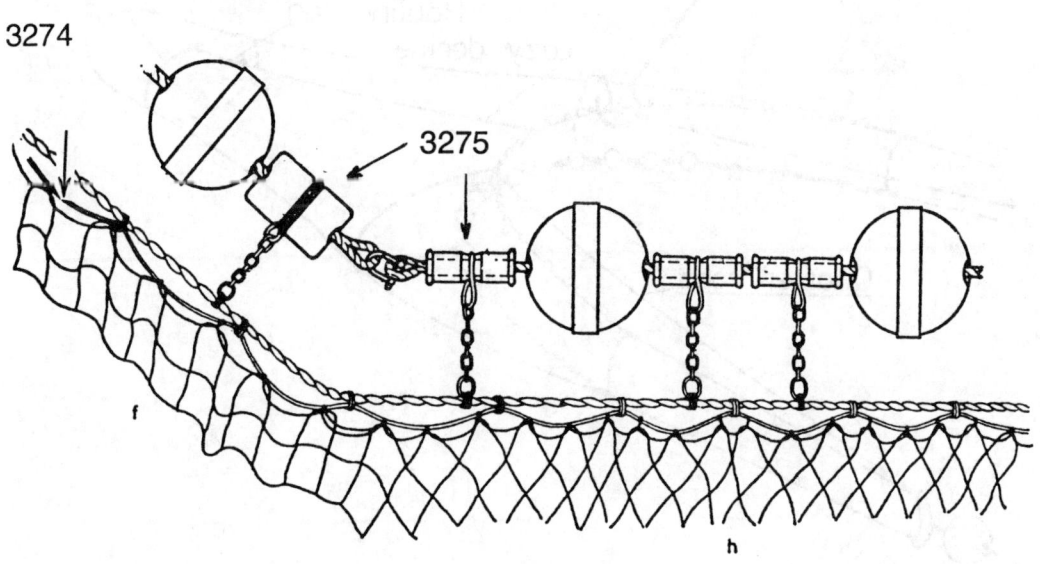

3275

f

h

3276

parpaillot; biribi

Filin servant à fermer l'ouverture du chalut au moment de l'apparition de la poche en surface. Les parpaillots, au nombre de deux, servent aussi à embarquer le bourrelet.

IT **cavo di chiusura**

Cavo che serve a chiudere la bocca della rete quando si sta salpando per evitare che il pesce fugga in avanti.
Rete da traino.

NL **buiktouw**

Op zijtrawlers: touw gebruikt voor het scheephalen van het midden van de onderpees.

PT **cabo de canto**

Cabo de fibra destinado a fechar a boca da rede de arrasto quando o saco chega à superfície; destina-se também a servir de auxiliar à alagem do arraçal.
São normalmente em número de dois.

3277

ES **red de nudos**

DA **knyttet fiskenet**

Net, der er fremstillet ved at knytte masker af garn med et enkelt eller dobbelt væverknob.

DE **geknotetes Fischnetz; geknotetes Fischernetz**

Handgeknotetes oder maschinell gestricktes Netz.

GR **πλεγμένο δίχτυ με κόμπους**

EN **knotted fishing net**

A net in which the netting is formed by knotting the twines using a single or double weavers' knot or reef knot.

FR **filet de pêche noué**

Filet de pêche dont les mailles sont formées par nouage.

IT **rete da pesca annodata**

Rete da pesca armata con pezze di rete annodate.

NL **geknoopt visnet**

Visnet, door middel van knopen, uit garens samengesteld.

PT **rede de pesca com nós**

3276 Headline float
Quarter rope
Headline becket
Headline
Halving becket
Hauling leg
Lazy deckie
Quarter rope

3278

ES estiraje de la red; estiramiento de la red

Cualquier operación tendente a apretar los nudos, o aquella operación que confiera una permanencia de formas por un tratamiento térmico o por la combinación de ambos procedimientos.

DA strækning af net

Proces, hvorved knuderne i nettet strækkes så formen stabiliseres.

Denne fiksering af nettet udføres ofte sammen med en varmebehandling eller anden påvirkning af nettet.

DE Streckung von Netztuchen

Unter Streckung werden die mechanischen, thermischen und sonstigen Behandlungen von Netztuchen verstanden, die eine Verfestigung der Knoten und Fixierung einer bestimmten Maschenform bewirken.

GR τέντωμα του διχτυού

EN stretching of netting

The operation of tightening of knots or of conferring a permanent shape by thermal or other means or a combination of both processes.

FR étirage de la nappe

Allongement sous tension de la nappe de filet dans le sens normal (N), destiné à serrer les nœuds afin de donner aux mailles leur forme et leur dimension.

IT stiro della pezza

Operazione necessaria per il serraggio dei nodi.

NL strekken van netwerk; fixeren van netwerk

Het mechanisch aantrekken van de knopen van een geknoopt net.

Deze behandeling wordt meestal gevolgd door een thermische en chemische behandeling die het materiaal stugger en slijtvaster maakt, en tevens de maaswijdte fixeert.

PT estiramento de um pano de rede

Operação destinada a apertar (socar) os nós dos panos de rede de pesca.

O estiramento executa-se segundo a direcção normal (N).

3279

ES arco; mediomundo

Velo o red embolsada como un canasto semiesférico empleado en el Cantábrico en la captura de peces pelágicos.

DA kassenet

En art vippenet.
Kendes ikke i Danmark.

DE Basnig

In der deutschen Fischerei nicht gebräuchlich.
In der Anwendung den Senknetzen vergleichbar.

GR

Δεν υπάρχει στην Ελλάδα.

EN basnig

Type of liftnet.

FR filet sac; basnig

Filet soulevé formé d'une poche parallélépipédique, relevé ou viré hors de l'eau à la main ou mécaniquement, à partir d'un bateau.
Engin peu utilisé dans les eaux européennes.

IT basnig

Non usato in Italia.

NL

Niet bekend in Nederland.

PT rede com saco

Qualquer tipo de arte de pesca fixa ou não e na qual os peixes capturados são acumulados num saco.

3280

ES eficacia de captura

De un arte.

DA fangsteffektivitet

Et mål for et fiskeredskabs fangstegenskaber.

DE Fängigkeit

Spezifische Fangleistung eines Fischfanggeräts.

GR αποτελεσματικότητα του αλιευτικού εργαλείου

EN catching efficiency

A measure used to compare the catching ability of fishing gears.

FR efficacité de capture

D'un engin de pêche.

IT efficienza di cattura; capacità di cattura

Di un attrezzo da pesca.

NL visnamigheid

Het vermogen van een vistuig om vis te vangen.
Een vistuig dat goed vangt heet visnamig.

PT eficiência de pesca; rendimento de pesca

Medida usada para comparar as capacidades de captura das artes de pesca.

3281

ES termofijado; tratamiento térmico

Operación que confiere una permanencia de formas.

DA varmestabilisering; termofiksering

Proces hvorved et nyt net udsættes for tør eller fugtig varme for at fiksere det inden ibrugtagning.

DE Thermofixierung; thermische Behandlung

Behandlung mit trockener oder feuchter Wärme zur Erhöhung der Haltbarkeit von Netztuch und Knotensitz.

GR θερμοστερέωση· θερμική σταθεροποίηση

EN stabilization by thermal means

The operation of setting the knots in a sheet of netting by baking in an oven and possibly applying tension.

FR fixation par la chaleur; thermofixation

Opération qui donne à la nappe de filet sa stabilité dimensionnelle définitive.

IT termofissaggio

Operazione per conferire la definitiva stabilità dimensionale delle maglie.

NL warmtebehandeling; thermische fixatie

Behandeling waarbij het netwerk wordt verhit om het materiaal te fixeren.

PT fixação pelo calor; termofixação

Operação que confere uma permanência de formas aos nós das redes de pesca através da utilização do calor.

3282

ES | corte K

Corte hecho en cualquiera de los sentidos N o T.
Símbolo K.

DA | K-skæring

Udskæring af net uden for knuderne.
a) Symbol K.
b)Samlebetegnelse for N- og T-skæring i tilfælde hvor nettets retning er uden betydning.

DE | K-Schnitt; Knotenschnitt

Schnitt außerhalb der Knoten.
a) Kennbuchstabe K.
b) Sammelbegriff für N-Schnitt und T-Schnitt in Fällen, in denen die Bezeichnung auf die Hauptlaufrichtung des Netzgarnes ohne Bedeutung ist.

GR | κοπή K· κόψιμο K

EN | K-cut; knot cut

Cut just beyond the knots.
(a) Symbol K.
(b) May be used instead of the N-cut or T-cut in cases where the relation to the general course of the netting yarn is insignificant.

FR | coupe K

Coupe N ou T, pratiquée au ras du nœud dans tous les cas où il n'est pas possible de faire référence au sens général d'avancement du fil pour filet.
Symbole K.

IT | taglio K

Nel caso in cui le direzioni N e T non siano determinabili si deve impiegare in luogo di «taglio normale» e «taglio parallelo» il termine «taglio K».
Simbolo K.

NL | K-snit

Snit overeenkomend met de N- of T-richting in netwerk waarbij geen specifieke richting van het garen is aan te geven.
a) Voor dit netwerk vervangt de K-snit de snitten N en T.
b) Symbool K.

PT | corte K

Corte N ou T, praticado junto ao nó, nos casos em que não é possível referenciar o sentido geral de avanço do fio na fabricação da rede.
Símbolo K.

3283

ES | corte

DA | udskæring; skæring

Tilpasning af et netstykkes form (trekant, parallellogram, trapez osv.).
Skæring bruges mest om siderne på en netsektion.

DE | Zuschneiden

Netztuche müssen durch Zuschneiden in die für das Fanggerät benötigte Form (Trapeze, Dreiecke, Parallelogramme oder andere Vielecke) gebracht werden.

GR | κοπή· κόψιμο

EN | cutting

Operation to obtain the desired final shape of the netting by tapering cuts in a suitable way.

FR | coupe

IT | taglio

Operazione per ottenere una pezza di rete di dimensioni e forma diverse da quella della pezza di rete ottenuta sul telaio.

NL | nettensnijden

Het snijden van netpanelen uit stukken netdoek.

PT | corte

Operação visando a obtenção de um pano de rede a partir de uma peça de rede.

3284

ES corte B; corte a pies

Corte hecho en el sentido AB.
Símbolo B.

DA B-skæring; stolperet skæring

Skæring lagt parallelt med en sammenhængende
række af stolper.
Symbol B.

DE B-Schnitt; Schenkelschnitt

Ein Schnitt parallel zu einer Linie von
aufeinanderfolgenden Maschenschenkeln, der
Maschenschenkel durchschneidet.
Kennbuchstabe B.

GR κοπή B· κόψιμο B· διαγώνιο κόψιμο· διαγώνια
κοπή

EN bar cut; B-cut

A cut parallel to a line of sequential mesh bars, each
from adjacent meshes, and severing one or more
bars.
Symbol B.

FR coupe biaise; coupe B

Coupe parallèle à une suite rectiligne de côtés de
maille, chacun d'eux appartenant à une maille
adjacente et tranchant un ou plusieurs côtés de
maille.
Symbole B.

IT taglio obliquo; taglio B

Taglio effettuato nella direzione AB su uno o più lati di
maglia.
Simbolo B.

NL snit B

B = been; dit zijn weggesneden benen, dus
driepoten.
Symbool B.

PT pernão; corte B

Corte paralelo ao lado da malha, isto é, à direcção AB
e sobre um ou mais lados de malha.
Símbolo B.

3285

ES corte T; corte transversal

Corte hecho en sentido T.
Símbolo T.

DA skæring tværs over; T-skæring

Skæring lagt parallelt med trådretningen.
Symbol T.

DE T-Schnitt

Ein Schnitt parallel zur Hauptlaufrichtung des
Netzgarnes außerhalb der Knoten.
Kennbuchstabe T.

GR οριζόντια κοπή· οριζόντιο κόψιμο· κοπή T·
κόψιμο T

EN horizontal cut; T-cut

A cut parallel to the general course of the netting yarn
just beyond the knots.
Symbol T.

FR coupe horizontale; coupe T

Coupe parallèle au sens général d'avancement du fil,
effectuée au ras des nœuds.
Symbole T.

IT taglio parallelo; taglio T

Taglio effettuato nella direzione T.
Simbolo T.

NL snit T

T = twijnrichting; dit is de loop van de mazen van
links naar rechts of omgekeerd; dit zijn dus opzet-
eindmazen of hangers.
Symbool T.

PT malha; corte T

Corte efectuado junto ao nó e segundo a direcção T.
Símbolo T.

3284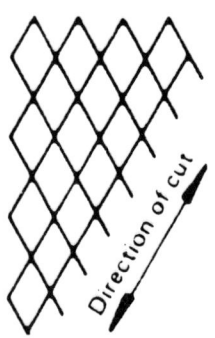

3286

ES corte N; corte normal

Corte hecho en el sentido N.
Símbolo N.

DA maskeret skæring; lige skæring; N-skæring

Skæring lagt vinkelret på trådretningen.
Symbol: N.

DE N-Schnitt

Ein Schnitt im rechten Winkel zur Hauptlaufrichtung des Netzgarnes außerhalb der Knoten.
Kennbuchstabe N.

GR κάθετη κοπή· κάθετο κόψιμο· κοπή N· κόψιμο N· κανονικό κόψιμο

EN vertical cut; N-cut

A cut at right angles to the general course of the netting yarn just beyond the knots.
Symbol N.

FR coupe verticale; coupe normale; coupe N

Coupe perpendiculaire au sens général d'avancement du fil, effectuée au ras des nœuds.
Symbole N.

IT taglio normale; taglio N

Taglio effettuato nella direzione N.
Simbolo N.

NL snit N

N = normaal; dit zijn normale mazen die langs de zijkant van het net van boven naar beneden of omgekeerd lopen (kantmazen).
Symbool N.

PT lombo; corte N

Corte efectuado ao nó e segundo a direcção N.
Símbolo N.

3287

ES corte AT

Corte completo en el sentido T (paralelo al sentido general de avance del hilo).

DA AT-skæring; maskeret skæring

Alle masker er skåret vinkelret på trådretningen.
Symbol AT.

DE AT-Schnitt

Glatter Schnitt in T-Richtung.
Parallel zur Hauptlaufrichtung des Netzgarnes.

GR μέθοδος κοπής AT· AT

EN AT-cut

All cuts entirely in the T-direction.
Parallel to the general course of netting yarn.

FR coupe AT

Coupe pratiquée entièrement dans le sens T.

IT taglio AT

Taglio nella sola direzione di taglio T.

NL snit AT

Snit waarbij alle mazen in de T-richting worden doorgesneden.
Dit is evenwijdig aan de richting van de netgarens.

PT corte direito; corte AT

Corte totalmente praticado na direcção T.
Símbolo AT.

3286

General course of the yarn

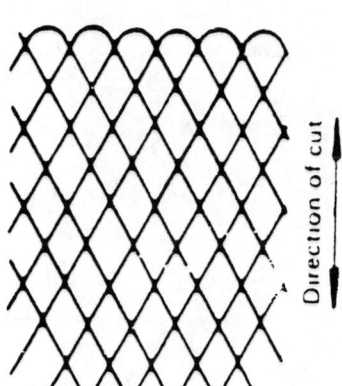

3285   General course of the yarn

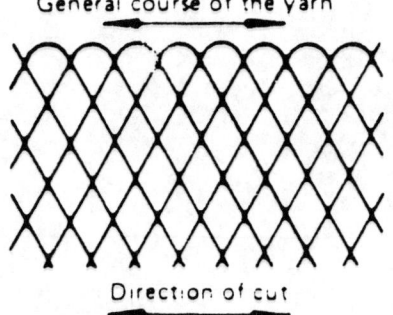

Direction of cut

3288

ES corte AN

Corte completo en el sentido N.
Perpendicularmente al sentido general de avance del hilo.

DA maskeret skæring; AN-skæring

Alle masker er skåret maskeret.
Symbol AN.

DE AN-Schnitt

Glatter Schnitt in N-Richtung.
Im rechten Winkel zur Hauptlaufrichtung des Netzgarnes.

GR μέθοδος κοπής AN· AN

EN AN-cut

All cuts entirely in the N-direction.
At right angles to the general course of the netting yarn.

FR coupe AN

Coupe pratiquée entièrement dans le sens N.

IT taglio AN

Taglio nella sola direzione di taglio N.

NL snit AN

Snit waarbij alle mazen in de N-richting worden gesneden.

PT corte ao lombo; corte AN

Corte totalmente praticado na direcção N.
Símbolo AN.

3289

ES corte AB

Corte completo sesgado.

DA stolperet skæring; AB-skæring

Alle masker er skåret stolperet.
Symbol AB.

DE AB-Schnitt

Alle Schenkel geschnitten.

GR μέθοδος κοπής AB· AB

EN AB-cut

All bars cut.

FR coupe AB

Coupe toute biaise.

IT taglio AB

Taglio nella sola direzione di taglio B.

NL snit AB

AB = allemaal weggesneden benen (driepoten).

PT corte em escada; corte AB

Corte totalmente praticado na direcção AB.
Símbolo AB.

3290

ES proceso de corte

Alternancia rítmica de los diferentes tipos de corte.
Símbolo C.

DA skæring

Skæringen udtrykker, hvordan maskerne er skåret for at give netsektionen den ønskede udformning.
Symbol C.

DE Schnittführung

Rhythmischer Wechsel der verschiedenen Schnittarten.
– Um eine gewünschte Form und Größe eines Netztuches durch Zuschneiden zu erreichen, müssen eine bestimmte Anzahl von N- oder T-Schnitten und B-Schnitten einander in rhythmischem Wechsel folgen.
– Kennbuchstabe C.

GR μέθοδος κοπής

EN cutting rate

Rhythmical alternation of the various types of cuts.
– To obtain a desired shape and area of netting by tapering, N- or T-cuts and B-cuts of a distinct length must follow each other in a rhythmical way.
– Symbol C.

FR processus de coupe

Alternance des différents types de coupe.
Symbole C.

IT schema di taglio

Modello che rappresenta l'alternarsi periodico di uno o più tagli N, T, B al fine di ottenere una pezza di rete di forma e area desiderata.
Simbolo C.

NL snit

Cyclische herhaling van combinaties van N, B en/of T, toegepast bij het snijden van netpanelen.
Symbool C.

PT processo de corte; método de corte

Alternância rítmica de vários tipos de corte.
Símbolo C.

3291

ES tipo de corte

DA skæring

DE Schnittart

GR τύπος κοπής

EN type of cut

Specification of cut such as N, T or B.

FR type de coupe

Désignation des éléments constitutifs de la coupe.

IT tipo di taglio

NL type snit; snijmethode

Wijze van vormen van stukken netwerk door het doorsnijden van de benen van een maas.
Men onderscheidt drie hoofdrichtingen: N-, B-, en T-snit.

PT tipo de corte

Designação dos elementos constituintes do corte.

3292

ES hilo acoplado; hilo reunido

DA paralleltrullet garn

To eller flere enkelttrådede, flertrådede eller kabelslåede garner oprullet ved siden af hinanden.

DE gefachtes Garn

Zwei oder mehr einfache Garne oder Zwirne, die zusammengespult, jedoch nicht miteinander verdreht sind.

GR πολύκλωνο κλώσμα

Κλώσμα που σχηματίζεται από την πλοκή δύο ή περισσοτέρων νημάτων ή κλωσμάτων. Τα τελευταία τοποθετούνται παράλληλα μεταξύ τους, χωρίς να στρίβονται.

EN multiple wound yarn; assembled yarn

Obtained by juxtaposition of two or more single, multiple or cabled yarns.

FR fil assemblé

Obtenu par juxtaposition de deux ou plusieurs fils simples retors ou câblés.

IT binato; accoppiato

Prodotto formato dall'unione, senza torsione, di due o più fili.

NL geassembleerd garen

Garen verkregen door het parallel naast elkaar groeperen van twee of meer eendraadsgarens, getwijnde garens of gekabelde garens.

PT fio multifilar

Obtido por justaposição de dois ou mais fios simples, torcidos ou entrançados.

3293

ES caja

Cuerpo o porción de madera que forma el motón y en el que hay practicada la cajera o hueco para la roldana.

DA blokhus

Den kappe af metal eller træ, som omgiver skiven i en blok.

DE Blockgehäuse; Gehäuse

GR θήκη τροχίλου

Μεταλλική ή ξύλινη θήκη (κάσα) ενός τροχίλου (κν. μακαρά), η οποία φέρει μεταξύ των παρειών της άνοιγμα για την τοποθέτηση του καρύου (τροχού).

EN shell; casing

The wooden or metal casing of a block in which the sheaves revolve.

FR caisse de poulie

Armature en bois ou en métal de la poulie, dans laquelle est fixé l'axe supportant le réa.

IT cassa del bozzello

Armatura in legno od in metallo, ad una o piú cavatoie, nella quale sono innestati gli assi delle pulegge del bozzello.

NL geraamte van een blok

PT caixa do moitão

Bloco de madeira ou metal que constitui o corpo do moitão.

3294

ES cuerda mixta

Cuerda, con o sin alma central, constituida por varios cordones cada uno de los cuales está combinado de la misma forma por fibras naturales o por hilos textiles sintéticos y por alambres de acero, pudiendo estar estos últimos galvanizados o no, según deseo del usuario.

DA kombinationstov; taifun

Tovværk, hvor kordelerne består af en kombination af ståltråde omgivet af syntetiske fibre.
Taifun er oprindeligt et handelsnavn.

DE Herkulestauwerk; gemischtes Seil

Seil aus Litzen mit Stahldrahtkern und Fasergarnumhüllung.

GR σχοινί ανάμεικτο

Σχοινί ανάμεικτο, σε αντιδιαστολή με τα (εξ ολοκλήρου μεταλλικά) σύρματα.

EN combination rope; combined rope

A rope, with or without a central core, consisting of several strands each of which is itself formed from a combination of natural fibres or synthetic fibre yarns and steel wires, the wires being either galvanized or ungalvanized as required by the customer.

FR câble mixte

Câble, avec ou sans âme centrale, constitué par plusieurs torons dont chacun est lui-même formé d'une âme de textile, d'une couche de fils d'acier galvanisé, en principe non jointifs, s'appuyant sur cette âme et d'une couverture de fils textiles.

IT corda mista; cavo misto

Corda con o senza anima, costituita da più legnoli, ciascuno dei quali è formato da una combinazione di fili di fibre naturali vegetali o fili di fibre sintetiche e di fili d'acciaio, questi ultimi zincati e/o plasticati o no a richiesta dell'utilizzatore.

NL compositiekabel

Tros, samengesteld uit drie staaldraadstrengen en drie vezelstrengen, alle van gelijke middellijn, geslagen om een vezelhart.

PT cabo misto

Cabo com ou sem alma central, constituído por diversos cordões formados cada um deles por uma alma têxtil e por arames de aço galvanizado e forrado exteriormente por cordões de fio têxtil.

3295

ES estacha

Cuerda que generalmente tiene un diámetro superior a los 40 mm (equivalente a una circunferencia de 5 pulgadas) empleada principalmente en la marina para el amarre de los buques.

DA trosse

Svært tovværk af naturlige eller syntetiske fibre eller stål. Anvendes til fortøjning, forhaling eller slæbning.

DE Trosse; Kabeltau

Starkes Stahlseil.

GR ρύμα· λατζάνα (κοινώς)

σχοινί ή συρματόσχοινο για ρυμούλκηση κλπ.

EN hawser

Man-made, natural fibre or wire rope used for mooring, warping or towing.

FR aussière

Filin en fibres synthétiques, en fibres naturelles ou en acier utilisé pour l'amarrage, le halage ou le remorquage.
Cordage mis à l'eau avec une tésure de filets dérivants au hareng et à l'extrémité duquel s'amarre le bateau durant l'opération de pêche.

IT cavo; gomena

Termine usato in marina sia per i prodotti tessili (corde) sia per i prodotti di acciaio (funi).

NL tros

Touwwerk van vrij grote dikte (omtrek van meer dan 4 cm), voornamelijk gebruikt voor het slepen, verhalen of afmeren van schepen.

PT amarra; cabo de amarra

Cabo de fibra ou de aço de grande diâmetro destinado à amarração e/ou reboque de navios.

3296

cuerda; cabo

Producto textil de diámetro superior a los 4 mm, constituido por cordones cableados o trenzados, con o sin alma.

En términos navales se utiliza la palabra cabo en lugar de cuerda.

DA **tov**

Fremstillet af fibre, ikke under 4 mm i diameter. Normalt spundet af 3 eller 4 kordeler eller flettet. Med eller uden hjerte i midten.

DE **Tau; Leine; Seil**

Seil aus Naturfaser, Kunstfaser, Draht oder deren Kombination gedreht, geschlagen oder geflochten.

GR **σχοινί**

EN **rope**

A textile product of not less than 4 mm diameter, generally consisting of three or four strands cabled or plaited together, with or without a core.

FR **cordage**

Produit textile n'ayant pas moins de 4 millimètres de diamètre, constitué de torons, câblés ou tressés, avec ou sans âme.

IT **corda**

Prodotto tessile con diametro non minore di 4 mm ottenuto per commettitura o trecciatura con o senza anima.

NL **lijn**

PT **cabo**

Conjunto de cordões de fibras têxteis ou de arames de aço, torcidos ou entrançados, com ou sem alma, com diâmetro superior a 4 milímetros.

3297

ES **hilo para red; hilo de red**

Todo hilo empleado en la fabricación de la red.

DA **tvist**

Alle liner egnet til fremstilling af net.

DE **Netzgarn**

Sammelbegriff für alle Arten von einfachen Garnen, Zwirnen, gedrehten oder geflochtenen Schnüren, Monofilen und Chemiedrähten sowie Kombinationen daraus, die für die Herstellung von Netztuchen verwendbar sind.

GR **νήμα για δίχτυα**

EN **netting yarn**

All yarns suitable for the manufacture of netting.

FR **fil pour filet**

Terme général servant à désigner toute espèce de fil utilisable pour la fabrication d'une nappe de filet.

IT **filo per rete**

Elemento tessile di grande lunghezza che può assumere diverse forme, designate specificamente con i termini: filato, filato voluminoso, filo continuo, filo continuo testurizzato, lamella, lamella fibrillata, binato, accoppiato, ritorto semplice, ritorto composto.

NL **netgaren**

Elk soort garen dat geschikt is voor het vervaardigen van netwerk.

PT **fio para rede de pesca**

Termo genérico utilizado para designar qualquer tipo de fio utilizável na fabricação de peças ou panos de rede de pesca.

3298

ES hilo continuo

DA filament garn

En eller flere kontinuerte fibre, evt. med en spinning. Udgør normalt første trin i fremstillingen af liner og tovværk.

DE Filamentgarn

Einfaches Garn, bestehend aus einem oder mehreren Filament(en), ohne Drehung oder mit Drehung hergestellt.

GR συνεχής ίνα

EN filament yarn

One or more continuous filaments, possibly twisted.

FR fil continu

Filament unique (monofilament) ou assemblage de plusieurs filaments continus (multifilament), sans torsion ou avec torsion (fil retors).

IT filo continuo

Filamento unico (monofilamento) o fascio di più filamenti (multifilamento) senza torsione o con torsione (torto).

Il termine «filo continuo» si deve intendere come contrazione dell'espressione «filo a filamenti continui».

NL filament; garenstreng

Streng gemaakt van doorlopende enkelvoudige garens (vezels) die bij het samenstellen worden getwijnd.

PT fio contínuo

Fio único ou feixe de vários filamentos de grande extensão.

3299

ES filástica retorcida

Filástica constituida por varios hilos retorcidos conjuntamente.

DA spundet garn; snoet garn

DE gedrehtes Garn

Mehrere gezwirnte Fäden zu einem (Netz-)Garn gedreht.

GR συστραμμένος σπάγγος

EN twisted twine

Twine composed of two or more yarns twisted together.

FR ficelle retordue

Ficelle reconstituée par plusieurs fils retordus ensemble.

IT ritorto semplice

Prodotto tessile ottenuto dall'unione di due o più fili singoli dei quali almeno uno con propria torsione e uniti mediante una sola operazione di torcitura.

NL getwijnde streng; geslagen streng

Streng opgebouwd uit verschillende in elkaar gedraaide garens.

PT fio torcido

Produto resultante da torção de vários fios.

3300

ES mecha; fibra discontinua; fibra cortada

DA stapelfibre

Korte fibre af passende længde til fremstilling af liner.
Liner af stapelfibre får et håret ydre, der gør dem behagelige at håndtere og giver en god knudefasthed.

DE Stapelfaser

Polyamidfaser, die durch Aufrauhung gute Dehnungseigenschaften erhält.
Diese Eigenschaft dient der erhöhten Knotenfestigkeit. Die Stapelfaser wird überwiegend als Anschlaggarn an Verbindungsstellen eingesetzt, die sicher fixiert werden müssen, z. B. Verbindung des Kopftaus mit der Bülschleine oder Anschlagleine des Netztuchs mit der Oberleine (Obersimm).

GR ασυνεχής ίνα

EN roving; rouving; staple fibre

Discontinuous fibres, usually prepared by cutting filaments into lengths suitable for the yarn spinning process. Staple fibres are twisted to form a spun yarn.

FR mèche; fibre discontinue; schappe

Assemblage de grande longueur de filaments discontinus, caractérisé par un aspect pelucheux et un allongement important sous une faible traction.

IT stoppino; fibra discontinua

Fascio di grande lunghezza di fibre discontinue, stirabile con debole torsione.

NL stapelvezel

Vezel met een korte lengte.
Garens samengesteld uit deze vezels vertonen een harig karakter, waardoor knopen minder snel doorslippen.

PT fibra descontínua

3301

ES torzal trenzado

Producto de trenzar en forma plana o tubular hilos o torzales (simples o cableados).

DA flettet garn; flettet tvist

Liner fremstillet ved at flette garnene.

DE geflochtenes Garn; Flechtgarn

GR σπάγγος για δίχτυα

Χρησιμοποιείται για την επιδιόρθωση ή την κατασκευή των διχτυών.

EN braided netting twine; plaited netting twine

The product of braiding or plaiting netting yarns and/or netting twines.

FR fil tressé pour filet; ficelle tressée pour filet

Produit résultant du tressage ou de l'entrelacement de plusieurs fils.

IT treccia per reti; filo trecciato per reti

Prodotto tessile con diametro minore di 4 mm ottenuto per trecciatura.

NL gevlochten netgaren

Door middel van een vlechttechniek samengesteld garen.
Bij netwerk worden tijdens het vlechtproces de mazen gevormd.

PT fio entrançado

Produto resultante do trançar ou entrelaçar de vários fios.

3302

ES cadeneta

DA garn; tråd

Her som udtryk for de tråde, der indgår i et knudeløst net.

DE Netzfaden; Netzgarn

Verbindung der Schenkel knotenloser Netze mittels Verflechtung verzwirnter Fäden entsprechend der Herstellungstechnik, z. B. Rascheltechnik.

GR κλώσμα διχτυού χωρίς κόμπους

κλώσμα κατασκευής διχτυών χωρίς κόμπους, η πλοκή του οποίου διαφέρει σύμφωνα με τη μέθοδο, π.χ. τεχνική Raschel.

EN thread

In knotless netting the individual filaments used to make the netting are sometimes called threads.

FR chaînette

Tresse entrant dans la formation de la maille dans le filet sans nœuds.

IT catenella

Intreccio che contribuisce alla formazione della maglia per la rete senza nodo.

NL vlechtdraad

Draad gebruikt bij het vlechten van knooploos netwerk.

PT

Entrelaçamento de que resulta a formação das malhas dos panos de rede sem nós.

Não tem equivalente em português.

3303

ES sentido AB

Sentido paralelo a una sucesión rectilínea de lados de malla, cada uno de ellos perteneciente a una malla adyacente.

DA stolperet retning; AB-retning

Retning parallel med sammenhængende stolper.

DE AB-Richtung

GR φορά AB

EN AB direction

Direction parallel to a rectilinear sequence of mesh bars, each from adjacent meshes.

FR sens AB

Sens parallèle à une suite rectiligne de côtés de maille, chacun d'eux appartenant à une maille adjacente.

IT direzione AB

Direzione parallela ad una serie rettilinea di lati di maglia ciascuno dei quali appartenente ad una maglia adiacente.

NL AB-richting

Richting langs de benen van de mazen in een netwerk.

PT direcção AB

Direcção paralela a uma sequência rectilínea de lados de malha de malhas adjacentes.

3303

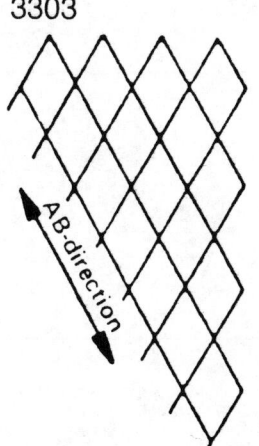
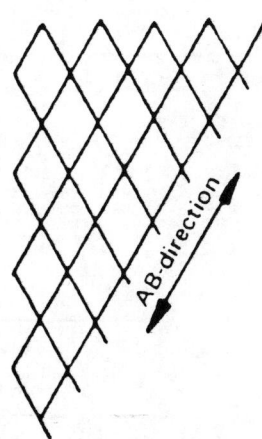

3304

ES dimensiones del paño de la red

La dimensión del paño de red está indicada:
— por el número de sus mallas en los sentidos T y N (estas dos indicaciones ligadas mediante el signo de la multiplicación);
— por el número de sus mallas en un sentido y por su dimensión expresada en una unidad de longitud universalmente admitida, por ejemplo el metro, en el otro sentido (estando la red completamente tensa en el momento de la medición).

DA størrelsen af en netsektion

Angives normalt som: antallet af masker (eller antallet af knuder = 2 gange antallet af masker) tværs over foroven, og antallet af masker i længden (eller netstykkets strakte længde) og skæringen i siden.

DE Netztuchabmessungen

Die Abmessungen von Netztuchen werden angegeben entweder durch die Anzahl der Maschen in T- und N-Richtung in Verbindung mit der Maschenlänge oder durch die Anzahl der Maschen in einer Richtung (z. B. T) und die Länge des Netztuchs (in Metern) in der anderen Richtung.

GR μέγεθος διχτυού· διαστάσεις διχτυού

EN size of netting

The size of netting is indicated either by the number of meshes in both the T- and N-directions (both indications are joined by a multiplication sign) or by the number of meshes in one direction and the length indicated in a recognized unit, for example metres, of the other direction, the netting being fully extended while the measurement is made.

FR dimensions de la nappe de filet

La dimension de la nappe de filet est indiquée par :
— le nombre de ses mailles dans le sens T et dans le sens N (ces deux indications étant reliées par le signe de multiplication);
— le nombre de ses mailles dans un sens et par sa dimension exprimée dans une unité de longueur universellement admise, le mètre par exemple, dans l'autre sens (la nappe étant complètement tendue au moment du mesurage).

IT dimensione della pezza di rete

Numero di maglie nella direzione T e nella direzione N unite dal segno di moltiplicazione; oppure il numero di maglie in una direzione e la dimensione della pezza (espressa in metri) nell'altra dimensione quando la pezza è completamente tesa.

NL afmetingen van een stuk netwerk

- Het aantal mazen in de T-richting en het aantal mazen in de N-richting.
- Het aantal mazen in de ene richting (T) en de gestrekte lengte in meter in de andere richting.

PT dimensões do pano de rede

— Número de malhas do pano de rede na direcção T e na direcção N (estes dois números são ligados pelo sinal de multiplicação).
— Número de malhas do pano de rede na direcção T e altura do pano de rede na direcção N expressa em metros e com o pano estirado.

3305

ES | sentido del estiramiento del paño

Sentido bajo el cual un paño de red puede estirarse.

DA | strækretning

Den retning et netstykke strækkes i brug. (Almindeligvis maskeret retning).

DE | Streckrichtung

GR | φορά τεντώματος του διχτυού

EN | direction of stretch; direction of netting stretch

Direction in which netting may be stretched: either N- or T-direction.

FR | sens d'étirage de la nappe

Direction selon laquelle la nappe de filet est étirée: sens N ou sens T.

IT | direzione di stiro della pezza

Direzione nella quale una pezza di rete viene stirata.

NL | richting waarin het netwerk wordt gestrekt.

De richting waarin het netwerk het gemakkelijkst kan worden gestrekt en waarin het beste kracht kan worden overgebracht.
Dit komt overeen met de N-richting.

PT | direcção de estiramento de um pano de rede

3306

3307

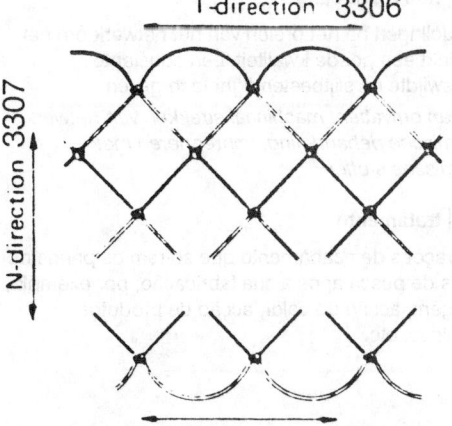

T-direction 3306

N-direction 3307

General course of the netting yarn

3306

ES | sentido T

a) En la red de nudos, sentido paralelo al de avance del hilo para fabricar el paño de la red.
b) En la red sin nudos, sentido perpendicular al sentido N.

DA | tværs over; T-retning

Retning angivet ved trådretningen i et net.

DE | T-Richtung

Richtung parallel zur Hauptlaufrichtung des Netzgarns.

GR | φορά T

EN | T-direction

(a) In knotted netting: the direction parallel to the general course of the netting yarn (Twinewise).
(b) In knotless netting: the direction at right angles to the N-direction.

FR | sens T

a) Pour la nappe de filet nouée: sens parallèle au sens général d'avancement du fil pour filet (sens du trajet du fil ou sens Transversal).
b) Pour la nappe sans nœuds: sens perpendiculaire au sens N.

IT | direzione T

a) Per la rete annodata: direzione parallela alla direzione di avanzamento del filo sul telaio.
b) Per la rete senza nodi: direzione perpendicolare (normale) alla direzione N.

NL | T-richting

De richting evenwijdig met de richting van het garen.

PT | direcção T; direcção transversal; largura

a) Para os panos de rede com nó é a direcção paralela à do avanço do fio durante a confecção.
b) Para os panos de rede sem nó é a direcção perpendicular à direcção N.

3307

ES sentido N

— Red de nudos: sentido perpendicular al general de avance del hilo para fabricar el paño.
— Red sin nudos: sentido correspondiente de la mayor longitud posible de la malla estirada.

DA maskeret retning; N-retning

Retningen vinkelret på trådretningen.

DE N-Richtung

Richtung im rechten Winkel zur Hauptlaufrichtung des Netzgarns.

GR φορά N

EN N-direction

— Knotted netting: the direction at right angles to the general course of the netting yarn (normal);
— In knotless netting: the direction of the longest possible meshaxis.

FR sens N

— Pour la nappe de filet nouée: sens perpendiculaire (Normal) au sens général d'avancement du fil pour filet.
— Pour la nappe sans nœuds: sens qui donne à la maille son extension maximale.

IT direzione N

— Per la rete annodata: direzione perpendicolare (normale) alla direzione di avanzamento del filo sul telaio.
— Per la rete senza nodi: direzione che fornisce alla maglia la massima estensione.

NL N-richting

De richting loodrecht op de richting van het garen.

PT direcção N; direcção normal; altura

— Para os panos de rede com nó é a direcção perpendicular à do avanço do fio durante a confecção.
— Para os panos da rede sem nó é a direcção segundo a qual se obtém, por estiramento, a dimensão máxima da malha.

3308

ES acabado de la red de pesca

Ejemplos de procedimientos de acabado de la red:
— crudo sin tratamiento
— crudo con impregnación
— teñido sin impregnación ni otro tratamiento
— teñido e impregnado.

DA færdigbehandling

DE Endbehandlung

Die Haltbarkeit des Netzes und der Knotensitz können nach der Herstellung durch Imprägnieren und Fixieren erhöht werden.
Des Netzes.

GR τελική επεξεργασία

EN finish of netting; final treatment

Operation on netting or netting yarn to impart particular physical properties and/or appearance.
Examples of possible processes: white (natural), untreated; white (natural), impregnated; dyed, without impregnation of other treatment; dyed and impregnated.

FR finition

Opérations de traitement intervenant après la fabrication; dans le cas d'une nappe de filet, désigne par exemple les opérations de teinture, d'imprégnation, de thermofixation, etc.

IT finissaggio

Operazione di nobilitazione del manufatto quale per esempio tintura, termofissaggio, polimerizzazione, impregnazione, ecc.
Della rete.

NL nabehandeling

Handelingen na het breien van het netwerk om het produkt een goede kwaliteit, een constante maaswijdte en slijtbestendigheid te geven.
Dit kan omvatten: machinaal strekken van netwerk, thermische behandeling, impregneren met chemische stoffen.

PT tratamento

Operações de acabamento que sofrem os panos de redes de pesca após a sua fabricação; por exemplo, tintagem, acção do calor, acção de produtos químicos, etc.

3309

ES cruce de cadenetes

DA Krydspunkt

Krydspunktet for maskerne i et knudeløst net.

DE Verbindungsstelle

GR κομβικό σημείο (διχτυού)

Σημείο συμβολής δύο κλωσμάτων ενός διχτυού χωρίς κόμπους (knotless netting).

EN cross-over point; joining point

Junction of two threads in knotless netting.

FR entrelacement des chaînettes

Jonction de 2 chaînettes dans le filet sans nœuds.

IT incrocio tra due catenelle

Unione tra 2 catenelle che si intrecciano tra di loro a formare, nella rete senza nodo, il corrispondente del nodo.

NL kruispunt

Punt waar de verschillende vlechtdraden elkaar kruisen bij de fabricage van knooploos netwerk.
Het kruispunt is vergelijkbaar met de knoop.

PT ponto de entrelaçamento

Redes sem nós.

3309

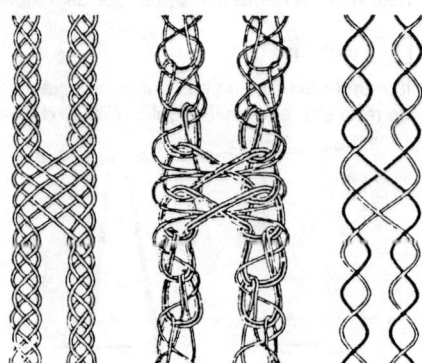

3310

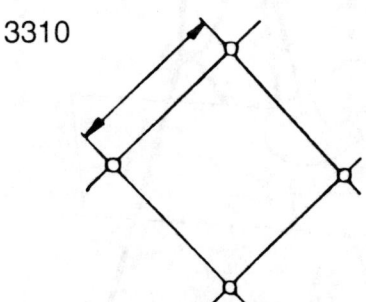

3310

ES longitud del lado de la malla

Distancia entre los centros de dos nudos a cruces consecutivos cuando el hilo está completamente tenso.

DA halvmaske

Længden af siden i en maske, målt fra midt-knude til midt-knude.
Danske fiskere og vodbindere anvender normalt denne længde som udtryk for et nets maskestørrelse.

DE Maschenweite

Abstand zwischen zwei aufeinanderfolgenden Knoten oder Verbindungsstellen, gemessen von Mitte bis Mitte Verbindungsstelle, wenn das Netzgarn bzw. anderes Netztuchmaterial zwischen diesen Stellen vollständig gestreckt ist.

GR μήκος πλευράς ματιού

EN length of mesh side

The distance between two sequential knots or joints, measured from centre to centre when the yarn between those points is fully extended.

FR longueur du côté de maille

Distance entre deux nœuds ou croisements consécutifs, mesurée de centre à centre alors que le fil, entre ces points, est complètement tendu.

IT lunghezza del lato di maglia

Distanza tra 2 nodi od intrecci consecutivi, misurata da centro a centro dei nodi o degli intrecci quando il filo, tra questi punti, risulta completamente teso.

NL lengte van de maaszijde

Gestrekte afstand tussen twee opeenvolgende knopen of verbindingen, gemeten van het hart van de knoop tot het hart van de naastliggende knoop, terwijl het netwerk zodanig is gestrekt, dat de mazen geheel open zijn.

PT comprimento do lado da malha

Distância entre dois nós (ou dos pontos de entrelaçamento) consecutivos, medida de meio do nó a meio do nó (ou do ponto de entrelaçamento), com a malha completamente estirada.

3311

longitud de corte

DA **skæring**

Her som angivelse af, hvor mange masker eller stolper, der skæres i en bestemt retning.

DE **Schnittlänge**

Zahl der zu schneidenden Maschen; sie ergibt sich aus der gewünschten geometrischen Form des herzustellenden Netzteils.

Anwendungstechnik zur Herstellung von Netzteilen aus vorgefertigtem Netztuch.

GR **μήκος κοπής· μήκος κοψίματος**

EN **length of cut; cutting length**

For N- and T-cuts it is the number of consecutive meshes cut. For B-cuts it is the number of consecutive bars cut along the edge excluding the bars on the preceding point.

FR **longueur de coupe**

IT **lunghezza di taglio**

Nel caso di tagli N o T: numero di maglie tagliate consecutivamente per ciascuna direzione. Nel caso di tagli B: numero di lati di maglia opposti e consecutivi tagliati, esclusi i lati uscenti dal nodo precedente.

NL **snitlengte; snitdiepte**

Voor N- en T-snitten: het aantal mazen dat aaneengesloten wordt gesneden. Voor B-snitten: het aantal lengten van de maaszijden (benen) dat wordt gesneden.

Wordt in Nederland niet gebruikt.

PT **comprimento do corte**

Para os cortes T e N trata-se do número de malhas consecutivas cortadas. Para o corte B trata-se do número de pernões cortados ao longo do pano de rede menos 1.

3312

ES **mallero**

Molde que se utiliza en la fabricación de paños de red a mano para conseguir mallas uniformes.

DA **skæl**

Fladt redskab, der anvendes ved håndbinding af masker for at opnå ensartet maskestørrelse.

DE **Strickholz; Maschenlehre**

GR

Εργαλείο με το οποίο εξασφαλίζεται η ομοιόμορφη κατασκευή των ματιών του διχτυού.

Δεν χρησιμοποιείται στην Ελλάδα.

EN **mesh stick; spool; mesh pin**

Gauge used in making nets by hand.

FR **moule; forme**

Pièce de forme et dimension appropriées, utilisée dans la confection à la main du filet afin d'obtenir des mailles uniformes.

IT **modulo**

Strumento che, usato nella costruzione della rete a mano, permette l'uniformità voluta delle maglie.

NL **breispaan; breischiel; schiel**

Langwerpig, plat stuk hout dat bij het breien van netten wordt gebruikt als mal voor de maaswijdte.

PT **malheiro**

Instrumento utilizado na confecção manual de panos de rede e que permite a uniformidade das malhas.

3312

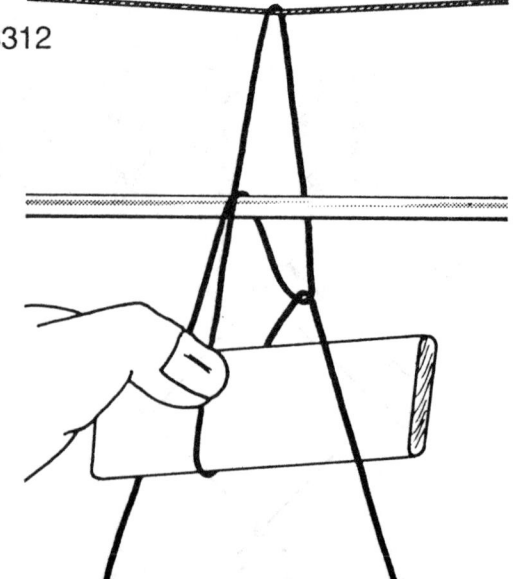

3313

ES nudo del paño de la red

DA knude

I net.

DE Knoten; Netzknoten

Netztuch.

GR κόμβος διχτυού· κόμβος

EN netting knot; knot

Netting knots are of two sorts, the single-sheet bend and the reef knot.

FR nœud de nappe de filet

IT nodo della pezza di rete

Legatura di un filo su se stesso, o di 2 insiemi di fili a due a due, atta a formare la maglia della pezza di rete.

NL knoop

Verbinding tussen twee of meer netgarens, gevormd met de garens zelf.

PT nó

Pano de rede de pesca.

3314

ES lado de la malla; barra; media malla; lado del cuadrado; pie; pata

DA stolpe

Siden i en maske.

DE Maschenseite

GR πλευρά ματιού· πλευρά βρόχου

EN mesh side; bar

The twine joining two knots and forming part of a mesh.

FR côté de maille

Portion du fil comprise entre deux nœuds consécutifs d'une même maille. La longueur du côté de maille est mesurée du milieu du premier nœud au milieu du nœud suivant.

IT lato di maglia

Porzione di filo compresa tra due nodi consecutivi di una stessa maglia.

NL maaszijde

PT lado de malha; meia malha

Porção de fio compreendida entre dois nós consecutivos de uma malha (rede com nós) ou entre dois pontos de entrelaçamento de uma malha (rede sem nós).

O termo meia malha também é empregue em português para designar metade de uma malha.

3315

ES | dimensión de la malla

Se expresa por la longitud del lado, longitud de la malla y abertura de la malla.

DA | maskestørrelse

Maskestørrelse kan defineres på 3 måder: indvendigt mål (se 314), helmaske (se 317), halvmaske (se 3310).
Se tillige 3008 og 3013.

DE | Maschengröße

Abmessung der Masche, entweder als Maschenweite, Maschenlänge oder Maschenöffnung. Die Maschengröße wird in jedem Falle in mm angegeben.

GR | μέγεθος ματιού· διαστάσεις ματιού

EN | size of mesh

Length of mesh size, length of mesh, opening of mesh.

FR | dimensions de la maille

Longueur de côté de maille, longueur de maille, ouverture de maille.
Il s'agit de trois dimensions différentes.

IT | dimensione di maglia

Proprietà che distingue la maglia ed espressa da lunghezza del lato di maglia, lunghezza di maglia e apertura di maglia.

NL | afmetingen van de maas

Lengte van de maaszijde, maaslengte, maaswijdte.

PT | dimensões da malha

Comprimento do lado da malha, malhagem e abertura da malha.

3316

ES | nudo doble de tejedor

DA | dobbeltknude; dobbelt væverknob

Knob, der bl.a. finder anvendelse ved samling af netstykker, hvor ekstra knudefasthed ønskes. Anvendes også til at binde liner af forskellig tykkelse sammen.

DE | doppelter Schotstek; doppelter Schotenstek

Verbindung von zwei ungleich starken Leinen; Befestigung von Leinen an Kauschen, Kettengliedern u. ä.

GR | διπλός κόμβος

EN | double weaver's knot

A type of knot used to form netting.

FR | nœud double

IT | nodo doppio

NL | vissersknoop; dubbele schootsteek

Type knoop dat veel gebruikt wordt bij het samenstellen van netwerk, en dat geschikt is voor het verbinden van touweinden van ongelijke dikte.

PT | nó duplo

Pano de rede de pesca.

3316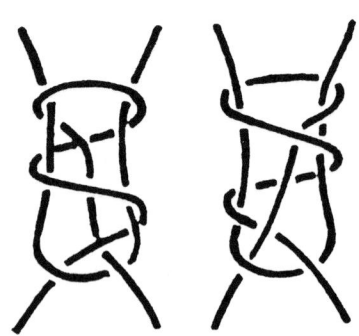

3317

ES hilo cableado

Dos o varios hilos retorcidos (o alternativamente hilos retorcidos e hilos simples) que vuelven a ser retorcidos conjuntamente por una o varias operaciones de retorsión.

En las industrias de redes de pesca y de cordelería un hilo cableado está generalmente constituido por hilos simples retorcidos conjuntamente, pero recibiendo cada uno un complemento de torsión durante el curso de la operación de retorcido que, en este caso, se llama cablear.

DA tovslået garn

Line spundet af tvistet garn (dvs. indeholder en spinding mere).

DE gedrehtes Netzgarn; geschlagenes Netzgarn

GR

Είδος κάβου που χρησιμοποιείται στο ψάρεμα.

EN cabled netting twine

The product of further twisting operations embracing two or more netting twines.

FR fil câblé pour filet; ficelle câblée pour filet

Produit résultant de la réunion par torsion de deux ou plusieurs fils retors pour filet.
Produit résultant de la réunion par torsion de deux ou plusieurs fils simples qui reçoivent chacun un complément de torsion pendant l'opération de retordage. L'ensemble de cette opération est appelé «câblage».

Le produit résultant est dit «câblé», mais est en fait un retors qu'il serait préférable de désigner par l'expression «retors indétordable». Ce «câblé» est en effet un produit équilibré; quand on le coupe, aucun décâblage ne se produit.

IT ritorto composto; tortiglia

Prodotto tessile ottenuto dell'unione di due o più fili dei quali almeno uno ritorto semplice, uniti mediante altra o altre operazioni di torcitura.

NL gekableerd netgaren

Twee of meer getwijnde garens, ineengedraaid in één of meer twijnbewerkingen.

PT fio torcido para rede de pesca

Produto resultante de torção acentuada de vários fios de modo a que, por corte, não destorçam.

3318

ES nudo plano; nudo llano; nudo de rizo

DA råbåndsknob

Knob brugt til sammenbinding af liner af samme diameter.

DE Kreuzknoten; Reffknoten

Knoten, bei dem die beiden Enden doppelt genommen und dann paarweise übers Kreuz verknotet werden.

Für die Verbindung von zwei annähernd gleich starken Leinen.

GR σταυρόκομπος

EN reef knot; square knot

A knot consisting of two successive overhand knots used for joining together two ropes of the same size.

The reef knot is chiefly used for small mesh nets such as shrimp trawls.

FR nœud droit; nœud plat

Nœud qui sert à réunir deux bouts de filin de même diamètre.

IT nodo piano

Nodo che serve per fare giunta a due cime.

NL platte knoop

Eenvoudige knoop voor het aan elkaar zetten van twee even dikke touwen.

PT nó direito

Nó utilizado para unir dois cabos ou linhas de igual diâmetro.

3318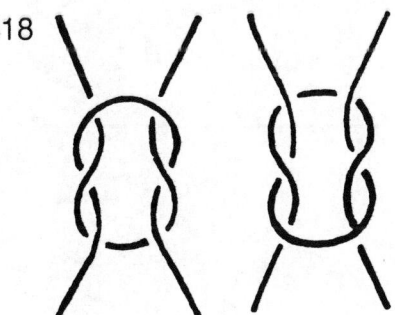

3319

ES filamento

DA filament

Meget lang fiber.

DE Filament

Synthetischer Endlosfaden von 0,10 mm Ø in feinen Abstufungen bis 5,00 mm Ø.

GR νημάτιο· ίνα

EN filament

Very long textile fibre.

FR filament

Fibre textile de grande longueur.

IT filamento

Fibra tessile di grande lunghezza.

NL vezel

PT filamento

Fibra sintética utilizada na confecção de filaças.

3320

ES hilo de torsión

DA skibsmandsgarn

Betegnelse for et garn med to eller tre kordeller af naturfibre brugt til at klæde tovværk (for at hindre skamfiling) eller wire (for at skabe fæste for bændsler).

DE Schiemannsgarn; gezwirntes Garn

Dünnes geteertes Tauwerk aus zwei oder mehreren Garnen gedreht.

— Umfang 10 – 30 mm.
— Schiemannsgarn wird zum Bekleiden von Drahttauwerk als Korrosionsschutz und zum Schutz des Tauwerks selbst eingesetzt. Die Bekleidung verhindert das Rutschen von Bändseln auf Drahttauwerk.

GR κλωσμένο νήμα

Σχοινί καμωμένο από κάναβι κατωτέρας ποιότητας.

EN spun yarn

A twine consisting of two or three twisted yarns, twisted together, each having a linear density of about 5 ktex, usually tarred, and used for tying.

FR bitord

Ficelle constituée de 2 ou 3 fils ayant chacun une masse linéique d'environ 5 kilotex câblés ensemble; le bitord est habituellement goudronné et utilisé comme ligature.

Il est confectionné avec du chanvre de qualité inférieure ou avec les fils de caret des torons de vieux cordages.

IT spago; spago catramato

Prodotto tessile con superficie esterna praticamente liscia ottenuto per torsione di due o più fili ritorti.

NL schiemansgaren

Twee- of driedraads, tegen de richting van de zon geslagen geteerd garen, vervaardigd van hennep-, sisal- of manillavezels, gebruikt voor bekleden van staand tuig en voor tijdelijke bindsels.

PT

Não tem equivalente em português.

3321

ES | torzal; torzal para red

Producto resultante de una operación de torsión uniendo dos o más hilos simples o dos o más monofilamentos.

DA | tvist; line

Udgangsmaterialet ved fremstilling og reparation af net. Produktet af en spinning af to eller flere garner.

DE | einstufiger Zwirn

Sammelbegriff für alle Zwirne, die in einem Zwirnvorgang aus zwei oder mehr einfachen Garnen hergestellt worden sind.

GR | νήμα δίκλωνο για δίχτυα

Ο όρος χρησιμοποιείται για περισσότερα από ένα νήματα στριμμένα μαζί.

EN | netting twine; folded yarn; plied yarn

The product of one twisting operation embracing two or more single yarns or monofilaments.

FR | ficelle pour filet; fil retors pour filet

Produit résultant d'une opération de torsion mettant en jeu deux ou plusieurs fils simples ou deux ou plusieurs monofilaments.

IT | tortiglia; filo ritorto

Prodotto tessile a forte torsione con diametro minore di 4 mm ottenuto per torsione o commettitura.

NL | getwijnd netgaren

Produkt vervaardigd door het ineendraaien van twee of meer enkelvoudige garens of monofilamenten in één twijnbeweging.

PT | fio torcido para rede de pesca

Produto resultante da torção de dois ou mais fios simples ou monofilamentos.

3322

ES | nudo de tejedor; nudo de escota

DA | væverknob

Alment brugt knob, især til at samle ender, der ikke har samme diameter. Også brugt ved fremstilling af net.

DE | einfacher Schotenstek; Schotstek; Schotenstek; Fischerknoten; Weberknoten

Verbindung von zwei ungleich starken Leinen an Kauschen, Kettengliedern u. ä.

GR | ποδόδεσμος· απλός κόμπος

EN | weaver's knot; sheet bend; becket bend; single bend; swab hitch; single sheet bend; signal halliard bend

A type of double hitch made by passing the end of one rope through the bight of another, around both parts of the other, and under its own part.

It is used for joining two ropes, especially when of different sizes, also in bending small sheets to the clews of sails, and in bending flags where snap hooks are not fitted. Also used to form netting.

FR | nœud de tisserand; nœud simple; nœud d'écoute

IT | nodo semplice

NL | schootsteek; enkele schootsteek; weversknoop; vissersknoop

Steek om twee lijnen of trossen van ongelijke dikte op elkaar te steken; daartoe wordt de tamp van de dikste dubbel genomen en in het aldus gevormde oog wordt met de dunste de schootsteek gemaakt. Steek om een lijn aan een oog te bevestigen.

PT | nó de escota; nó simples

Nó muito simples, com várias aplicações a bordo. Pode ser singelo ou dobrado.

3322

Weaver's knot – Z-type   Weaver's knot – S-type

3323

ES red de nudos de un solo hilo

La red de nudos constituida por un hilo único se fabrica generalmente a mano. El hilo bobinado sobre una aguja especial y todas las mallas de una misma fila quedan anudadas una después de la otra. La lazada uniforme de las mallas puede conseguirse con ayuda de un mallero, para mallas durante el anudado. Si la red de pesca se fabrica plana, el hilo cambia de sentido alternativamente de izquierda a derecha y de derecha a izquierda. Si las mallas de la red se anudan en redondo (en forma de tubo o de cilindro), el hilo progresa siempre en el mismo sentido.

DA håndbundet net

Ved håndbinding fremstilles nettet som række efter række af enkelte masker under anvendelse af en bødenål til tvisten og evt. et skæl for at opnå ensartet maskestørrelse. Tvisten går ubrudt gennem hele nettet.

DE handgeknotetes Netztuch

Handgeknotete Netztuche bestehen aus einem einzigen Fadensystem. Der Faden wird für die Herstellung des Netztuches auf Netznadeln gewickelt. Die Maschen einer Reihe werden einzeln nacheinander geknüpft. Das Verwenden einer Maschenlehre (Strickholz) beim Knoten erleichtert es, eine einheitliche Maschengröße zu erreichen. In flachen Netztuchen verläuft das Netzgarn abwechselnd von links nach rechts und von rechts nach links, in rundgeknoteten Netztuchen stets in derselben Richtung.

GR δικτύωμα με κόμπους, πλεγμένο με ένα μόνο νήμα

EN knotted netting with single yarn

Knotted netting consisting of a single-yarn system is mostly hand made. The yarn is wound on a netting needle and all the meshes in the same row are knotted individually one after another. A uniform mesh size may be achieved by the use of a mesh gauge during knotting. If the netting is made as a flat panel, then the netting yarn runs alternately from left to right and from right to left. If the netting is knotted round and round (as a 'tube' or 'cylinder'), then the yarn proceeds continuously in the same direction.

FR nappe de filet nouée à un seul fil

La nappe de filet nouée constituée d'un seul fil est, le plus souvent, fabriquée à la main. Le fil est enmagasiné sur une navette et toutes les mailles de la même rangée sont nouées individuellement l'une après l'autre. Au laçage, l'uniformité des mailles peut être obtenue à l'aide d'un moule.

IT pezza di rete annodata formata da un solo filo

Pezza di rete annodata formata da un solo filo e, nella maggior parte dei casi, fabbricata a mano.

NL handgeknoopt netwerk uit één garen

Hierbij wordt gebruik gemaakt van boetnaalden waarop het garen wordt gewikkeld. Alle mazen worden met de hand gebreid, de een na de ander. Gelijke maaswijdte kan worden verkregen door een maaswijdtemalletje.

PT rede de pesca de um só fio com nós

A rede com nós construída com um só fio é, a maior parte das vezes, feita à mão. O fio é enrolado numa agulha de coser rede e todas a malhas da mesma carreira são efectuadas individual e sucessivamente. A uniformidade das malhas é obtida com o auxílio de um malheiro.

3323

General course of the netting yarn

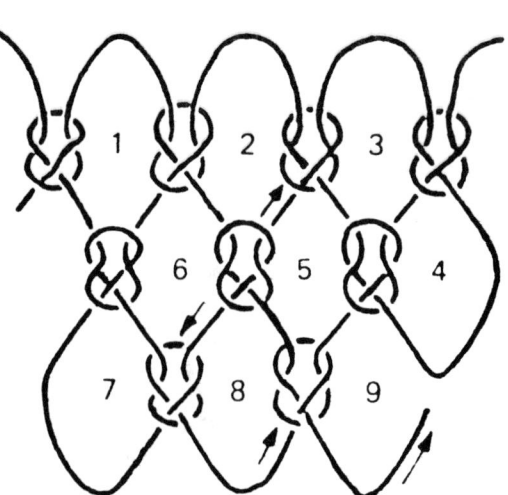

Single-yarn system

3324

paño de red de nudos

DA **knyttet net**

Til forskel fra knudeløst net.

DE **geknotetes Netztuch**

GR **πλεγμένο δικτύωμα με κόμπους**

EN **knotted netting**

Netting made by hand or machine from twines which are knotted together at intervals to form meshes.

FR **nappe de filet nouée**

IT **pezza di rete con nodo**

Pezza di rete con maglie ottenute mediante nodi.

NL **geknoopt netwerk**

Netwerk dat samengesteld is met behulp van knoopverbindingen.

PT **pano de rede com nós**

3325

Two-yarn system

3325

ES **red de pesca con dos series de hilos anudados**

La red de nudos constituida por dos series de hilos se fabrica generalmente con ayuda de una máquina de anudar redes. Una de las dos series de hilos se desplaza, como en el caso de la urdimbre para tejer, a partir de bobinas, mientras que la otra serie de hilos son encanillados para las lanzaderas que los transportan hacia el gancho o hacia las agujas de un dispositivo anudador. Todos los nudos de una misma fila quedan anudados simultáneamente.

DA **maskinknyttet net**

Net knyttet på maskine under anvendelse af to garnsystemer. Garnet i det ene af de to systemer løber ligesom vævens kædegarn fra spoler, mens garnet i det andet system er vundet op på en skyttel, som leder garnet med en krogformet eller retformet knytteanordning.

DE **maschinengeknotetes Netztuch**

Maschinengeknotete Netztuche werden auf der Netzknüpfmaschine hergestellt und bestehen aus zwei Fadensystemen. Die Fäden des einen Systems, Kettfäden (oder Obergarn) genannt, laufen in der Art eine Webkette von Spulen ab, während die Fäden des zweiten Systems, Schußfäden (oder Untergarn) genannt, einzeln auf Schiffchen gewickelt sind und von diesen einem hakenförmigen Knoten zugeführt werden. Alle Knoten einer Maschenreihe werden gleichzeitig geknüpft.

GR **πλεγμένο δικτύωμα με κόμπους αποτελούμενο από δύο είδη νημάτων**

EN **knotted netting consisting of two systems of yarns**

Knotted netting consisting of two systems of yarns is mostly manufactured on a knotting machine. The yarn of one of the two systems runs like a weaving warp from bobbins, while the yarn of the other system is wound on shuttles that guide it towards a hook-shaped or needle-type knotting device. All the knots in one row are knotted simultaneously.

3325

FR nappe de filet nouée à deux ensembles de fils

La nappe de filet nouée à deux ensembles de fils est le plus souvent fabriquée à l'aide d'une machine à lacer. Les fils du premier ensemble courent à partir de bobines, comme dans une chaîne de tissage, tandis que les fils du second ensemble sont emmagasinés sur des navettes qui les amènent sur les crochets ou les aiguilles d'un dispositif noueur. Tous les nœuds d'une même rangée de mailles sont noués simultanément.

IT pezza di rete con nodo formata dalla annodatura di due insiemi di fili

Pezza di rete annodata, costituita da due insiemi di fili, e fabbricata nella maggior parte dei casi mediante telaio per annodare.

I fili del primo insieme si svolgono come in un ordito di tessitura, a partire da bobine, mentre i fili del secondo insieme sono immagazzinati su spole contenute in navette che le trasportano verso l'uncino o verso gli aghi di un dispositivo di annodatura. Tutti i nodi di un medesimo rango vengono annodati contemporaneamente.

NL geknoopt netwerk samengesteld uit twee garens

Machinaal geknoopt netwerk samengesteld uit twee garensystemen. De garens van het ene systeem zijn afkomstig van spoelen, terwijl de garens van het andere systeem naar het knoopvormend mechanisme worden geleid. Alle knopen in een rij worden tegelijkertijd gemaakt.

PT pano de rede com nós com dois sistemas de fornecimento de fio

Pano de rede com nós de confecção mecânica com dois sistemas de fornecimento de fio; os nós de uma mesma carreira de malhas são efectuados simultaneamente.

3326

ES paño de red sin nudo

DA knudeløst net

Net fremstillet ved vævning eller fletning af garner, der med faste intervaller bliver samlet til masker. Der findes en række metoder til fremstilling af sådanne masker.

DE knotenloses Netztuch

GR δίχτυ δίχως κόμπους

EN knotless netting

Netting made by machine from yarns which are interlaced at intervals to form meshes. Yarns which comprise one side of a mesh at one point in the netting may not be associated together in succeeding meshes.

FR nappe de filet sans nœuds

IT pezza di rete senza nodo

Pezza di rete con maglie ottenute mediante l'incrocio tra catenelle.

NL knooploos netwerk

Netwerk dat samengesteld is zonder het leggen van knopen.

PT pano de rede sem nós

3327

ES sonar; ecogoniómetro

Equipo acústico para la localización, escucha y comunicación submarinas. Consta de un emisor o proyector submarino en el que las emisiones se efectúan por impulsos de muy corta duración y si en su trayectoria hallan un obstáculo se reflejan en forma de eco y vuelven al proyector. Por el tiempo entre la emisión de un impulso y la recepción del eco se obtiene la distancia al obstáculo o blanco, mientras que la dirección de éste se determina por la del haz de radiación.

DA sonar

Hydroakustisk instrument til at bestemme afstand og retning til forhindringer under vand. Princippet bygger på måling af den tid, det tager for et ekko at blive kastet tilbage. Anvendes til opsporing af fiskestimer på lang afstand.

DE Sonar

Sammelbegriff für Schallortungsverfahren. Prinzip ist die Laufzeitmessung ausgestrahlter und reflektierter Ultraschallwellen, die gerichtete Schallausbreitung sowie der Doppeleffekt zur Bestimmung der Geschwindigkeit des Ziels.

GR ηχοβολιστικό

EN sonar

Method or equipment for determining, by underwater sound, the presence, location or nature of objects in the sea.

Sound navigation and ranging.

FR sonar

Appareil servant à la détection et à la localisation horizontale des poissons; permet également d'apprécier les changements du profil et de la nature du fond autour du bateau.

IT sonar

Apparecchio destinato a localizzare un oggetto situato sotto la superficie del mare. Esso è basato sulla conoscenza della velocità di propagazione nell'acqua delle vibrazioni sonore ed ultrasonore e sulla misura del tempo intercorso tra l'istante in cui ha luogo l'emissione del suono e dell'ultrasuono e quello della ricezione dell'eco riflesso.

NL sonar

Verzamelnaam voor instrumenten die door middel van ultrasone trillingen objecten onder water opsporen.

PT sonar

Equipamento hidroacústico utilizado para a detecção e/ou localização horizontal, oblíqua e/ou vertical de peixes e para a determinação da natureza e topografia dos fundos.

3327

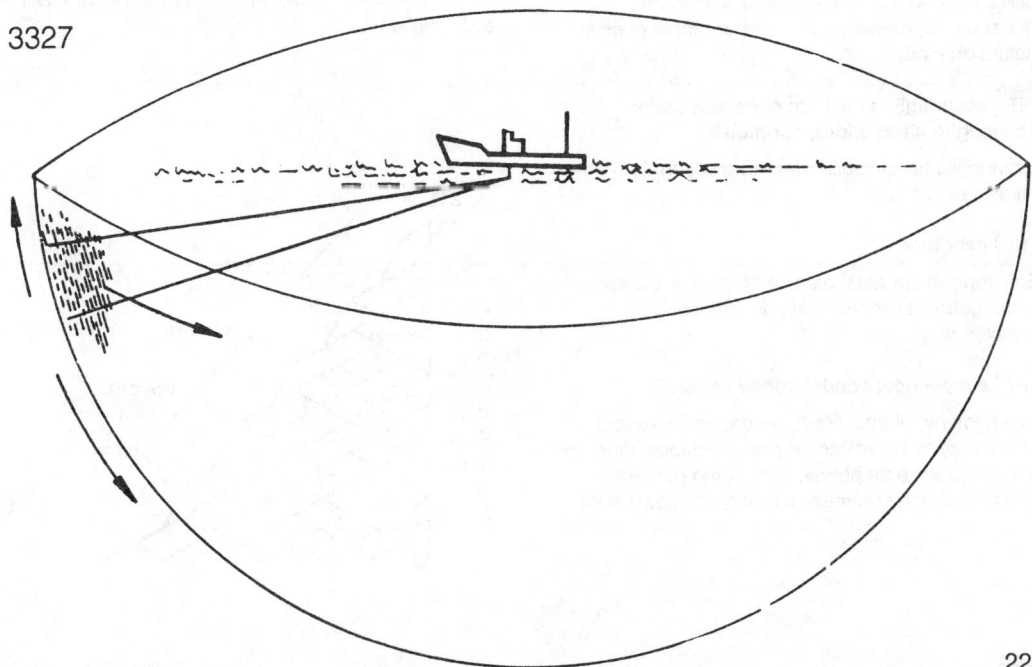

3328

ES ecosonda; ecómetro

Sondador por eco.

DA ekkolod

Hydroakustisk instrument til at bestemme afstanden til og naturen af forhindringer under fartøjet, f.eks. bunden eller fiskeforekomster. Princippet bygger på måling af den tid, det tager for et ekko at blive kastet tilbage.

DE Echolot; akustisches Lot

Gerät zur Messung der Wassertiefe auf indirektem Wege, indem die Laufzeit von Ultraschall-Impulsen gemessen wird, die vom Schiffsboden aus abgestrahlt, am Meeresboden reflektiert und am Schiffsboden wieder empfangen werden. Bei bekannter Ausbreitungsgeschwindigkeit des Schalls im Seewasser läßt sich damit die Wassertiefe unter dem Kiel bestimmen.

GR βυθόμετρο· ηχοβολιστικός ενδείκτης ψαριών· ηχητικός βυθομετρητής

'Όργανο που δείχνει το βάθος του νερού κάτω από το πλοίο.

EN echo sounder; acoustic sounder

An apparatus used on a fishing boat for the detection and identification of fish and the determination of depth of water and nature of the seabed.

FR sondeur; écho-sondeur

Instrument installé à bord d'un navire de pêche, utilisé pour la détection et l'identification des poissons, l'appréciation de la topographie et de la nature du fond.

IT scandaglio acustico; ecoscandaglio; scandaglio ultrasonoro; ecometro

Scandaglio basato sulla riflessione di suoni o ultrasuoni in acqua.

NL echolood

Elektronisch apparaat dat aan boord van schepen wordt gebruikt voor verticale dieptemeting en visopsporing.

PT ecossonda; sonda; sonda acústica

Equipamento hidroacústico de detecção vertical utilizado para a medição de profundidades, detecção e indentificação de peixes, bem como para a determinação da natureza e topografia dos fundos.

3328

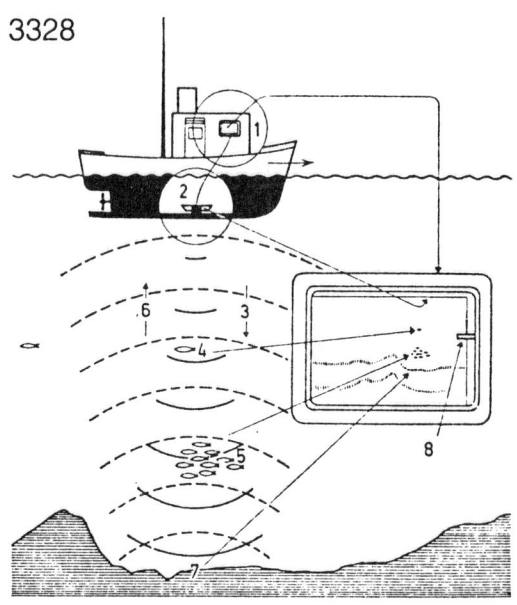

De werking van het echolood. Het registreertoestel (1), geplaatst nabij de stuurstand, geeft de echo's weer in een getekend diagram. De transducer (2), die gemonteerd is tegen de scheepsbodem, gewoonlijk op een derde van de scheepslengte verwijderd van de voorsteven, zendt de geluidstrillingen (3) uit en ontvangt de echo's (6). Als de geluidstrillingen een enkele vis (4) of een school vissen (5) tegenkomen, wordt een deel van het geluid weerkaatst naar de transducer, die dan werkt als een microfoon en de impulsen doorgeeft naar de ontvanginrichting. Het signaal wordt overgebracht naar de schrijfpen (8), die het op de bewegende papierstrook aftekent. De echo's die van de zeebodem (7) terugkeren, vormen daaronder een doorlopende weergave van het bodemprofiel.

3329

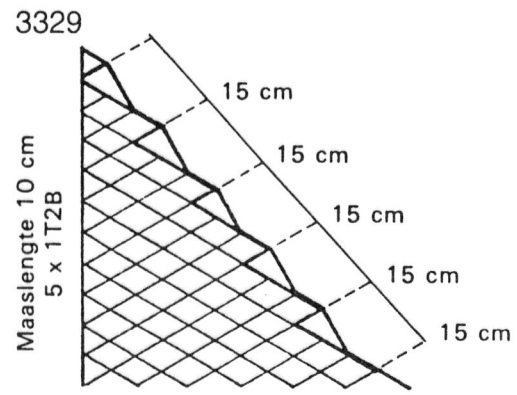

3329

ES armado

Operación consistente en el montaje de una red de acuerdo con una relación específica entre la longitud de la red y de la última porción de cordel o del marco comprendido entre los dos puntos extremos de fijación.

DA føring

Udtryk for monteringen af nettet på en line, f.eks. en tælle, med et bestemt forhold mellem nettets længde og linens længde, eller med en bestemt afstand mellem maskerne.

DE Einstellen

Anschlagen eines Netztuches in einem Verhältnis von Netztuchlänge zur Leinenlänge.

GR ανάρτηση δικτύου (καθ.)· κρέμασμα διχτυού

EN hanging

The mounting of netting according to a specific relationship between the length of that part of the final rope or frame on which the netting is mounted and the length of the netting.

FR armement

Montage d'une nappe de filet selon une relation spécifique entre la longueur de la nappe et la longueur de la portion du dernier cordage ou du cadre de support sur laquelle la nappe est montée.

IT armamento

Operazione consistente nell'attaccare la rete direttamente ad una corda od indirettamente alla stessa per mezzo di una corda di sopporto. Ciò viene effettuato secondo un rapporto tra la lunghezza di ogni singola pezza costituente la rete e la lunghezza del tratto di corda compresa tra i 2 punti estremi di fissaggio di ogni singola pezza di rete.

NL aanslaan; verdeling

Het bevestigen van het netwerk aan de pees gewoonlijk aan de hand van een bepaalde verhouding tussen de lengte van het netwerk (al dan niet gestrekt) en de lengte van de pees, uitgedrukt in de zgn. verdelingsverhouding.

PT armar

Montagem de um pano de rede num cabo segundo uma relação específica entre o comprimento do pano e o comprimento do cabo sobre o qual é montado.

3330

ES longitud de la red; longitud del paño de la red

La dimensión de la red montada paralelamente al terminal del cordón o marco, medida cuando la red está extendida por completo, antes de ser montada, en uno de los dos sentidos N o T.

DA nettets længde

Angivelse ved fastlæggelsen af føringen af net på en line eller tælle. Nettet måles forinden monteringen i dets fuldt strakte længde parallelt med tællen.

DE Netztuchlänge

Die Länge des anzuschlagenden Netztuches, gemessen völlig gestreckt, parallel zur Leine.

GR μήκος δικτυώματος

EN length of netting

That dimension of the netting to be mounted which is parallel to the final rope or frame, measured when the netting is fully extended, prior to being hung in the direction N or T.

FR longueur de nappe de filet

Dimension du bord de la nappe qui est parallèle au cordage ou au cadre de support, mesurée alors que la nappe est complètement étirée dans ce même sens avant d'être montée.

IT lunghezza della pezza di rete da armare

Dimensione del bordo della pezza da armare parallelo alla corda o alla corda di sopporto misurata quando la pezza è completamente tesa prima di essere armata in una delle due direzioni N o T.

NL lengte van het netwerk

Afstand tussen de uiteinden van het gestrekte netwerk, gemeten over de mazen, die aan de pees zijn of zullen worden aangeslagen.

PT comprimento de um pano de rede

Dimensão do bordo do pano de rede paralelo ao cabo onde vai ser armado, medida com a rede completamente estirada na mesma direcção.

3331

ES longitud de la cuerda

Longitud de la porción de cuerda o de marco comprendido entre los puntos extremos de fijación de la red.

DA længden af tællen

Angivelse ved fastlæggelsen af føringen. Målet tages mellem yderpunkterne af den del af tovet, som nettet skal føres på.

DE Leinenlänge

Die Länge des Abschnittes der Leine, an die das Netztuch unmittelbar angeschlagen wird.

GR μήκος σχοινιού ανάρτησης δικτύου

EN length of rope

The length of the section of the rope or frame between the extreme points of mounting of the netting.

FR longueur de cordage

Longueur de la portion du dernier cordage monté, comprise entre les points extrêmes de fixation de la nappe (le cordage est parfois remplacé par un cadre de support).

IT lunghezza del tratto di corda d'armamento

Lunghezza della parte di corda compresa tra i punti estremi di fissaggio di ogni singola pezza.

NL lengte van de pees

Afstand tussen de uiteinden van het gedeelte van de pees waartussen het netwerk of de dunnepees is bevestigd, of bevestigd zal worden.

PT comprimento de um cabo para armar

Comprimento do cabo sobre o qual vai ser armado um pano de rede, medido entre os dois extremos do pano de rede.

3332

ES cuerda para armar; cordel

Cable de acero o cuerda sobre la que se monta el paño de red, directamente o por medio de trencilla.

DA tælle

Den line eller det tov, som nettet skal monteres på.

DE Einstell-Leine

Die Leine, an die das Netztuch angeschlagen wird.

GR σχοινί ανάρτησης

EN rope; hanging rope

The rope on which the netting is fixed.

FR cordage; cordage d'armement

Cordage sur lequel la nappe de filet est fixée.

IT corda d'armamento

Corda sulla quale la pezza di rete è fissata.

NL pees

Touw, staaldraad of anderszins, waaraan het netwerk uiteindelijk is bevestigd.
Deze bevestiging kan zowel rechtstreeks, als door middel van een dunnepees gebeuren.

PT cabo para armar

Cabo sobre o qual um pano de rede é armado (montado).

3333

ES relación del montaje

Relación entre la longitud del cordel final y la longitud de la red.
a) Símbolo E;
b) la relación del montaje se podrá escribir en forma de fracción ordinaria, de fracción decimal o en porcentaje.

DA føring

Udtryk for forholdet mellem nettets længde og tællens længde.

DE Einstellungsfaktor E

Verhältnis von Leinenlänge zu Netztuchlänge.
Der Einstellungsfaktor kann als echter Bruch, als Dezimalbruch oder in Prozent angegeben werden.

GR λόγος ανάρτησης

EN hanging ratio

The ratio between the length of final rope and the length of netting.
(a) Symbol E;
(b) the hanging ratio may be written as a vulgar fraction, as a decimal fraction or as a percentage.

FR rapport d'armement

Rapport de la longueur de cordage à la longueur de nappe de filet.
a) Symbole E;
b) le rapport d'armement peut être exprimé sous trois formes: fraction ordinaire, fraction décimale, pourcentage.

IT rapporto di armamento

Rapporto tra la lunghezza del tratto di corda e la lunghezza della pezza di rete.
Simbolo E.

NL verdelingsverhouding

De lengte van de pees gedeeld door de lengte van het netwerk.
a) Symbool E;
b) de verdelingsverhouding mag worden opgegeven als een breukgetal, als een decimaalgetal of als een percentage.

PT coeficiente de montagem

Relação entre o comprimento do cabo para armar e o comprimento do pano de rede.
Símbolo E.

3334

ES estiramiento en porcentaje

Incremento de la longitud de una muestra en el transcurso de un ensayo de tracción expresado en porcentaje sobre la longitud inicial.

DA forlængelsesprocent

En angivelse ved trækprøver for forlængelsen af emnet i forhold til den oprindelige længde.

DE prozentuale Dehnung

Produkt aus Dehnung x 100.

GR ποσοστιαία επιμήκυνση

EN elongation per cent

The increase in length of a specimen expressed as a percentage of the original length.
Ropes and yarn.

FR allongement en pour cent

Augmentation de longueur d'une éprouvette au cours d'un essai de traction exprimée en pour cent de la longueur initiale.

IT allungamento percentuale

Aumento di lunghezza di una provetta espressa come percentuale della lunghezza originale.

NL verlenging; specifieke verlenging; rek

Lengtetoename gedeeld door de oorspronkelijke lengte, vermenigvuldigd met 100 %.
Wordt gebruikt bij het beschrijven van trekproeven.

PT alongamento percentual

Aumento de comprimento de uma amostra verificado no decurso de um ensaio de tracção; exprime-se em percentagem do comprimento inicial.
Fios e cabos.

3335

ES atmósfera normal

Temperatura: 20°C ± 2°C; humedad relativa: 65% ± 2%.

DA standardatmosfære

Den temperatur og luftfugtighed, som skal overholdes ved forskellige afprøvninger: 20° ± 2° C, 65 ± 2% relativ luftfugtighed.

DE Normalklima

20 °C ± 2 °C und 65 % ± 2 % relative Luftfeuchte.

GR κανονική ατμόσφαιρα· πρότυπη ατμόσφαιρα

EN standard atmosphere

Standard temperature atmosphere for testing. An atmosphere which has a relative humidity of 65 ± 2% and a temperature of 20 ± 2°C.

FR atmosphère normale

Température: 20 °C ± 2 °C, humidité relative: 65 % ± 2 %.

IT atmosfera normale

Atmosfera avente umidità relativa dal 65 ± 2% e temperatura di 20 ± 2 °C.

NL standaard atmosfeer

Atmosferische condities overeenkomend met een temperatuur van 20° ± 2°C en een relatieve vochtigheid van 65% ± 2%.

PT atmosfera normal

Atmosfera caracterizada por 20° C ± 2° C de temperatura et 65% ± 2% de humidade relativa.

3336

ES carga a la rotura

Es la carga final observada en el momento en que la probeta o el primer elemento de ella se rompe, y se produzca en la carga de rotura o después de que esta carga haya sido alcanzada.

Habitualmente, pero no siempre, la carga a la rotura es idéntica a la carga de rotura.

DA belastning ved brud

Den belastning, der måles under trækprøver, når den første del af prøven brister.

DE Bruchkraft

Kraft, unter der beim Zugversuch in der Zerreißmaschine der Bruch eintritt.

GR τελικό φορτίο θραύσης· τελικό φορτίο ρήξης· όριο θραύσης του δικτυού

EN load at rupture

The final load at the moment that the specimen or the first component of the specimen breaks at, or after, attainment of the breaking load.

The load at rupture is usually, but not always, identical to the breaking load.

FR force à la rupture

Force finale notée au moment où l'éprouvette, ou bien le premier élément de celle-ci, se rompt.

IT forza a rottura

Forza che si rileva al momento nel quale la provetta o una parte dei suoi elementi costituenti si rompe in modo rilevabile dallo strumento di misura.

NL belasting bij breek; breeksterkte

De waarde van de trekkracht op het moment waarop het materiaal breekt.

Uitdrukking gebruikt bij trekproeven.

PT carga à ruptura; tensão à ruptura

Força máxima necessária, anotada durante um ensaio de tracção, para o primeiro elemento de uma amostra se romper.

Fios e cabos.

3337

ES carga de rotura

Es la carga máxima desarrollada en el transcurso del ensayo de rotura. Se distinguirán los casos siguientes:
a) carga de rotura del hilo seco;
b) carga de rotura del hilo mojado;
c) carga de rotura en el nudo en estado seco;
d) carga de rotura en el nudo en estado mojado.

DA brudstyrke

Den maksimale belastning, som måles ved en trækprøve.

DE Bruchkraft

GR φορτίο θραύσης· φορτίο ρήξης

EN breaking load

The maximum load observed during a breaking test.
Distinction is made between:
(a) the dry yarn breaking load;
(b) the wet yarn breaking load;
(c) the dry knot breaking load;
(d) the wet knot breaking load.
Ropes and yarn.

FR force de rupture

Force maximale notée au cours d'un essai de traction conduit jusqu'à la rupture de l'éprouvette.

IT forza di rottura; carico di rottura

Forza massima nel corso della prova di rottura.
Si distingue in:
a) forza di rottura del filo ambientato;
b) forza di rottura del filo bagnato;
c) forza di rottura al nodo allo stato ambientato;
d) forza di rottura al nodo allo stato bagnato.

NL breeksterkte; treksterkte

De maximale trekkracht optredend gedurende een trekproef.
Vaak loopt de trekkracht vlak voor het moment van breuk terug.

PT carga de ruptura; tensão de ruptura

Força máxima necessária, anotada durante um ensaio de tracção, para se obter o rompimento total de uma amostra.
Fios e cabos.

3338

ES carga de rotura de la malla

Fuerza máxima aplicada a la malla de la red que causa su rotura durante el ensayo de rotura por tracción.

DA maskebrudstyrke

Den maksimale brudstyrke, der måles ved en trækprøve af en maske.

DE Maschen-Höchstzugkraft; Maschenreiß-kraft *

Die beim Zugversuch an Netzmaschen im Normalklima gemessene Höchstzugkraft.
Naß-Maschen-Höchstzugkraft ist die beim Zugversuch an Netzmaschen im nassen Zustand gemessene Höchstzugkraft.
* Veralteter Begriff.

GR φορτίο ρήξης του ματιού· φορτίο θραύσης βροχίδας

EN mesh breaking load

The maximum load applied to a mesh, as observed during a breaking test.
Distinction is made between: the dry mesh breaking load, the wet mesh breaking load.

FR force de rupture de la maille

Force maximale appliquée à la maille de filet et causant sa rupture, notée au cours de l'essai de rupture par traction.

IT forza di rottura della maglia

Forza massima applicata alla maglia rilevata nel corso della prova di rottura.
Si distingue in:
a) forza di rottura della maglia ambientata;
b) forza di rottura della maglia bagnata.

NL treksterkte van de maas

Maximale trekkracht optredend bij een trekproef aan een maas van een stuk netwerk.
Hierbij dient een voorgeschreven atmosferische conditie te gelden.

PT carga de ruptura da malha; tensão de ruptura da malha

Força máxima que, aplicada à malha de um pano de rede, provoca o seu rompimento, anotada durante um ensaio de tracção.

3339

ES | longitud de rotura

Es la longitud calculada de una probeta acondicionada cuya masa ejerce una carga igual a su carga de rotura. Se expresa en kilómetros cuando está calculada en kilogramos fuerza y es numéricamente igual a la tenacidad calculada en gramos fuerza.

DA | brudlængde

Den længde af torvet, hvis vægt svarer til brudkraften.

DE | Reißlänge; Garnreißlänge

Die Garnreißlänge (eigentlich Reißkraftlänge) ist der Quotient aus Garnreißkraft und Feinheit. Sie ist die Länge (gewöhnlich in km ausgedrückt) des Seilgarnes, bei der der Zahlenwert des Gewichtes (Masse) gleich dem der Garnreißkraft ist.

Die bisherigen Rechnungsgrößen Garnreißlänge, Seil-Nennreißkraft und Seilreißkraft sind weggefallen. Die Rechnungsgrößen sind nunmehr:
- Garn-Höchstzugkraft (bisherige Bezeichnung: Garnreißkraft);
- rechnerische Seil-Höchstzugkraft;
- Seil-Vorspannkraft;
- Seil-Lieferlänge.

GR | μήκος ρήξης· μήκος θραύσης· χιλιομετρική αντοχή

EN | breaking length

The calculated length of a specimen whose conditioned weight exercises a force equal to its breaking load.

It is expressed in kilometres and, when calculated in kgf units, is numerically equal to the tenacity calculated in gf units.
Ropes and yarn.

FR | résistance kilométrique; longueur de rupture

Longueur calculée d'un échantillon conditionné en atmosphère normale dont la masse exerce une force égale à la force maximale de rupture de ce même échantillon.

IT | lunghezza di rottura

Lunghezza calcolata di un filo la cui massa ambientata esercita una forza uguale alla sua forza massima di rottura allo stato ambientato.

NL | breeklengte; lengte bij breuk

Lengte van het proefstuk op het moment van breken.

PT | comprimento de ruptura

É o comprimento calculado de uma amostra cujo peso exerce uma força igual à carga de ruptura; é expressa em quilómetros. Quando calculado em quilogramas força (Kgf) é numericamente igual à tenacidade calculada em gramas força (gf).

3340

ES | masa lineal

Masa por unidad de longitud.

Para las cuerdas la masa lineal se expresará generalmente en kilotex (masa en kilogramos por cada 1 000 m, o masa en gramos por metro); se determina bajo une tensión inicial fijada para cada tipo de cuerda.

DA | lineær massefylde; løbelængde

Massen i kilogram for 1 000 m (af garnet/linen); enheden benævnes tex.

DE | Feinheit; längenbezogene Masse; Lineardichte

Quotient aus Gewicht und Länge eines Garnes oder Zwirnes, angegeben in Einheiten des Tex-Systems.

In der Praxis wird die längenbezogene Masse von textilen Fasern, Garnen und dgl. „Feinheit" genannt.

GR | γραμμική πυκνότητα

EN | linear density

Mass per unit of length.

For ropes, the linear density is generally expressed in kilotex (mass in kilograms per 1 000 m, or mass in grams per metre); it is measured under a tension defined for each type of rope.

FR | masse linéique

Masse par unité de longueur.

Pour les cordages, la masse linéique est généralement exprimée en kilotex (masse en kilogrammes par 1 000 m, ou masse en grammes par mètre); elle est mesurée sous une tension fixée pour chaque type de cordage.

IT | massa per unità di lunghezza

Massa per unità di lunghezza espressa in tex.

NL | lineaire dichtheid; lineïeke massa

Massa per lengte-eenheid.

PT | massa por unidade de comprimento

Para cabos é normalmente expressa em kilotex. É medida sob uma tensão fixada para cada tipo de cabo.

3341

ES ES sentido de la torsión

DA snoningsretning

Den retning et tov eller en lines bestanddele er snoet. To på hinanden følgende snoninger vil som oftest gå hver sin vej. Man taler om S-snoning og Z-snoning, hvor disse bogstavbilleder viser snoningen.

DE Drehungsrichtung

Die schraubenlinienförmige Steigungsrichtung:
- der Fasern im Seilgarn
oder der Garne im Zwirn eines Seilgarnes;
- der Seilgarne in einer Seil-Litze;
- der Seil-Litzen im Trossenschlag-Seil
oder im Kardeel;
- der Kardeele im Kabelschlag-Seil.
Für die Drehungsrichtung sind die Buchstaben S und Z festgelegt.

GR φορά συστροφής· φορά σύστρεψης

EN twist direction; direction of twist

The product has an S twist if, when it is held vertically, the spirals formed by the filaments around its axis incline in the same direction as the central portion of the letter S. The opposite direction of twist is denoted by Z twist.

FR sens de torsion

Direction des spires des éléments constitutifs d'un fil ou cordage.
On distingue deux directions de câblage: S ou Z.

IT senso di torsione

Direzione che presentano le spire dei vari elementi costituenti la corda.

NL twijnrichting

PT sentido da torção; direcção da torção

Direcção das espiras dos elementos constituintes de um fio ou cabo.
Distinguem-se dois sentidos de torção: S e Z.

3342

ES colchar

DA slå; spinne

Slå et tov, spinne en line.

DE schlagen

GR πλέκω
σχοινί

EN lay (verb)

A rope.

FR commettre

Un cordage, un fil, un câble, etc.

IT commettere

NL touw slaan

PT cochar

Cabos.

3343

ES | masa lineal resultante; R

DA | resulterende lineær massefylde

Den lineære massefylde af det færdige produkt, en line, et tov.
Symbol R eller R-tex.

DE | resultierende Feinheit; R

GR | τελική γραμμική πυκνότητα· R

EN | resultant linear density; R

The linear density of a yarn resulting from twisting, folding or cabling operations.

FR | masse linéique résultante; R

Masse par unité de longueur du produit fini, après les différentes opérations de fabrication.

IT | massa risultante per unità di lunghezza; R

Massa per unità di lunghezza del prodotto finito risultante dalla torcitura, doppiatura, ritorcitura o, in generale, dalla lavorazione.

NL | resulterende lineïeke massa; resulterende lineaire dichtheid

De lineïeke massa van het eindprodukt, verkregen na twisten, twijnen of kableren en eventuele nabehandelingen.

PT | massa por unidade de comprimento do produto acabado

Normalmente expressa em R-Tex.

3344

ES | tiempo de rotura

Es el tiempo, expresado en segundos, necesario para alcanzar la carga de rotura y contado a partir del momento en que se aplique la carga.

DA | tid indtil brud indtræffer

Den tid, der ved trækprøver måles (i sek.), fra prøven belastes, til den bryder.

DE | Reißdauer

GR | διάρκεια θραύσης· διάρκεια ρήξης

EN | time-to-break

The time, in seconds, taken ro reach the breaking load, measured from the moment of application of the load.
Ropes and yarn.

FR | durée de rupture; temps d'essai de rupture

Durée nécessaire pour atteindre la force de rupture. S'exprime en secondes et est comptée à partir du moment où la force est appliquée.

IT | tempo di rottura

Tempo, espresso in secondi, che intercorre tra l'inizio dell'applicazione della forza ed il momento nel quale si raggiunge la rottura.

NL | breektijd; tijd benodigd voor breuk

Tijdsinterval tussen het moment van het aanbrengen van een kracht en de uiteindelijke breuk van het proefstuk.

PT | tempo de ruptura

Espaço de tempo necessário para obter a carga de ruptura, contado a partir do momento em que a força é aplicada.
Normalmente expresso em segundos.

3345

ES tenacidad

Es la carga de rotura por unidad de masa lineal resultante de la probeta sin forzar en estado acondicionado.

DA tenacitet; brudspænding

Mål for brudstyrken pr. enhed lineær massefylde.
Angives i N/tex.

DE Zugfestigkeit; Feinheit-Festigkeit

GR συνεκτικότητα· ανθεκτικότητα

EN tenacity

The breaking load per unit resultant linear density of the unstrained specimen in the conditioned state.
Ropes and yarn.

FR ténacité

Force de rupture par unité de masse linéique du fil sans nœud.
-Il s'agit ici de la masse linéique résultante déterminée sur l'éprouvette non tendue, conditionnée en atmosphère normale d'essai des textiles.
- S'exprime en centinewtons par tex.

IT tenacità

Forza massima di rottura rilevata allo stato ambientato, riferita all'unità di massa risultante per unità di lunghezza dalla provetta prima che questa venga sottoposta a trazione.

NL

Breeksterkte per eenheid van lineaire dichtheid.

PT tenacidade

Propriedade que os materiais apresentam para contrariar a ruptura.

3346

ES torsión

La torsión de una cuerda está caracterizada por el sentido de la torsión de la cuerda acabada y por el número de espiras por metro.

DA snoning; slåning; spinning

Bestemmes for en line eller et tov ved snoningsretningen (S- eller Z-) og antallet af omdrejninger pr. meter.

DE Drehung

Drehungsrichtung und Drehungszahl von Garnen und Zwirnen.
Die Drehung und die Drehungsrichtung bei Faserseilen sind durch die Steigungsrichtung der Fasern im einfachen Garn, der einfachen Garne im Zwirn, der Garne in der Litze und der Litzen im Seil oder im Kardeel gekennzeichnet. Für die Drehrichtung sind die Buchstaben S für S-Drehung und Z für Z-Drehung festgelegt. Genormt gedrehte Seile sind in Z-Drehung hergestellt. Nach deutscher Norm wird die Zahl der Schläge nicht pro Meter angegeben.
Beispiel: Ein Drahtseil von 20 mm Ø, 6 Litzen mit 7-fachem Schlag hat eine Schlaglänge von 20 x 7 = 140 mm.

GR συστροφή· σύστρεψη

EN twist; amount of twist

The number of turns per metre of a twisted yarn.
The twist of a yarn is characterized by the direction of twist of the finished yarn and by the number of turns per metre.

FR torsion

La torsion d'un cordage ou d'un fil est caractérisée par sa direction (S ou Z) et par le nombre de tours par mètre.

IT torsione

Operazione che consiste nell'avvolgere tra di loro gli elementi che costituiscono il prodotto considerato.
La torsione di una corda è caratterizzata dal senso di torsione della corda finita e dal numero di giri al metro.

NL twist; twijning

Aantal slagen per lengte-eenheid, verdraaiing van garen.
Twijning komt voor in de Z-richting en in de S-richting (resp. rechtsom en linksom).

PT torção

A torção de um fio ou de um cabo é caracterizada pela sua direcção (S ou Z) e pelo número de voltas por metro.

3347

ES puerta Süberkrüb

Puerta para arte pelágica de profundidad regulable.

DA suberkrub-skovl

En krum trawlskovl med et stort forhold mellem højde og længde.
Anvendes kun pelagisk.

DE Süberkrübscherbrett; Süberkrüb-Scherbrett

GR πόρτα τράτας τύπου «Süberkrüb»

EN suberkrub otter board

All-steel cambered midwater otter board with vertical aspect greater than its horizontal aspect.
Named after its inventor.

FR panneau süberkrüb

Panneau de chalut à grand allongement vertical et profil creux, employé principalement pour la pêche pélagique.
Des panneaux similaires, avec allongement un peu plus faible et munis d'une semelle renforcée, peuvent être utilisés pour la pêche au fond (panneaux japonais, par exemple).

IT divergente süberkrüb

Divergente pelagico.

NL süberkrübbord

In de pelagische visserij: hydrodynamisch gebogen stalen bord, met de korte zijde van de rechthoek horizontaal, dat niet over de bodem gaat.
Süberkrüb is de naam van de uitvinder.

PT porta de arrasto Suberkrüb

Porta para arrasto pelágico, de aço e com forma hidrodinâmica.

3348

ES driza

DA

En line, der anvendes i forbindelse med indhalingen af bobbinsgearet på en sidetrawler.
Kendes ikke i Danmark.

DE Hißtau

GR υπέρα· μαντάρι
Σχοινί σημαίας

EN quarter strop

Wire rope fitted with swivels, two in number. They are used in conjunction with the quarter ropes for heaving the mouth of the trawl inboard.

FR drisse

Du parpaillot.

IT drizza; cavo di recupero

Del cavo di chiusura.

NL val

PT adriça

3347

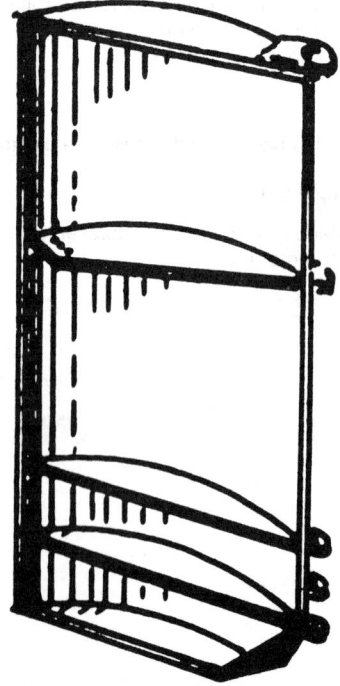

3349

ES trasbordo

DA omladning

Overførsel af fangst fra et fartøj til et andet. F.eks. for forarbejdning eller transport.

DE Fangübergabe; Fangübernahme

Übergabe des Fangs in der Flottillenfischerei vom Fangschiff (Zubringerschiff) an das Transport- oder Verarbeitungsschiff, wobei die Qualität des Fisches möglichst wenig beeinträchtigt werden soll.

GR μεταφόρτωση

EN transhipment; transhipping of catch; handing over of catch

Transfer of fish from a fishing vessel to a larger ship for processing and/or transportation.

FR transbordement

Transfert de la capture d'un bateau au navire-mère.

IT trasbordo

Trasferimento del pesce catturato da una nave all'altra.

NL vangstovername

Het overnemen van door andere vissersvaartuigen gevangen vis door een fabrieksschip voor verdere verwerking.

PT transbordo

Transferência de capturas de um barco de pesca para um navio de maior tonelagem para processamento e/ou transporte.

3350

ES chigre de espiar

DA forhalingsspil

Spil til at hale trosser under forhalingen af et fartøj. Trossen hales over spillet, men opbevares ikke derpå.

DE Verholspill

Deckshilfsmaschine zum Einholen von Trossen.
Beim Verholspill werden mit der Trosse um den Spillkopf Törns gelegt und dann steifgehievt, um sie danach am Poller zu belegen.

GR βαρούλκο πρόσδεσης

EN warping winch

A winch used solely for warping, on which a rope may be wound under power but not stored.

FR treuil de touage

Treuil utilisé uniquement pour le touage, sur lequel un câble peut être enroulé sous tension mais non stocké.

IT verricello di tonneggio

NL verhaallier; verhaalspil

Windas voor het behandelen van trossen aan boord.

PT alador de cabeço

Guincho utilizado para alar cabos e no qual um cabo pode ser enrolado sob tensão, mas não armazenado.

3351

ES chigre

DA spil

Dækmaskineri til haling af liner eller tove med en tromle, hvorpå tovet vindes op.

DE Winde

Deckshilfsmaschine zum Heben von Lasten, die das Lastseil im Gegensatz zum Spill aufwickelt.

GR βαρούλκο· βίντσι

EN winch

A powered, or unpowered machine, having one or more horizontally mounted drums and/or warping ends, on which a rope may be wound.

FR treuil

Appareil motorisé ou non permettant d'enrouler un câble sous tension au moyen d'un ou de plusieurs tambour(s) à axe horizontal et, éventuellement, d'une ou plusieurs poupée(s) à axe horizontal.

IT verricello

Piccolo argano ad asse verticale od orizzontale che serve ad alare le cime per il tonneggio della nave, a salpare àncore di peso limitato e le reti nei grossi pescherecci.

NL lier; windas

Hijswerktuig met horizontale, draaibare spil, voorzien van een cilindrische trommel waarop een touw, staaldraad of tros kan worden gewonden.

PT guincho

Aparelho de força mecânico ou manual equipado com um ou mais tambores horizontais que permitem enrolar e armazenar um cabo sob tensão e que eventualmente pode ser dotado com um ou mais cabeços de eixo horizontal.

3352

ES aparejo

Sistema o máquina compuesta por dos motones, dos cuadernales o un motón y un cuadernal con un cabo guarnido entre ambos, y que permite realizar un trabajo con menos fuerza de la necesaria sin su utilización.

DA talje

Et system af blokke og et tov. Blokkene kan have en eller flere skiver, bestemt af det mekaniske arbejdes størrelse. En talje reducerer de kræfter, der er nødvendige ved hal og løft.

DE Talje; Takel

Kräftesparende Anordnung von ein- und mehrscheibigen Blöcken und Tauwerk. Die festen Blöcke dienen darin der Richtungsänderung der Kraft, die losen der Kraftersparnis. Das Ende des Läufers, an dem die Kraft angreift, ist die holende Part, und die anderen sind die tragenden Parten. Im Schiffsbetrieb finden Taljen z. B. am Ladegeschirr, an den Aussetzvorrichtungen der Rettungsboote sowie als Hilfsmittel bei verschiedenen Arbeiten Verwendung. Die Taljen mit fünf und mehr Seilscheiben werden als Gien bezeichnet.

GR παλάγκο· σύσπαστο

Ανυψωτήρας με δύο τροχαλίες και ένα σχοινί

EN tackle; purchase

A combination of ropes and blocks working together, or any similar contrivance affording a mechanical advantage to assist in lifting or controlling a weight or applying tension on board ship. Tackles vary in design according to their different uses, every form or adaptation having its own specific name.

FR palan

Appareil composé de deux poulies et d'un cordage, le garant. Les palans servent à multiplier la force exercée sur le garant; ils permettent de raidir sans secousse et de retenir plus aisément un cordage qui a déjà subi une certaine tension.

3352

IT paranco

Attrezzo molto usato a bordo. È formato da due bozzelli, l'uno fisso e l'altro mobile, e da un cavo che passa per le pulegge di entrambi. Un'estremità del cavo è fissata allo stroppo di uno dei due bozzelli e viene chiamata dormiente o arricavo; l'altra estremità rimane libera ed a questa viene applicata la forza che mette in azione il paranco; viene detta tirante. I paranchi vengono utilizzati per ridurre la forza necessaria a vincere una resistenza.

A seconda del numero delle pulegge nei bozzelli, i paranchi sono semplici, doppi o caliorne.

NL takel

Takel bestaande uit twee blokken, elk met een of meer schijven, waardoor een touw is geschoren.

PT talha; estralheira

Aparelho de força composto por um cadernal de dois gornes e um moitão (talha singela), ou por dois cadernais de dois gornes (talha dobrada).

3352

3353

ES maquinilla para el lanteón

DA gilsonspil; takkelspil

Et spil, der hjælper med at hale trawlposen om bord på fartøjet. Gilsonspillet haler posen op gennem rampen på en hæktrawler; takkelspillet tager løftet ind over siden på en sidetrawler.

DE Beihieverwinde

Winde zum Einholen des Beihievers.

GR βαρούλκο παλάγκου

Βοηθητικό βαρούλκο που χρησιμοποιείται για την ανέλκυση της τράτας, όταν αυτή έχει μεγάλο βάρος.

EN gilson winch

Winch which holds the gilson wire.

FR treuil de caliorne

IT verricello per ghia; verricello per trinca; verricello per caliorna

Verricello per recuperare la ghia (trinca; caliorna).

NL jomperlier

Lier voor de bediening van de jomper.

PT guincho para uma talha

3354

En los buques pesqueros, aparejo formado por un cable que pasa por una pasteca firme al palo.
Para cargas mayores se usa el aparejo real.

DA takkelwire; gilsonwire

Wire anvendt til at hale fangstposen om bord på en trawler. Takkelwire: på en sidetrawler en wire gennem en blok på formasten. Gilsonwire: på en hæktrawler en wire gennem en blok placeret bag ved styrehuset.

DE Beihiever

Seil mit Haken zum Beihieven von Schleppnetzen.

GR σχοινί του βαρούλκου

EN gilson; gilson wire; jilson

Wire tackle used for hauling onboard and emptying the codend.
For heavy bags a stronger purchase called fish tackle is used.

FR caliorne

Palan servant à hisser à bord le cul de chalut; formé d'un câble d'acier muni d'un croc à son extrémité.

IT ghia; trinca; caliorna

Cavo con o senza paranco per il recupero della saccata.

NL jomper

Strop voor het inhalen van de kuil van een trawl.

PT

Talha para alar os sacos das redes de arrasto.
Não tem equivalente em português.

3355

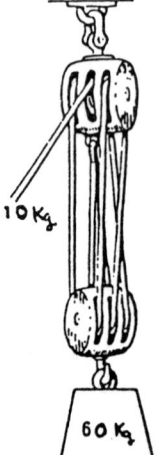

3355

ES aparejo real

Aparejo de grandes dimensiones y numerosos guarnes empleado para izar grandes pesos.

DA gie

Talje, der er fem-eller seksskåret.

DE Gien; Gientalje

Talje aus Blöcken mit fünf und mehr Scheiben zum Anheben schwerer Lasten.

GR πολύσπαστο· μάντος (κοινώς)

EN winding tackle

Large purchase, comprising three-fold block aloft and double block in lower end. Secured at lower masthead and used for lifting heavy weights.

FR caliorne; palan de caliorne

Palan de fort échantillonnage, composé de deux poulies triples, ou d'une poulie double et d'une poulie triple, utilisé pour soulever des poids importants.

IT paranco triplo; caliorna

Paranco con un bozzello triplo e uno doppio, usato per sollevare grossi pesi o mettere in forza cime d'ormeggio.

NL jijn; gijn

Zwaar takelgestel bestaande uit twee blokken die samen vijf of meer schijven hebben, waardoor een loper is geschoren.

PT talha tripla; estralheira tripla; aparelho real

Aparelho de força constituído por dois cadernais de três gornes, ou por um cadernal de dois gornes e um de três gornes.

3356

ES paño interior del trasmallo

DA indergarn

Det midterste småmaskede net i et toggegarn.

DE Inngarn

Inneres Netzteil eines Spiegelnetzes.

GR μεσαίο δίχτυ των μανωμένων διχτυών

EN lint; inner net

The middle panel of a trammel net made of small mesh netting.

FR flue

Nappe intérieure d'un trémail.

IT pezza interna del tremaglio

NL flouw; fluwe; vluwe, vlouw

Binnennet van een schakel.

PT miúdo

Pano de rede interno de um tresmalho, caracterizado por reduzida malhagem.
Redes de tresmalho.

3357

ES cebo vivo

DA levende agn; levende madding

Agn, der består af fisk, skaldyr eller lignende. I modsætning til kunstig agn.

DE lebender Köder; Lebendköder

GR ζωντανό δόλωμα

EN live bait

Bait comprising fish, shellfish or other animal which is, or has recently been, alive and is still attractive to the prey which it is used to catch.

FR appât vivant

Poisson gardé à bord en vivier et utilisé comme appât pour la pêche aux lignes.

IT esca viva

NL levend aas

PT isco vivo

Peixe mantido vivo a bordo em viveiros e que é utilizado como isco na pesca à linha.

3356

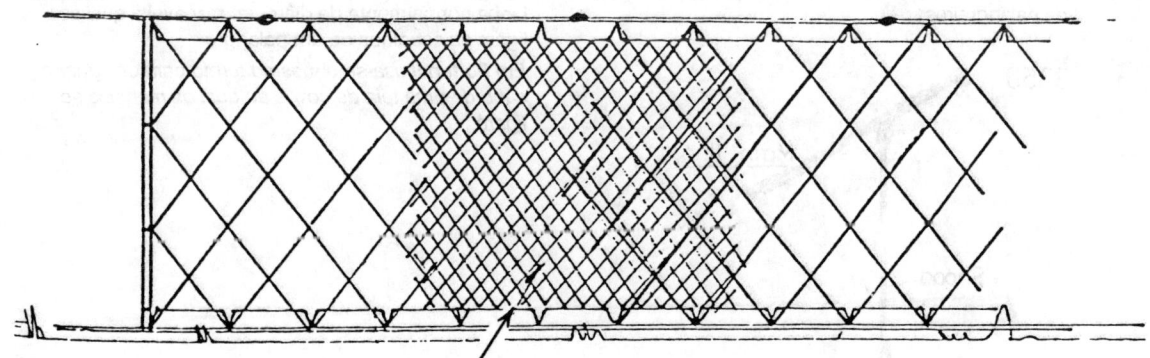

3358

ES palangre automático

DA automatiseret langline

System, hvor større eller mindre dele af operationen er mekaniseret: haling, afpilning af fiskene fra krogene, opbevaring af liner og kroge i magasin, agning af kroge og udsætning.

DE automatisierte Langleine

GR αυτοματοποιημένο παραγάδι

EN automatic longline

Automated longline system in which the shooting (including baiting the hooks) and hauling (including removal of fish) of the lines are completed without the need for significant manual work.

FR palangre automatisée

Dispositifs permettant l'opération entièrement mécanisée (automatisée) d'une palangre à bord d'un bateau.

IT palangaro automatico

NL geautomatiseerde beug

PT palangre automático

Equipamento que permite uma operação totalmente mecanizada e automatizada dos palangres a bordo de palangreiros.

3359

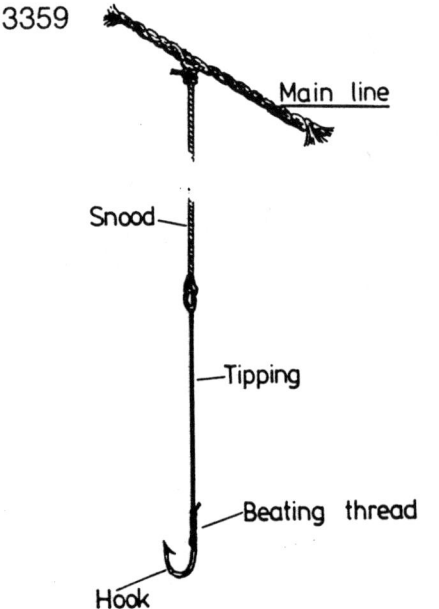

Main line

Snood

Tipping

Beating thread

Hook

3359

ES madre; línea madre

Cordel principal del palangre.

DA line; hovedline

Line, hvortil tavserne er fastgjort.
Langline.

DE Hauptleine

Langleine.

GR μάνα

EN main line

The line of a string to which the snoods are attached.
Line fishing.

FR ligne principale; ligne-mère

Ligne de plus fort diamètre, à laquelle sont fixés les avançons.
Palangre.

IT trave; madre; letto; maestra; stesa

Palangaro.

NL lijn

Hoofdlijn van een beug waaraan de sneuen zijn vastgemaakt.

PT madre

Linha normalmente de diâmetro razoável à qual se ligam os estralhos de um palangre.

Em Portugal usa-se o mesmo termo para designar o cabo de uma teia de covos ao qual os mesmos se ligam.

3360

ES contramalla *; esmais **

Paños exteriores de malla grande de un trasmallo.
* Norte de España. ** Alicante.

DA stormasket net; spejlmasker

De to lag af stormasket net yderst i et toggegarn.

DE Spiegelmaschen

Außenwand eines Trammelnetzes (Dreiwandnetzes).

GR εξωτερικό φύλλο (μανωμένου διχτυού)

Εξωτερικό φύλλο ενός μανωμένου διχτυού με τρία φύλλα. Διακρίνεται για το μεγάλο μάτι σε σχέση με το μεσαίο κυρίως φύλλο του διχτυού.

EN armouring; armoring

Outer walls of larger mesh netting of a trammel net.

FR aumée

L'une des nappes extérieures d'un trémail.

IT parete; maglione

Pezze esterne del tramaglio (pareti); sono a maglie molto grandi che vengono chiamate maglioni.

NL laddernet

Één van de twee buitenste netten van een schakel.

PT albitana; alvitana

Panos externos de maior malhagem das redes de tresmalho.

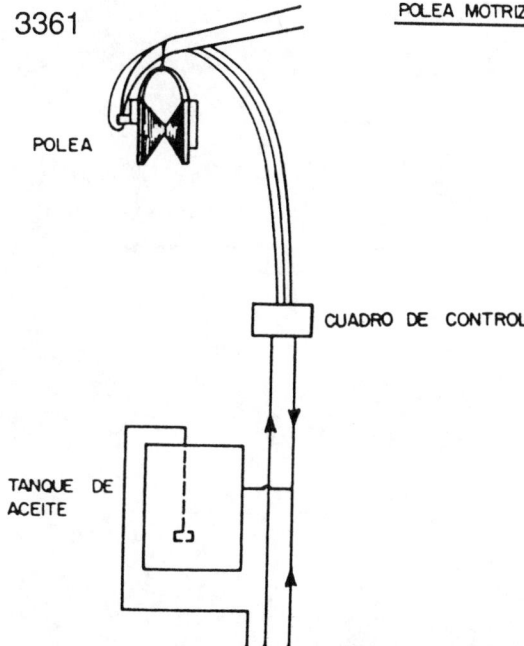

3361

POLEA MOTRIZ

POLEA

CUADRO DE CONTROL

TANQUE DE ACEITE

BOMBA HIDRÁULICA

3361

ES polea motriz; power block; halador mecánico

Consiste en una roldana que gira por acción hidráulica sobre un bastidor. La forma de la roldana permite el paso de la red completa con sus respectivas relingas. Está forrada de goma vulcanizada para que la tracción sea mayor, de tal suerte que la tripulación solamente tiene que recoger el paño conforme va saliendo de la misma e ir estibándolo directamente sobre cubierta, ahorrando tiempo y esfuerzo.

DA kraftblok

En hydraulisk drevet blok med gummiskive (med V-formet rille) eller system af tre gummiruller. Anvendes i snurrevods- og notfiskeriet til at hale redskabet indenbords.

DE Kraftblock; Powerblock

Frei hängende angetriebene Rolle mit V-förmiger Rille. Der Kraftblock ist speziell für die Ringwadenfischerei entwickelt und dient zum Einholen der Ringwade. Dabei wird die Ringwade durch die V-förmige Rille der Rolle als Netzstrang gebündelt. Durch die zwischen Rillenwand und Netzstrang entstehenden Reibkräfte kann durch Drehen der Rolle der Netzstrang eingeholt werden.

GR μηχανικό παλάγκο· μηχανική τροχαλία

EN power block

A mechanized pulley used to haul in nets, purse seine, etc.

FR poulie mécanique; poulie motrice; power-block

Appareil utilisé pour remonter les filets, la senne coulissante, etc.

IT bozzello motorizzato; bozzello salparete; salpacianciolo; power-block

Puleggia mossa da un motore generalmente di tipo idraulico. Essa è montata alla sommità di un albero e serve ad agevolare il trasferimento, dal mare a bordo, di grosse reti a circuizione.

NL power-block

Hydraulisch aangedreven schijf, gebruikt bij de visserij met de purse-seine, om het inhalen van dit net te vergemakkelijken.

PT alador de redes mecânico

Aparelho de força mecânico, normalmente hidráulico, dotado de uma gola e utilizado para alar redes de cerco (power block) e redes de emalhar e tresmalhos.

3362

ES segundo barco; compañero

En la pesca a la pareja, barco que no tiene la responsabilidad de la pesca.

DA

Det fartøj ved dobbeltslæbning, der ikke sætter og haler trawlen.

I dansk fiskeri vil de to fartøjer, der udgør et makkerpar ved dobbeltslæbning skiftes til at sætte trawlen. De er derfor ligeværdige.

DE

In der deutschen Gespannfischerei wird der Fang auf Verabredung der Partner im Wechsel übernommen. Es gibt daher keine spezielle Bezeichnung für ein an der Fangübernahme nicht beteiligtes Fahrzeug.

GR βοηθητικό σκάφος· δευτεροκάικο

EN second boat

In pair trawling, the boat which does not shoot or haul the net.
See also 3181 and 3372.

FR veau

Dans le chalutage à bœuf, bateau qui ne vire pas le filet à bord.
Voir aussi 3181 et 3372.

IT barca secondaria

Barca che effettua solo il traino senza calare e salpare la rete.
Nella pesca al traino a coppia.

NL maat; spanmaat

In de Nederlandse spanvisserij hebben beide schepen doorgaans een vistuig op de nettenrol, zodat de rol kan wisselen; men spreekt wel van maat of spanmaat als aanduiding van het andere schip van het span.

PT

Embarcação que no arrasto de parelha não ala a rede para bordo.
Não tem equivalente em português.

3363

ES marea; campaña; turno

DA rejse

Den tid et fartøj er på fiskeri, fra afsejling til hjemkomst.

DE Einsatzfahrt; Fahrt; Reise

GR αλιευτικό ταξίδι

EN trip

The period when a fishing boat is away from port.

FR marée

Sortie en mer d'un bateau de pêche.

IT campagna di pesca; bordata

Permanenza in mare dei pescherecci.

NL reis; vaart

Alle handelingen van een vissersschip vanaf het vertrek uit de haven tot aan het afmeren.

PT campanha de pesca; saída de pesca; viagem de pesca; maré

Período durante o qual um barco de pesca se encontra fora do porto em faina de pesca.
Designa-se por maré quando não ultrapassa 24 horas.

3364

ES abertura; entrada

La boca u orificio por el cual se accede al interior del artilugio. Tiene especial importancia en las nasas en las que adopta forma de tubo o embudo. En los reteles dicha entrada es la superficie que delimita el aro armazón.

Nasa.

DA tragt; indgang

Åbningen ind til en ruse eller tejne, udformet således, at den er vanskelig at komme ud af igen.

DE Einkehlung

Reuse.

GR είσοδος ιχθυοπαγίδας

Άνοιγμα εισόδου ιχθυοπαγίδας (π.χ. κοφινέλου) διαμορφωμένη κατάλληλα ώστε να εμποδίζεται η διαφυγή των εγκλωβισμένων αλιευμάτων.

EN opening; entrance

The mouth of a trap or pot usually designed so that escape is difficult.

FR goulotte; entrée de casier

Ouverture par laquelle les animaux entrent dans le casier ou la nasse; sa forme est spécialement étudiée pour empêcher la sortie des animaux capturés.

IT apertura; bocca

Nassa.

NL keel; enkel

Trechtervormig deel van een fuik dat de vis gemakkelijk toegang geeft tot het uiteinde van de fuik, en belet terug te gaan.

PT abertura; entrada; boca; endiche; andiche

Abertura através da qual os peixes, crustáceos ou moluscos entram nas armadilhas; a sua forma é especialmente estudada para impedir a saída das capturas.

Nassa.

3365

ES colchado

DA slåning

Proces, der fører frem til dannelsen af et tov ved at sno, (tvinde, slå) fibre, garner og kordeler.

DE Schlag; Kabelschlag

Herstellung eines Faserseils aus pflanzlichen und/oder synthetischen Fasern, das entweder durch Seilformung oder Drehung (Legen oder Verkleben), durch zwei- oder mehrstufiges Verseilen, durch Ummanteln oder durch Verflechten entsteht.

GR πλέξιμο σχοινιού

EN lay

The length of one complete turn or between two successive plaiting points of the same strand, measured parallel to the axis of the rope.

FR commettage

Opération consistant à confectionner un cordage en réunissant par une torsion convenable les éléments qui le composent.

IT commettitura

Torsione di due o più legnoli attorno ad un asse comune, di regola con senso di torsione contrario a quello dei componenti stessi, ai quali si dà contemporaneamente una sovratorsione in modo che il complesso risulti stabile per l'equilibrio delle torsioni dei singoli elementi.

NL slag

Het samenstellen van een touw uit strengen door deze in elkaar te draaien.

PT cochar; torcer

Operação que consiste em fabricar um fio ou cabo reunindo por torção adequada os elementos que o compõem.

3366

ES pegadura

Popular.

DA samling af net

Proces hvorved de enkelte sektioner i et redskab slås eller bindes sammen.

DE Zusammensetzen von Netztuchen

GR συναρμογή διχτυών· συνένωση διχτυών

EN joining of netting

The process of connecting, by means of a thread, the edges of netting panels which can differ in the number of meshes, mesh size and types of cut.

There are two methods of joining netting: sewing and seaming.

FR assemblage des nappes de filet

Procédé qui consiste à unir à l'aide d'un fil les bords de pièces de filet, ces pièces pouvant différer quant à leur nombre de mailles et à la dimension de celles-ci et quant au processus de coupe de bords.

On distingue deux méthodes d'assemblage des pièces de filet: par couture et par abouture.

IT assemblaggio delle pezze di rete

Procedimento che consiste nell'unire, con un filo, i bordi di pezze di rete. Queste ultime possono differire sia come numero di maglie, sia come dimensioni delle medesime, sia come tecnica di taglio sui bordi.

NL aanslaan van netwerk; aanslaan van netstukken

Het door middel van een rijgdraad aan elkaar verbinden van stukken netwerk.

De rijgdraad vormt de zgn. aanslag.

PT porfio *; pegamento **

* Junção, por intermédio de um fio, dos bordos laterais de panos de rede que podem diferir quer quanto à dimensão e número de malhas, quer quanto ao tipo de corte.
** Junção, por intermédio de um fio, dos bordos transversais de panos de rede que diferem apenas quanto à dimensão e número de malhas.

3367

ES cordón grueso

DA kordel

Bestandel af et tov eller line. Der indgår normalt heri tre eller fire kordeler.

DE Kardeel

Durch Verseilen von Seil-Litzen hergestelltes Halberzeugnis für die Weiterverarbeitung zu einem Faserseil im Kabelschlag.

Ein Kardeel entspricht im Aufbau einem Seil im Trossenschlag.

GR δευτερεύον έμβολο· έμβολο

EN strand; secondary strand

That produced in the second laying process by the twisting of strands.

FR toron secondaire; toron

Produit obtenu par la réunion de plusieurs fils ou duites retordus ensemble.

Les torons sont assemblés par commettage.

IT legnolo; trefolo

NL kardeel

PT cordão secundário

3366

248

3368

ES banda de plomo; banda de abajo; ala inferior

Pieza de la parte baja del arte de arrastre, situada bajo la banda de corcho, por delante del vientre.

DA undervinge

Den del af forparten af en trawl, der strækker sig frem sammen med over- og sidevinger.

DE Unterflügel

GR κατώτερο φτερό

Τμήμα της κάτω πλευράς της τράτας που βρίσκεται μπροστά από την κοιλιά.

EN lower wing

Net section extending forward from one side of the belly and usually joined to the adjacent top wing (two panel trawls) or adjacent side wing (four panel trawls).

FR aile inférieure

Pièce de la face inférieure du chalut située en avant du ventre.

IT braccio inferiore

NL ondervlerk

Puntvormig gedeelte tussen de onderpees en de ondernaad (4-bladig net) of tussen de onderpees en de naad (2-bladig net) tot aan de middeling.

PT asa inferior

Secção das redes de arrasto que se estende para a parte anterior da rede a partir de cada um dos lados da primeira barriga inferior e que se liga à asa superior adjacente (redes com duas faces) ou à face lateral adjacente (redes com quatro faces).

3368

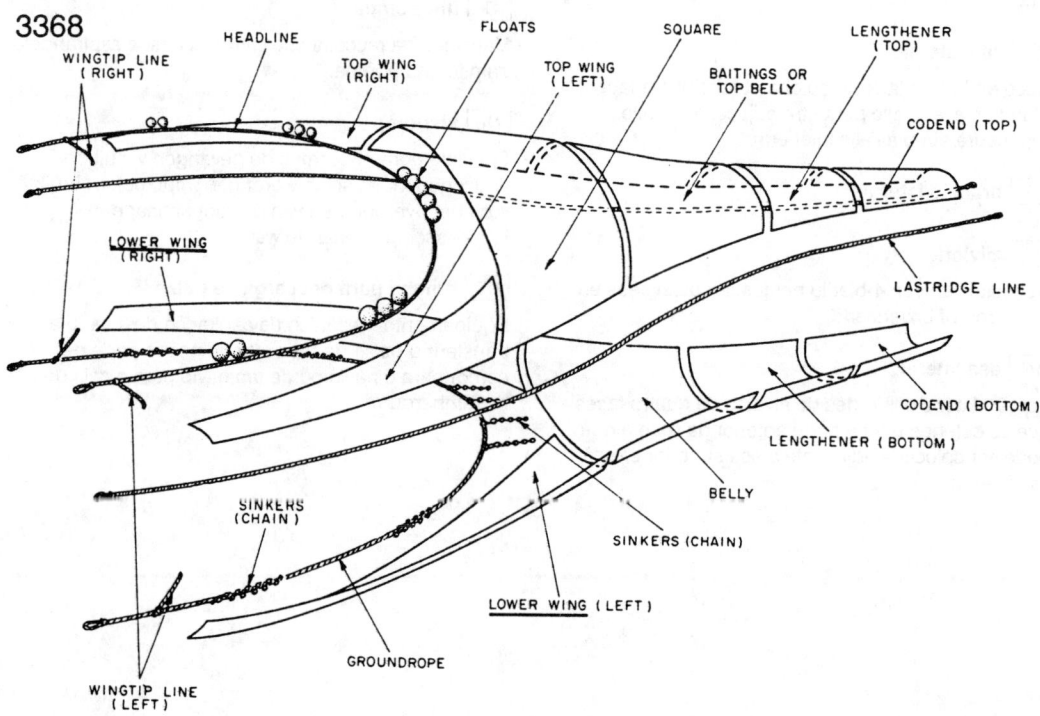

WINGTIP LINE (RIGHT) — HEADLINE — TOP WING (RIGHT) — FLOATS — TOP WING (LEFT) — SQUARE — BAITINGS OR TOP BELLY — LENGTHENER (TOP) — CODEND (TOP) — LOWER WING (RIGHT) — LASTRIDGE LINE — SINKERS (CHAIN) — CODEND (BOTTOM) — LENGTHENER (BOTTOM) — BELLY — SINKERS (CHAIN) — LOWER WING (LEFT) — GROUNDROPE — WINGTIP LINE (LEFT)

3369

ES banda lateral

En la red de cuatro caras, parte situada en la zona delantera de las caras laterales, generalmente adyacente a las bandas de corcho y plomo.

DA sidevinge

Den del af forparten af en 4-panels trawl, der strækker sig frem sammen med over- og undervinger.

DE Seitenflügel

GR πλαϊνό φτερό

Τμήμα της πλάγιας πλευράς της τράτας (τράτα τεσσάρων φύλλων) τοποθετημένο μπροστά. Μπορεί να υπόκειται του ανώτερου φτερού και/ή του κατώτερου φτερού.

EN side wing

Lower or upper wing of side panel of a four panel trawl.

FR aile latérale

Pièce de la face latérale du chalut (chalut à 4 faces), située en avant; elle peut être adjacente à l'aile supérieure et/ou à l'aile inférieure.

IT braccio laterale

NL zijvlerk

Gedeelte van een 4-bladig net tussen de zijpees en de onder- of bovennaad.

PT asa lateral

Secção lateral das redes de arrasto de quatro faces que se estende para a parte anterior da rede e que pode ser colocada adjacente à asa superior e/ou à asa inferior.

3370

ES bomba para descarga de pescado

Red de cerco.

DA fiskepumpe

Maskineri til at overføre fangsten fra redskabet i vandet (not eller trawl) eller fra lasten under losning.

DE Fischpumpe

GR αντλία για ψάρια

EN fish pump

Apparatus to transfer fish from net to boat, or from boat to dock.

FR pompe à poisson

Appareil servant au transfert ou au déchargement du poisson.
Senne coulissante.

IT ittiopompa

Macchina da raccolta che cattura il pesce aspirandolo tramite una pompa.

NL vispomp

Speciale pomp waarmee de gevangen vis uit de ringzegen in het schip wordt gepompt; ook gebruikt voor het overpompen van de vangst naar een fabrieksschip of naar de wal.

PT bomba para descarga de peixe

Equipamento mecânico de aspiração destinado a transferir o peixe de uma arte de pesca (rede de cerco) para o navio ou de um navio para o cais de desembarque.

3371

ES boya

Objeto flotante destinado a indicar la presencia de un arte de pesca, al que se une por un cabo llamado orinque.

DA bøje; gaj

Stort flydelegeme, der anvendes til at afmærke positionen af et fiskeredskab. Kan være forsynet med et flag.
En gaj vil normalt være stor, uden flag og af blødt plastik.

DE Boje

Tonnen-, kugel- oder kegelförmiges schwimmendes Seezeichen.

GR πλωτήρας· σημαδούρα

Επιπλέον σώμα που χρησιμοποιείται για να επισημαίνει τη θέση ή να προσθέτει άνωση σ' ένα εργαλείο στον πυθμένα ή στα μεσόνερα με τη χρήση ενός σχοινιού σημαδούρας.

EN buoy

Large float used to mark or support part of the gear.

FR bouée

Corps flottant servant à signaler ou à soutenir un engin au fond ou entre deux eaux par l'intermédiaire d'un orin.

IT boa; gavitello

Galleggiante, in genere pitturato a colori vivaci, di forma sferica o a doppio cono, di legno, sughero o lamierino, comunque atto a stare a galla e a sorreggere una cima o una catenella che lo unisce con un peso adagiato sul fondo. Si usa per indicare la posizione di un oggetto che si trovi sul fondo.

NL boei; joon; breel

Drijvend, verankerd voorwerp voor het aangeven van visnetten, of van een positie in het algemeen.

PT bóia; flutuador

Corpo flutuante utilizado para dar elevação vertical a uma rede de pesca, para indicar a sua posição, ou, simultaneamente, para ambas as situações.

3372

ES primer barco

Buque de pareja.
En España los buques de la pareja son llamados «primero» y «segundo». Primero es el que lleva a bordo el técnico que manda ambas unidades y dirige la pesca. El embarcar el arte se hace alternativamente por ambos buques, no determinando denominación especial a uno u otro.

DA

Det fartøj, der sætter og haler trawlen ved dobbeltslæbning.
I dansk fiskeri vil de to fartøjer, der udgør et makkerpar ved dobbeltslæbning, skiftes til at sætte trawlen. De er derfor ligeværdige.

DE

In der Gespannfischerei eines der beiden beteiligten Boote.

GR κυρίως σκάφος

Κατά το καλάρισμα τράτας με ζευγαρωτά σκάφη, ο όρος «κυρίως σκάφος» χαρακτηρίζει το σκάφος στο οποίο φορτώνεται η τράτα. Το άλλο σκάφος ονομάζεται «βοηθητικό σκάφος».
Στην Ελλάδα οι όροι «κυρίως σκάφος» και «βοηθητικό σκάφος» ή «δευτεροκάικο» χρησιμοποιούνται κυρίως από τα γρι-γρι.

EN bull

In pair trawling, the boat which shoots or hauls the net.
See also 3181 and 3362.

FR bœuf

En chalutage à bœufs, le terme «bœuf» désigne le bateau sur lequel on embarque le chalut.
Voir aussi 3181 et 3362.

IT barca principale

Nella pesca con rete da traino a coppia, il peschereccio che dirige le azioni.

NL maat; spanmaat

In de Nederlandse spanvisserij hebben beide schepen doorgaans een vistuig op de nettenrol, zodat de rol kan wisselen; men spreekt van maat of spanmaat als aanduiding van het andere schip van het span.

PT

Embarcação que no arrasto de parelha ala a rede para bordo.
Não tem equivalente em português.

3373

ES embolsamiento; cierre de la jareta

Operación de cerrar un arte de cerco de jareta, por su parte inferior.

DA snurpning

Den proces hvorved noten lukkes i bunden ved haling på snurpewiren.

DE Schnüren

Schließen der an Luv des Schiffs schwimmenden Ringwade.

GR στιγγάρισμα· κλείσιμο

Διαδικασία κλεισίματος του κάτω μέρους ενός γρίπου με τη βοήθεια ενός στίγγου και αφού έχουν περικυκλωθεί τα ψάρια.

EN pursing

The operation of drawing in the purse line of a purse seine to complete the enclosure.

FR boursage; coulissage

Opération de fermeture d'une senne par en bas, après encerclement du poisson, effectuée au moyen de la coulisse.

IT chiusura

NL

Dichttrekken van een omringend vistuig d.m.v. de sluitlijn.

PT fechar a rede; fechar a retenida

Operação que consiste no fecho de uma rede de cerco com retenida na sua parte inferior, através da alagem da retenida, após o cerco do cardume.

3374

ES enganchón; embarre

Obstáculo por accidente natural, restos de naufragio u objetos sumergidos, en los que puede enganchar un arte de pesca.

DA hold

Forhindring på bunden.
Man taler i trawlfiskeriet om at få hold eller gå i et hold, når trawlen sætter sig fast.

DE Haker

GR εμπόδιο

Εμπόδιο στον πυθμένα, στο οποίο μπορεί να μπλεχτεί ή να γαντζωθεί ένα αλιευτικό εργαλείο.

EN fastener; obstruction

An obstruction on the sea bed which can foul fishing gear.

FR croche

Obstacle sur le fond dans lequel peut s'accrocher l'engin de pêche.

IT afferratura; presura

NL obstakel; vastloper

Object op de zeebodem dat de voortgang van een gesleept vistuig belet.

PT peguilho

Obstáculo ou irregularidade no fundo do mar e no qual se pode prender uma arte de pesca.

3375

ES embarrar

Enganchar el arte de pesca en un embarre.

DA få hold; gå i et hold

Redskabet sætter sig fast i bunden.

DE einen Haker haben

Festfahren des Grundgeschirrs am Grund.

GR σκαλώνω· μπλέκω

Ακούσιο μπλέξιμο ενός αλιευτικού εργαλείου σε ένα εμπόδιο ή σε μία ανωμαλία του πυθμένα.

EN get fast (verb)

The unintentional fouling of fishing gear on an obstruction on the sea bed.

FR crocher

Accrocher involontairement l'engin de pêche sur un obstacle ou une irrégularité de fond.

IT afferrare; fare presura

NL vastlopen

Het onvrijwillig raken van een obstakel op de zeebodem met een gesleept vistuig, waardoor de visserij-operatie moet worden gestopt.

PT pegar; prender

Acção resultante do facto de uma arte de pesca ficar presa em algum obstáculo ou irregularidade existente no fundo do mar.

3374

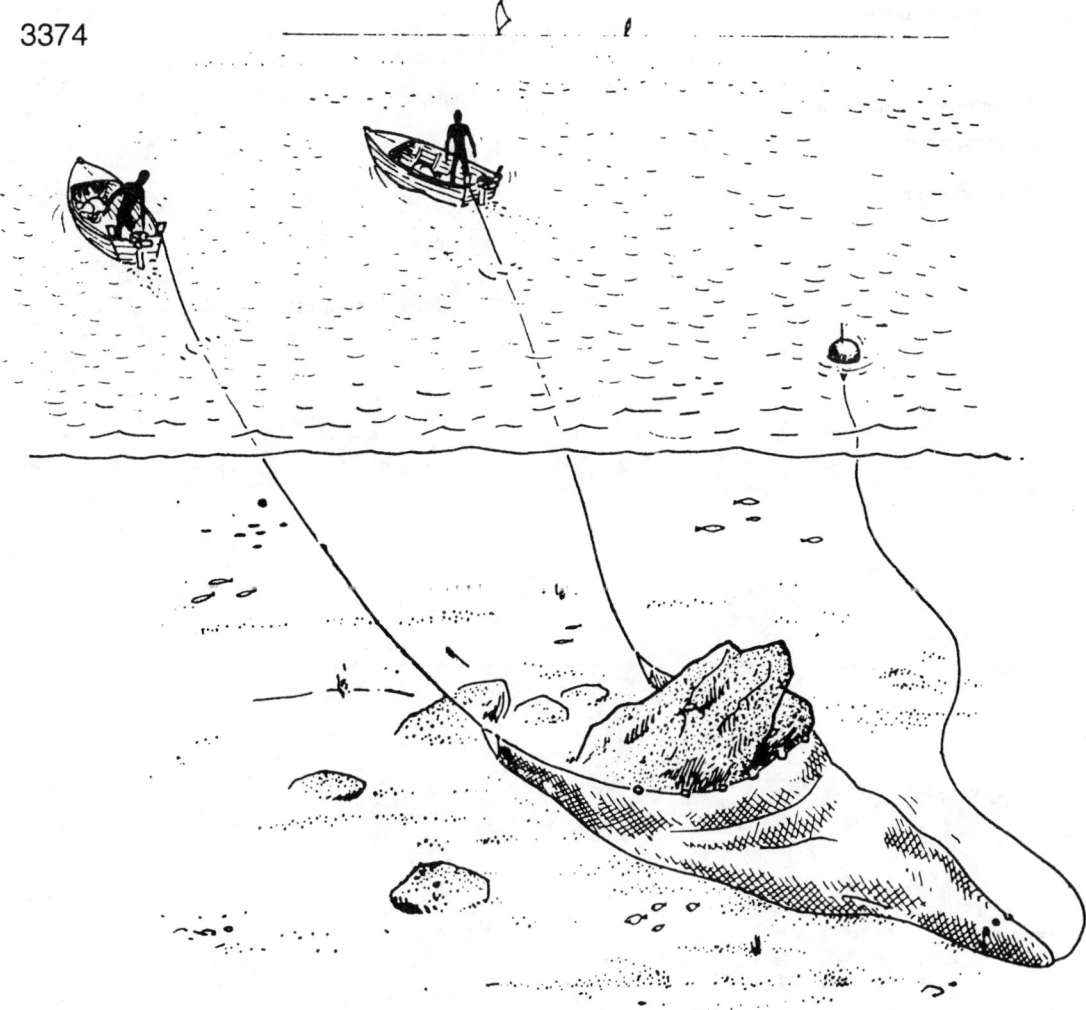

3376

ES boca

Red.

DA åbning

Indgangen til en trawl.

DE Netzmaul; Maulöffnung

Schleppnetz.

GR στόμιο διχτυού τράτας

EN mouth; opening

The open end through which fish enter.
Trawl.

FR ouverture; gueule du chalut

IT apertura; bocca

Rete.

NL netmond; netopening

Opening waardoor de vis het net binnenzwemt.

PT abertura; boca

Rede.

3377

ES trenzar

Hilo.

DA flette

Samlebetegnelse for flere fremstillingsprocesser for liner og tove, der har til fælles, at linen/tovet ikke tvindes eller slås.

DE Seilflechten

Sammelbegriff für unterschiedliche Arten des Verflechtens (z. B. Verkreuzen, Verschlingen) von Seilgarnen oder Seil-Litzen.

GR πλέκω

νήμα, σπάγγο

EN braid (verb); plait (verb)

Twine; yarn.

FR tresser

Fil; ficelle.

IT trecciare

Filo.

NL vlechten

PT trançar; entrançar

Fios.

3378

ES bote auxiliar; panga

Embarcación auxiliar empleada en las maniobras de los buques de pesca al cerco.

Los grandes cerqueros atuneros emplean embarcaciones con grandes potencias (400/600 CV) a las que se llama «pangas».

DA dory

Jolle med en kraftig motor, som bruges i visse notfiskerier (ikke det danske) til at trække i notfartøjet under snurpningen. Doryen skal udøve et træk svarende til trækket i snurpewiren for at forhindre notfartøjet i at blive trukket ind i noten.

DE Skiff

Kleines motorgetriebenes Hilfsboot, das die Fischsuche und das Aussetzen der Ringwade unterstützt.

GR βοηθητική λέμβος

Βοηθητική λέμβος για τον εντοπισμό των ψαριών και το άπλωμα των δικτύων κατά την αλιεία με κυκλικά δίκτυα (γρι-γρι).

Είναι σήμερα κυρίως μηχανοκίνητη, παλαιότερα κωπήλατη, από πλαστικό υλικό, ξύλο ή αλουμίνιο και μήκους έως 8 μέτρα περίπου, εγκατεστημένη ισχύς μηχανής έως 150HP περίπου, ενίοτε υψηλότερη.

EN skiff

Light rowing or sculling boat.

FR annexe de senne; skiff

Embarcation motorisée mise à l'eau avant le filet et servant de point fixe pour le filage de la senne coulissante.

Elle intervient également au cours d'opérations de virage de la coulisse et du filet, en particulier pour maintenir le senneur en dehors du filet, ainsi que durant le salabardage.

IT stazza

Natante ausiliario messo in acqua prima della rete e che serve da punto fisso per calare il cianciolo.

Interviene inoltre in operazioni quali il salpamento del cavo di chiusura e della rete, in particolar modo per mantenere il peschereccio al di fuori della rete, nonché nel corso di operazioni con il coppo.

NL skiff

Kleine open boot, gebruikt bij het vissen met de purse-seine, voor het nauwkeurig lokaliseren van de school en het leiden van de omsingelingsmanoeuvre, zonder de vis te verstoren.

PT chalandra; chata; enviada; embarcação auxiliar; «skiff»

Embarcação motorizada que é lançada à água no início da largada das redes de cerco com retenida e que serve de ponto fixo para a largada. A bordo da chalandra encontra-se a cuba de vante da rede de cerco e a maçariça, que é passada paro o cercador no final do cerco a fim de se iniciar o fecho e a viragem da rede. A chalandra intervém igualmente no decurso das operações de viragem da retenida e da rede, não só para manter o cercador fora da rede como também durante o desenvasar do peixe.

Embarcação auxiliar do cerco com retenida.

3379

ES | **guía para el saco**

Dispositivo construido de tubos de acero para inmovilizar el saco al ser izado a bordo.

Cumple el papel de las «contras», utilizadas con el mismo fín en los buques de costado.

DA |

Anordning til at fastholde posen og hindre den i at svinge faretruende under tømning. Udformningen varierer.

Ikke noget dansk udtryk.

DE | **Steerthalter**

Am Heckportal befestigte bogenförmige Halterung, die das Ausschwingen des Steerts beim Einhieven des Schleppnetzes verhindern soll.

GR | **οδηγός άκρης σάκου τράτας· οδηγός άκρης πετσαλίου τράτας**

Σωληνοειδές μεταλλικό πλαίσιο του τελικού άκρου του σάκου μιας τράτας που επιτρέπει τη διατήρηση της σωληνοειδούς μορφής του κατά την ανέλκυσή του στο σκάφος.

EN | **codend guide**

A curved, tubular device fixed to the stern gantry and used to stabilize the codend when this is hauled on board.

FR | **banane; guide-poche**

Élément tubulaire de forme échancrée, fixé au portique et permettant de stabiliser le cul du chalut lorsque celui-ci a été hissé à bord.

IT | **guida sacco**

Elemento tubolare incavato fissato all'arcone, che consente di stabilizzare il sacco quando questo viene issato a bordo.

Poco usato nella pesca mediterranea.

NL | **kuilvang**

Buisvormige constructie op het dek van een vissersschip, waarin de kuil bij het naar binnen halen wordt opgevangen.

Bij oudere hektrawlers was de kuilvang ook wel onder het visluik geplaatst en kwam dan in de juiste stand na het openen van dit luik.

PT | **forca do pórtico de arrasto**

Estrutura tubular com forma de chanfro, fixada ao pórtico de arrasto e que permite a estabilização do saco da rede de arrasto no momento em que é içado para bordo.

3380

ES | **cosquillera**

Cadena colocada entre las puertas de los buques que pescan gamba o langostino, con objeto de remover el fondo y facilitar la captura de los crustáceos.

También se emplean cadenas similares en la pesquería de peces planos y en la de cefalópodos.

DA | **skrabekæde**

Kæde fastgjort i hver ende til undertællen af en trawl eller et snurrevod. Længden og fastgørelsen er således, at den under fiskeri slæbes i kort afstand foran midten (kværken) af undertællen.

Skrabekæder anvendes i visse bundfiskerier, f.eks. efter fladfisk, jomfruhummer o.l., for at skræmme fangsten op fra bunden.

DE | **Scheuchkette**

Vor dem Grundtau befestigte Kette zum Aufscheuchen von Plattfischen, Garnelen u. ä.

GR | **αλυσίδα παρενόχλησης**

Ελεύθερα κρεμάμενη αλυσίδα που τοποθετείται μπροστά από το άνοιγμα μιας τράτας βυθού για την παρενόχληση των αλιευμάτων βυθού (ψάρια βυθού, γαριδοειδή κλπ.) και διευκόλυνση της αλίευσής τους.

EN | **tickler chain**

Chain towed in front of groundrope to disturb flatfish.

3380

FR chaîne gratteuse

Chaîne libre, remorquée en avant du bourrelet et ayant pour but de soulever du fond les poissons plats ou les crevettes afin de faciliter leur capture.

La chaîne gratteuse peut être simple (cas des chaluts à panneaux) ou multiple (cas des chaluts à perche).

IT catena per la pesca a strascico

Catena situata davanti alla lima da piombo e trainata allo scopo di sollevare dal fondo i pesci piatti o i gamberi onde agevolarne la cattura.

NL wekkerketting

Ketting, bevestigd tussen de sloffen van een boomkor, die tot doel heeft de vis van de bodem op te schrikken. De ketting sleept over de bodem; vaak wordt er een aantal tegelijk gebruikt.

Daarnaast kent het vistuig een reeks soortgelijke kettingen, die aan de onderpees zijn bevestigd, de zgn. kietelaars.

PT duplo arraçal

Corrente(s) de ferro livre(s), rebocada(s) à frente do arraçal com a finalidade de fazer levantar do fundo os peixes chatos ou os camarões e, assim, facilitar a respectiva entrada na rede.

Podem ser simples, como no caso das redes de arrasto com portas, ou múltiplos, como no caso das redes de arrasto de vara.

3381

ES red de tiras; red de cabos; red «spaghetti» *

Arte, generalmente pelágico, en la que el paño de la parte anterior y las bandas se substituyen por cabos o tiras de red enrollada.

* Término familiar.

DA stræbertrawl

Trawl hvori forpartens masker er erstattet af liner, stræbere, der løber mere eller mindre parallelt med slæberetningen. Forekommer i flydetrawl og visse høje bundtrawl.

DE Tauwerk-Netz; Spaghettinetz

Meist pelagisches Schleppnetz, bei dem das Vornetz aus Leinen besteht.

GR σχοινοειδής τράτα· τράτα σπαγγέτι

Τύπος πελαγικής τράτας, ενίοτε και τράτας βυθού, που χαρακτηρίζεται από τη διαμόρφωση του ανοίγματος της εισόδου της τράτας που εδώ αποτελείται από πλεγμένα, ισχυρά σχοινιά, σχεδόν παράλληλα μεταξύ τους, έτσι ώστε να μειώνεται κατά το δυνατόν η αντίσταση του συρόμενου δικτύου σε σχέση με συμβατικές τράτες.

EN rope trawl; rope wing trawl; spaghetti trawl

With the object of saving energy required for towing by decreasing the resistance of the gear, the netting of the front part of the four-seam trawl consists of plaited warps running parallel for a few metres.

3381

FR | **chalut à cordes**

Type de chalut souvent pélagique, mais parfois de fond, dont la partie antérieure et les ailes sont constituées de cordages plus ou moins parallèles reliant la partie en filet aux ralingues d'ouverture.

IT | **rete a corde; rete spaghetti**

Tipo di rete generalmente pelagica, a volte anche a strascico, la cui parte anteriore ed i bracci sono costituiti da corde più o meno parallele che collegano la pezza alle relinghe di apertura.

Poco usata nella pesca italiana.

NL | **lijnennet; lijnentrawl**

Trawl waarin de mazen van het voorste deel, alsmede de vlerken, zijn vervangen door min of meer parallelle lijnen. Deze lijnen hebben tot doel de netweerstand te verminderen en schade bij vissen aan de bodem te beperken.

Dit type vistuig is in Nederland veel beproefd, maar niet commercieel ingezet.

PT | **rede de arrasto de cabos**

Tipo de rede de arrasto pelágico, algumas vezes de arrasto pelo fundo, cuja parte anterior e as asas são constituídas por cabos mais ou menos paralelos, os quais ligam os malheiros da parte posterior da rede com os cabos de abertura (cabo de pana e arraçal).

3382

ES | **red marisquera; red camaronera**

Arte utilizado para la captura de gambas, langostinos y especies afines. El tamaño de sus mallas es adecuado para estas especies. Puede ser una red de arrastre de fondo con puertas, de redes gemelas, de redes remolcadas desde tangones o de redes de vara.

DA | **rejetrawl**

Trawltype tilpasset fangst af rejer.

DE | **Garnelenschleppnetz**

GR | **γαριδότρατα**

Ειδικό δίκτυ-τράτα για την αλίευση γαριδοειδών αλιευμάτων.

EN | **shrimp trawl**

A trawl with a shape and mesh size suitable for catching shrimp and other crustaceans.

FR | **chalut à crevettes**

Type de chalut avec une forme et un maillage adaptés à la capture de ces crustacés.

IT | **rete da gamberi; rete da traino per gamberi**

Rete da traino con i divergenti attaccati direttamente alla rete.

Ogni peschereccio ha due reti.

NL | **garnalennet**

Trawl ontworpen voor de vangst van garnalen.

In Nederland meestal boomkorren.

PT | **rede de arrasto para camarões**

Tipo de rede de arrasto pelo fundo com forma e malhagem especialmente adaptadas à captura de crustáceos.

3383

ES red de prueba; red de muestra

Pequeño arte, utilizado en los buques que emplean redes camaroneras manejadas desde tangones, similar a los dos artes principales.
Manejada desde la popa, se vira durante el arrastre, sin interrumpir el lance, para tener una idea de la calidad y la cantidad capturable.

DA »trynet« *

En ganske lille trawl, der især anvendes i forbindelse med bomtrawlsfiskeri i troperne. Den sættes sammen med de egentlige trawl og hales hyppigt. Fangsten heri gør det muligt at bestemme, hvornår de egentlige net skal tømmes.
* Ikke noget dansk udtryk. Almindeligvis anvendes det engelske »trynet«.

DE Suchnetz; Probenetz

Kleines Netz, das in der Garnelenfischerei verwendet wird, um größere Garnelenansammlungen ausfindig zu machen, indem es in regelmäßigen kurzen Abständen vom Heck des Fahrzeuges aus eingeholt wird.

GR δοκιμαστικό δίκτυ

Μικρό δοκιμαστικό δίκτυ που σύρεται πρύμνηθεν αλιευτικού σκάφους και ανελκύεται περιοδικά ώστε να διαπιστώνεται η σύνθεση και συγκέντρωση των αλιευμάτων, ιδιαίτερα μικροαλιευμάτων και γαριδοειδών.

EN trynet

A small net used to locate large concentrations of shrimp. It is hung from the stern and pulled in at frequent intervals.

FR chalut d'essai

Petit filet manœuvré de l'arrière d'un crevettier, relevé périodiquement à courts intervalles et servant à localiser les concentrations de crevettes.

IT rete di prova

Piccola rete manovrata dalla poppa di un peschereccio e che viene issata periodicamente, a brevi intervalli, per localizzare importanti concentrazioni di gamberi.

NL proefnetje

Klein net dat wordt gebruikt door een garnalenvisser om concentraties garnalen op te sporen.

PT rede de amostrar; rede de prova

Pequena rede de arrasto para camarões, manobrada pela popa dos arrastões para crustáceos com tangões, que é virada com frequência visando a localização de concentrações importantes de camarões.

3384

ES arte de arrastre selectivo

Red de arrastre que, utilizando paneles interiores de separación o mallas de forma o tamaño determinado, permite, bien separar especies diferentes a copos independientes, aprovechando su diferente comportamiento ante el arte, bien dejar escapar los individuos de pequeña talla.

DA selektiv trawl

Trawltype der på forskellig vis er i stand til at sortere fangsten. Udformningen bygger på en viden om fiskenes adfærd og indbefatter ofte indsættelsen af et sorteringspanel eller en sorteringsrist på et eller andet sted i trawlen. Hvis de forskellige fiskearter eller -størrelser opfører sig forskelligt i fangstfasen, kan det evt. lade sig gøre at skille dem i trawlen og lede dem til hver sin pose, eller lede en evt. uønsket fangst ud af posen igen.
Princippet med at kunne sortere fangsten, mens den endnu er levende, forventes at få betydning som redskab i forvaltningen af ressourcerne.

DE Schleppnetz für Artenselektion

GR τράτα (δίκτυ) επιλογής-διαλογής

Εξειδικευμένο δίκτυ επιλογής-διαλογής συγκεκριμένων αλιευμάτων ορισμένου ελάχιστου μεγέθους και είδους, έτσι ώστε να διευκολύνεται η αλίευση επιλεγμένων ειδών αλιευμάτων και να προστατεύονται άλλα είδη.

EN selective trawl

Used to catch certain small species, while allowing larger ones to escape, thus facilitating sorting and protecting species.

FR chalut sélectif

Chalut conçu spécialement pour capturer certaines espèces de petite taille et permettre l'échappement d'oopòooo do pluo grando taillo, pour faoilitor lo tri ot/ ou protéger une espèce.
Cette sélection est basée sur les comportements différents des espèces, grâce à une forme particulière du chalut ou par le montage d'un dispositif séparateur. Si l'espèce rejetée était de petite taille, la sélectivité du cul suffirait dans la plupart des cas.

3384

IT rete da traino selettiva

Particolare tipo di rete concepita per catturare talune specie di piccole dimensioni, lasciando libere quelle di dimensioni maggiori, allo scopo di agevolare la cernita e/o proteggere una specie.

La selezione si basa sui diversi comportamenti delle specie, grazie ad una particolare forma della rete o all'aggiunta di un dispositivo di separazione. Se la specie è di piccole dimensioni, la selettività del sacco è sufficiente nella maggior parte dei casi.

NL selectieve trawl

Trawl met voorzieningen die de vangst van bepaalde soorten of afmetingen bevordert boven andere.

De selectie kan worden veroorzaakt door verschil in gedrag van vis, t.a.v. het naderende vistuig en/of door de grootte of vorm van mazen voor of in de kuil, in relatie tot de afmetingen van de vis.

PT rede de arrasto selectiva

Rede de arrasto especialmente concebida para apenas capturar certas espécies e/ou classes de comprimento de uma dada espécie, com a finalidade de proteger uma espécie e/ou de facilitar a escolha das capturas.

Este selecção tem por base os diferentes comportamentos das espécies alvo e é conseguida graças a uma forma particular da rede de arrasto ou à montagem de painéis separadores. Se a espécie rejeitada é de pequeno tamanho, a selectividade no saco da rede será, na maior parte dos casos, suficiente.

3385

ES factor de selectividad

Cociente entre la longitud de un pez, de cuya talla el arte retiene el 50% de los capturados, y la medida de la abertura de la malla, expresados en milímetros.

DA selektionsfaktor

Forholdet mellem den længde af fisken, hvor sandsynligheden er 50% for at blive tilbageholdt i en trawlpose - og maskestørrelsen (indvendigt mål) i posen.

DE Selektionsfaktor

GR συντελεστής επιλογής-διαλογής

Συντελεστής που εκφράζει το ελάχιστο μήκος ενός είδους αλιευμάτων για τη διατήρηση μέσα στο δίκτυ τουλάχιστον 50 % των εγκλωβισμένων αλιευμάτων και το αντίστοιχο μέγεθος των ανοιγμάτων (ματιών) του δικτύου, εκφρασμένο στις ίδιες μονάδες (συνήθως MM).

EN selectivity factor

FR facteur de sélectivité

Relation entre une certaine longueur des poissons d'une espèce, pour laquelle 50% des individus sont retenus à l'intérieur du cul du chalut, et la dimension correspondante de l'ouverture des mailles de la poche, exprimées dans les mêmes unités.

IT fattore di selettività

Rapporto tra una determinata lunghezza dei pesci di una data specie, della quale il 50% della cattura è mantenuta nel sacco delle rete, e la dimensione corrispondente dell'apertura delle maglie del sacco, espresse nelle stesse unità di misura.

NL selectiefaktor

Het quotiënt van de 50%-retentielengte en de maaswijdte van de kuil.

De 50%-retentielengte is de volgens een bepaalde wiskundige functie (b.v. logistische) berekende lengte, waarbij 50% van de vis door de mazen van de kuil wordt tegengehouden.

PT factor de selectividade

Relação entre um certo comprimento dos peixes de uma dada espécie, para o qual 50% dos indivíduos são retidos no interior do saco da rede de arrasto e a dimensão correspondente do vazio das malhas do saco, expressos nas mesmas unidades.

3386

ES enganche de pie de gallo

Pieza metálica colocada en la parte trasera de la cara de fuga de la puerta. Puede ser una anilla o una pieza con 3 o 4 orificios, que permiten variar el ángulo de ataque de la puerta, según cual sea usado para afirmar el pie de gallo.

DA beslag; fastgørelsespunkt for kædetrækket på en trawlskovl

Øjer eller beslag med flere huller på bagkanten eller bagsiden af en trawlskovl. Tjener til fastgørelse af agtertrækket. Placeringen er medbestemmende for skovlens angrebsvinkel med slæberetningen.

DE Fixierpunkt für Scherbretthahnepoten

GR δακτύλιος πόρτας· σκουλαρίκι τράτας

Χαλύβδινος δακτύλιος επί της εξωτερικής πλευράς της πόρτας μιας τράτας για τη σύνδεση των ολκών της πόρτας με το κυρίως δίκτυ της τράτας.

Η θέση του δακτυλίου καθορίζει τη γωνία κλίσης της πόρτας ως προς την κατεύθυνση έλξης της τράτας.

EN backstrop ring

Steel ring on back of otter board for attachment of backstrop.

FR fixation de patte de panneau

Ferrure, comportant habituellement un œil, placée sur la face externe ou au bord arrière du panneau de chalut et servant à la fixation d'une patte de panneau.

Son emplacement détermine l'angle d'incidence du panneau.

IT punto di attacco della braga del divergente

Elemento metallico, generalmente dotato di un occhiello, posto sul lato esterno o sul bordo posteriore del divergente e usato per fissare una braga del divergente.

La sua ubicazione determina l'angolo di incidenza del divergente. Pesca al traino.

NL bevestigingsoog voor de bordstrop

Een aan de buitenzijde of achterkant van een visbord bevestigd oog, waaraan de bordstrop wordt bevestigd.

Het kan ook een stalen plaatconstructie met ingeboorde gaten zijn.

PT ponto de fixação do brinco da porta; régua de fixação

Estrutura ou ferragem, normalmente dotada de um orifício, implantada na face externa ou no bordo posterior da porta de arrasto, e que se destina à fixação do brinco da porta.

O seu posicionamento determina o ângulo de ataque da porta de arrasto.

3387

ES estibador

Sistema formado por dos rolines verticales que, al ser movidos en sincronización con el tambor de la maquinilla, permiten un correcto enrollamiento del cable.

El movimiento del estibador puede ser automático o manual.

DA styrerulle

Består af to parallelle ruller af jern monteret lige foran wiretromlen på et spil, vinkelret på tromlens aksel. Wiren lægges imellem rullerne under indhalingen og ved at flytte rullerne fra side til side styres pålægningen af wiren.

DE Leitgeschirr; Aufleitgeschirr; Aufleitvorrichtung

Mechanische Vorrichtung an der Kurrleinentrommel, die ein gleichmäßiges Aufwickeln der Leinen gewährt.
Siehe auch 3126.

GR οδηγός σχοινιών

Αυτόματος ή χειροκίνητος μηχανισμός που επιτρέπει την ορθή οδήγηση των σχοινιών προς το τύμπανο ενός εργάτη.

EN guiding-on gear

A mechanical system enabling cables to be wound evenly on the drum.

FR guide-câble

Système mécanique permettant de régulariser l'enroulement des câbles sur un tambour de treuil.

IT guida cavo

Sistema meccanico che consente un avvolgimento uniforme dei cavi sul tamburo del verricello.

NL kabelgeleider

Mechanische constructie die ervoor zorgt dat een kabel gelijkmatig over een trommel wordt opgewonden.
Deze bestaat doorgaans uit geleiderollen, waardoor de kabel loopt, die zijdelings heen en weer worden bewogen terwijl de trommel draait.

PT espalha-cabos

Sistema mecânico dos guinchos que permite distribuir uniformemente o enrolamento de cabos sobre os repectivos tambores.

3388

ES sisga; tirador

Usado en los buques de pesca de pareja, para pasar elementos del uno al otro. También se emplea para enviar cabos a tierra. Consta de un cabo fino con un peso en su extremo que lanzado a distancia permite establecer comunicación con el otro buque o con tierra.

DA kasteline

Smækker line med en tung genstand i den ene ende, f.eks. en lille pose med sand. Anvendes til at skabe en forbindelse med et andet fartøj, således at en kraftigere line, tov eller wire kan overføres, f.eks. under partrawling.

DE Wurfleine

GR κάβος έλξης· ρυμουλκείο

Ισχυρό σχοινί (κάβος) για τη ρυμούλκηση άλλου σκάφους ή τη διατήρηση μιας σταθερής απόστασης μεταξύ δύο κινούμενων σκαφών, π.χ. κατά τη ζευγαρωτή αλιεία.

EN hauling line

FR lance-amarre

Filin muni d'une boule tressée à une extrémité. Sert à passer un cordage (bras, fune ou rapporteur) d'un bateau à l'autre, en particulier dans la pêche en bœufs.

IT sacchetto

Cavo destinato a far passare una corda (calamento, cavo di traino o penzolo) da un peschereccio all'altro, particolarmente nella pesca a coppia.

NL werplijn

Lijn die in de spanvisserij wordt gebruikt om één van de voorlopers van het vistuig van het ene schip naar het andere te verplaatsen.

PT pinha; mensageiro com pinha; retenida com pinha

Fio ou cabo munido, numa das extremidades, com uma bola (pinha) entrançada. Utiliza-se para passar um cabo (malheta, cabo real ou brinco) de um barco para outro, em particular no arrasto de parelha.

3389

ES | eslabón de patente

Malla formada por dos partes unidas por piezas machinembradas, que permiten zafar ambas con facilidad, dejando libres los elementos empalmados por la misma.

DA | samleled

Kædeled, der består af to halvdele, der hver især er åbne, men til hver sin side. Samles ved hjælp af nitter og tillader derved en hurtig samling af en kæde eller lignende.

Betegnelsen samleled omfatter flere typer led med samme funktion.

DE | Steckglied; Kettenschloß

Verbindungselement aus zwei offenen Teilen, das zum rasch lösbaren Verbinden von Ketten und Seilen dient.

GR | κρίκος λυόμενος

Λυόμενος κρίκος που επιτρέπει την εύκολη σύνδεση και αποσύνδεση δύο στοιχείων. Αποτελείται από δύο μέρη που συνδέονται μεταξύ τους μέσω ενός μηχανισμού ασφάλισης, π.χ. ένα αγκύλιο με κοχλιωτό πείρο στα άκρα.

EN | false link; split link

A link consisting of two open parts; used to quickly connect two sections of gear.

FR | maille brisée; maille brisée à river

Maille formée de deux pièces ouvertes dont l'assemblage par rivetage permet de relier rapidement deux éléments du gréement.

IT | maglia falsa

Maglia costituita da due parti aperte, la cui rivettatura consente di collegare rapidamente due elementi dell'attrezzatura.

NL | verbindingsschalm

Schalm voorzien van een v-vormige opening, waardoor een soortgelijke schalm snel kan worden gekoppeld.

PT | elo patente

Elo formado por duas peças abertas cuja união, através de rebites, permite ligar rapidamente dois elementos de um armamento.

3390

ES | malla abierta

Malla con un corte o rebaje que facilita su empalme con otra similar.

DA | tovled

Kædeled med en åbning i den ene side. To sådanne led kan glide ind i hinanden, når de holdes vinkelret på hverandre. Herved kan foretages en hurtig samling af kæder eller tove. Anvendes således ved samlingen af tovene i snurrevodfiskeriet. Denne form for samling er imidlertid ikke så pålidelig som visse andre samlinger.

DE | offenes Kettenglied

GR | κρίκος ανοικτός

Κρίκος με άνοιγμα πλαγίως για την εύκολη (χρονικά περιορισμένη) σύνδεση και αποσύνδεση δύο στοιχείων.

EN | open link; C-hook

A link with an opening on one side; used to quickly connect with another link as a temporary measure.

FR | maille coupée; maille ouverte

Maille dans laquelle une ouverture a été ménagée sur le côté. Permet de réaliser rapidement un assemblage temporaire avec une autre maille.

La tenue d'une maille coupée est moins sûre que celle d'une maille à méplat sur croc en G ou de deux mailles à sifflet.

IT | maglia aperta

Maglia dotata di un'apertura laterale. Consente di effettuare rapidamente una giunzione temporanea con un'altra maglia.

NL | C-haak

Haakvormig verbindingselement met een open kant, geschikt voor snelle koppeling of loskoppeling. Het open deel kan voorzien zijn van een verend sluitelement (veiligheidshaak).

Vaak wordt een snelle verbinding gemaakt d.m.v. een G-haak met bijpassende schalm (= patent schalm). Deze kan slechts in één stand in elkaar schuiven en is daarom veiliger.

PT | elo aberto

Elo dotado de uma abertura lateral (não biselada). Permite a realização rápida de uma ligação temporária com um outro elo.

A eficiência dos elos abertos entre si é menor que a de um elo de ligação num gato ou de dois elos abertos em bisel.

3391

ES metros por kilo; runnage

Número de metros de un determinado hilo que pesan un kilogramo. Este número se emplea como unidad para clasificar hilos empleados en la construcción de artes de pesca. Actualmente se emplea también el sistema Denier y se recomienda el uso del Rtex para la clasificación de hilos por su grosor.

DA løbelængde (m/kg)

Den længde af en line som vejer 1 kg. Den tilsvarende R-tex er givet ved formlen:

$$R\text{-}tex = \frac{1\ 000\ 000}{m/kg}$$

DE Lauflänge (m/kg)

GR τρέχον μέτρο

Τα τρέχοντα μέτρα ανά μονάδα βάρους (m/kg) των κλωσμάτων ενός δικτύου.

Χαμηλός αριθμός τρεχόντων μέτρων, ανά μονάδα βάρους, σημαίνει ισχυρή αντίσταση έναντι θραύσης του κλώσματος, π.χ. ένα κλώσμα από πολυαμίδιο (POLYAMIDE), χωρίς κόμβους, των 2 500 m/Kg αντέχει σε φορτίο θραύσης 21 KGF, ενώ για 1 000 m/kg ισχύει φορτίο θραύσης 49 KGF. Ο αριθμός τρεχόντων μέτρων ανά kg συνδέεται με τον αριθμό RTEX μέσω του τύπου: RTEX = 1 000 000/ (m/kg), π.χ. RTEX = 400 σημαίνει 2 500 m/kg.

EN runnage; m/kg

Length in metres of a kilogram of yarn. The equivalent in tex is obtained using the formula

$$R\text{-}tex = \frac{1\ 000\ 000}{m/kg}$$

FR métrage au kg; m/kg

Longueur en mètres d'un écheveau de fil pesant 1 kg. Par exemple: 2000 m/kg. L'équivalence en tex est donnée par la formule $tex = \dfrac{1\ 000\ 000}{m/kg}$

IT titolo metrico

Lunghezza in metri di un kg di filo.

L'equivalente in R tex è dato dalla formula :

$$R\ tex = \frac{1\ 000\ 000}{m/kg}$$

NL looplengte (m/kg)

Aantal meters van een garen per kg. De conversie naar de texwaarde wordt gegeven door de formule:

$$tex = \frac{1\ 000\ 000}{m/kg}$$

Er is een verschil tussen tex en R-tex (de resulterende tex-waarde van een geslagen of gevlochten touw). R-tex volgt uit de methode van samenstellen.

PT metragem por quilo; m/kg

Comprimento em metros de uma amostra de fio com 1 kg de peso. Por exemplo: 2000 m/kg. O equivalente em R-tex é dado pelo fórmula

$$R\text{-}tex = \frac{1\ 000\ 000}{m/kg}$$

3392

ES carrete

Aparato colocado en la regala del buque o en una caña, que sirve para recoger el hilo de los aparejos de anzuelo y almacenarlo enrollado.

DA fiskehjul

Mindre hjul til håndtering af en fiskeline. Anvendes ved krogfiskeri og sættes på skibssiden eller på en fiskestang.

DE Rolle

GR έντροχον· μπαστέκα

Οδηγητικός, σταθερός τρόχιλος για το χειρισμό σχοινιών ή δικτύων. Ευρίσκεται συχνά επί της κουπαστής αλιευτικών σκαφών.

EN reel

A device to haul lines.

FR moulinet

Appareil utilisé pour la manœuvre des lignes et servant à stocker le fil. Entraîné à la main ou mécaniquement, il peut être fixé sur une canne ou sur le plat-bord du bateau.

IT mulinello

Attrezzo usato per manovrare le lenze e per avvolgere il filo. Azionato a mano o meccanicamente, può essere fissato ad una canna o sul capodibanda del peschereccio.

NL haspel; trommel

Cilindrisch werktuig ter bediening en opslag van kabels of lijnen.

PT molinete; carreto

Equipamento utilizado para a manobra de linhas de pesca e que se destina ao respectivo armazenamento. Operado à mão ou mecanicamente, pode ser fixado numa cana de pesca ou sobre a borda falsa de uma embarcação.

3393

ES paño principal

Paño que forma el cuerpo principal de una red de cerco. Suele estar constituido por varios paños unidos entre sí, del mismo mallero y grosor de hilo.

DA hovednot

Midterste og mest omfattende del af en not.

DE Mittelstück einer Wade

GR κύριο τμήμα δικτύου

Κυρίως τμήμα ενός κυκλικού δικτύου (γρι-γρι), αποτελούμενο από δίκτυ μεγέθους και μορφής ανάλογα με την εκάστοτε μέθοδο αλιείας και τον τύπο των αλιευμάτων. Οριοθετείται στο άνω μέρος από τη γραμμή των φελλών και στο κάτω/πλάγιο μέρος από τη γραμμή των βαριδίων και των σχοινιών σφιξίματος και έλξης του δικτύου.

EN main netting body

FR nappe principale

Nappe de filet dont le maillage et la force de fil s'appliquent à la majeure partie du corps de la senne.

IT pezza centrale di una rete a circuizione

Termine poco usato in Italia.

NL hoofdnet van de zegen

Grootste gedeelte van een zegennet, dat de vis geleidt naar de zak, waar ze gevangen wordt

PT corpo; talhões

Pano de rede cuja malhagem e titulação do fio é comum à maior parte do corpo da rede de cerco.
Redes de cerco com retenida.

3394

corte e izado del copo; saco

Método utilizado en caso de mucha captura. Se ata la sereta y se arroja el copo al agua cobrando de la manga para que el pescado corra hacia el copo. Cuando está lleno se iza de nuevo a bordo y así se repite la operación hasta terminar de meter la captura.

DA **bjærgning i løft; takle ind i løft**

Den operation at tage fangsten ind i portioner ved afsnøring af løftet i fangstposen.

DE **Steertunterteilung**

Unterteilen des Steertfanges in hievfähige Portionen.

GR **εκκένωση σάκου τράτας**

Τμηματική εκκένωση του τελικού άκρου (κν. πετσάλι) του σάκου μιας τράτας όταν παρουσιάζεται υπερπλήρωση του σάκου με ψάρια.

EN **codend lift; lift of codend**

Emptying the codend in several steps when there is a large catch.

FR **palanquée**

Vidage du chalut par étapes successives lorsque la prise est grosse.

IT **taglio della saccata**

Operazione che si effettua quando la cattura è troppo grande per essere salpata in un'unica operazione.

Rete da traino.

NL **halen in delen; jojoën; «pakken»**

Methode van binnenhalen van de vangst door het op en neer bewegen van de jomperdraad, vastgemaakt aan de verdeelstrop van de kuil en de zgn. jojo, die om het netwerk van de tunnel wordt geslagen.

Deze methode is standaard op Nederlandse hektrawlers die geen «slipway» hebben, en dus niet de gehele kuil in eenmaal kunnen scheephalen. Het voorkomt kwaliteitsverlies van de vangst (met name haring) door grote druk. Tegenwoordig wordt de vangst steeds meer binnengehaald met behulp van een vispomp.

PT **sacada; saco**

Acção de vazar um saco de rede de arrasto no caso de grandes volumes de capturas obrigarem a efectuar várias sacadas.

A expressão utilizada é «fazer um dado número de sacadas» ou «fazer um dado número de sacos».

3395

ES **saco**

Cada una de las viradas del copo lleno de pescado de un arrastrero convencional.

La captura de un lance puede componerse de varios sacos, es decir, que para vaciar el arte son necesarias varias viradas del copo completo.

DA **løft**

Betegnelse for den del af fangsten, som løftes ind på en gang under trawlfiskeriet. Ved store fangster kan hele posen ikke tømmes på en gang, og det er nødvendigt at tage den ind i flere løft.

Da det enkelte fartøj tager nogenlunde den samme mængde ind i hvert løft, bliver betegnelsen nærmest en enhed, hvorefter fangstens størrelse opgøres.

DE **Teilsteert**

GR **αλιεία με πυροφάνι**

Μέθοδος αλιείας με τη βοήθεια τεχνητού φωτός που προσελκύει τα ψάρια στην εστία του φωτός όταν τριγύρω επικρατεί σκοτάδι.

Εφαρμόζεται ως αλιεία με δίκτυα (γρι-γρι νύχτας), καθετή-πετονιά, καμάκι ή απόχη.

EN **codend lift**

The part of a trawl net containing a certain quantity of fish which is hauled on board in a single operation.

FR **palanquée**

Cul de chalut contenant une certaine quantité de poissons mise à bord en même temps.

Dans le cas de grosses captures, sur certains chalutiers, il est nécessaire de faire plusieurs palanquées.

IT **taglio della saccata**

Parte della saccata che viene salpata in una singola operazione.

NL **pak; zak**

PT **sacada**

Parte do saco da rede de arrasto, contendo uma certa quantidade de peixe, que é virada para bordo de uma só vez.

No caso de grandes volumes de capturas, obtidas por certos arrastões, é necessário efectuar várias sacadas.

3396

potencia; pescante

Elemento que sostiene las pastecas por las que se guían a la maquinilla los extremos de la jareta y de los cabos de un arte de cerco. Tiene forma de T y puede ser abatible hacia el interior del buque.

DA notgalge

T-formet opstander på siden af et notfartøj til montering af notblokke, som snurpewiren løber i.

DE Schnürleinengalgen

GR επωτίδα δικτύων γρι-γρι· καπόνι δικτύων γρι-γρι

Επωτίδα για την ανέλκυση δικτύων γρι-γρι. Επ' αυτής στηρίζονται διάφοροι τρόχιλοι για την έλξη σχοινιών και δικτύων. Ορισμένοι τρόχιλοι μπορεί να έχουν υδραυλική κίνηση.

EN purse line davit

FR potence

Sorte de bossoir en T, rabattable ou non, supportant les poulies dans lesquelles passent la coulisse et, dans certains cas, le câble de remorque.
De senne.

IT archetto del cianciolo

Porta pulegge, abbattibile o no, a forma di T che serve da supporto ai cavi nelle operazioni di pesca con reti a circuizione.

NL sluitlijndavit

T-vormige galg met twee schijven, waardoor de sluitlijn van een ringzegen door de sluitlijnlier wordt ingehaald. Deze davit bevindt zich aan de verschansing van het schip (meestal op het voorschip).

PT cornuda; cruzeta

Estrutura com a forma de T, rebatível ou não, que serve de suporte às patesgas por onde passa a retenida da rede de cerco e, em certos casos, a maçarica e o cerrador, durante a viragem da rede.

3397

ES cañón para anillas de jareta

DA ringnål

Kraftig jernstang, der monteres vandret, pegende agterud på siden af et notfartøj ud for notringen. Notringene trækkes på ringnålen, hvorved det gøres lettere at få snurpewiren gennem inden udsætningen af noten.

DE Wadenringstange

GR ράβδος δακτυλίων γρι-γρι

Χαλύβδινη ράβδος στην πρύμνη αλιευτικών γρι-γρι, κυρίως κατά μήκος της κουπαστής, για την αποθήκευση των δακτυλίων του σάκου δικτύων γρι-γρι.

EN shooting bar; ring stand

A device placed along the gunwale to hold the purse seine rings.

FR ratelier à anneaux

Barre d'acier placée sur le plat-bord d'un senneur à l'arrière et sur laquelle sont enfilés les anneaux de coulisse. Le ratelier est fixe ou peut être débordé pour le filage de la senne.
Ce dispositif facilite le passage de la coulisse dans les anneaux, lors du filage de la senne.

IT barra guida anelli

Barra in acciaio situata sul capodibanda di poppa di un peschereccio a cianciolo e sulla quale sono infilati gli anelli del cavo di chiusura. La barra è fissa o può essere fatta ruotare fuori bordo per calare il cianciolo.

NL schietstang

Stalen stang waarop de sluitingen van een ringzegennet worden geschoven.

PT ferro das argolas; calha das argolas

Barra de aço colocada sobre a borda falsa e para a popa dos cercadores de rede com retenida, na qual são enfiadas as argolas da retenida.
Esta barra pode ser fixa no convés, ou móvel de modo a poder ser rodada para fora de borda durante a largada da rede de cerco, o que facilita a passagem da retenida nas argolas. Pode também tratar-se de uma calha onde as referidas argolas são colocadas. Cercadores de rede com retenida.

3398

ES zapata; zapatilla

DA køl; sko

Kraftigt stykke jern monteret på underkanten af en trawlskovl. Skal dels fungere som vægt, dels tage noget af sliddet på skovlen. Består ofte af to dele: selve kølen (vægten) og slidkølen af meget hårdt jern. På visse trawlskovle er den delt op i sektioner, der hver for sig kan udskiftes.

DE Scherbrettsohle

GR τρόπιδα πόρτας τράτας· σόλα πόρτας τράτας

Κάτω άκρο της πόρτας μιας τράτας βυθού υπό τη μορφή οριζόντιου πτερυγίου, έτσι ώστε να διευκολύνεται η σύρση της τράτας σε σταθερή απόσταση από τον βυθό της θάλασσας.

EN shoe plate; keel

Of otterboard.

FR semelle

Pièce d'acier, de largeur et d'épaisseur variables, placée à la partie inférieure du panneau et assurant le contact avec le fond. Peut être constituée d'un seul morceau, ou de plusieurs éléments démontables.

IT scarpa del divergente

Elemento in acciaio, di larghezza e spessore variabili, posto nella parte inferiore del divergente per assicurare il contatto con il fondo. Può essere costituito da un solo pezzo o da più pezzi smontabili.

NL slee; schaats; loper

Zware metalen strip bevestigd aan de onderkant van visborden.

PT sapata; arrasto; rastro

Peça de aço, com largura e espessura variáveis, colocada na parte inferior das portas de arrasto e que assegura o respectivo contacto com o fundo. Pode ser constituída por uma única peça ou por vários elementos desmontáveis.

Portas de arrasto.

3399

ES numeración del hilo

DA garnnummer

Betegnelse for tykkelsen (og typen) af et garn. Flere systemer med forskelligt udgangspunkt finder anvendelse ved denne angivelse, men anbefalingerne fra ISO foreskriver anvendelsen af R-tex (den resulterende lineære massefylde, se 3343.

DE Garnnummerierung

Numerierung zur Bezeichnung der Feinheit von Netzgarnen.

GR κωδικός μεγέθους κλώσματος

Αριθμητική κωδικοποίηση του μεγέθους (διαμέτρου) των κλωσμάτων ενός σχοινιού.

Η συνήθης κωδικοποίηση γίνεται με αναφορά στα τρέχοντα μέτρα ανά κιλό κλώσματος (m/Kg) ή κατά το διεθνές σύστημα μέτρησης ISO σε R-TEX.

EN numbering of yarn

Numbering to designate the thickness of a yarn.

FR titre du fil

Désignation numérique exprimant la grosseur d'un fil.

En France, on utilise surtout le métrage par kg (m/kg), mais le système recommandé par l'ISO est le tex résultant (R-tex).

IT titolo di un filo

Grandezza numerica che definisce lo spessore di un filo.

In Italia si usa prevalentemente il titolo in tex.

NL garennummer; titer

Getal dat het verband tussen de lengte en het gewicht van een garen weergeeft. Volgens de ISO-normen uitgedrukt in R-tex, vaak in m/kg.

In Nederland wordt de dikte van visserijgarens vaak aangegeven in Denier. Veel gebruikt wordt garen in veelvoud van 210 Denier, aangeduid als 210/3, 210/10, ..., 210/132 enz. De eenheid 1 Denier is het gewicht in gram van een filament van 9000 m lengte.

PT titulação; título

Designação numérica que exprime a espessura de um fio.

Em Portugal utiliza-se normalmente o diâmetro e a metragem por quilo (m/kg), mas o sistema recomendado pela ISO é o tex resultante (R-tex).

3400

ES tracción

DA belastning; træk

Kraften som påvirker en belastet line, et tov eller en wire. F.eks. i wirerne på en trawl.

DE Zugkraft

GR δύναμη έλξης

Η απαιτούμενη δύναμη για την έλξη ενός αντικειμένου, π.χ. η εξασκούμενη στα σχοινιά δύναμη για την έλξη μιας τράτας.

EN pull; traction; machine effort

Force exerted on a trailer rope or by an engine.

FR traction

Force exercée sur un câble de remorque ou de manœuvre d'un engin. Par exemple: traction exercée sur les funes d'un chalut de pêche.

IT forza di traino; forza di tiro

Forza esercitata su un cavo di rimorchio o di manovra di un attrezzo. Ad esempio: la trazione esercitata sui cavi di una rete da traino.

NL trekkracht

Kracht die door middel van lijnen of kabels wordt overgebracht, b.v. op een vistuig via de vislijnen.

PT tracção

Força que é exercida sobre um cabo de reboque ou de manobra de uma arte de pesca. Exemplo: tracção exercida em pesca sobre os cabos reais de uma rede de arrasto.

3401

ES resistencia

Del aparejo o de la red.

DA modstand

Den ene af de to kræfter, der udgør den hydrodynamiske modstand. Den er rettet bagud og er parallel med slæberetningen.

DE Widerstand; Zug

GR αντίσταση (υδροδυναμική)

Υδροδυναμική δύναμη που εξασκείται σε ένα κινούμενο σώμα ενάντια στην κατεύθυνση κίνησής του. Εφόσον το σώμα κινείται εξ ολοκλήρου μέσα σε ένα ρευστό, χωρίς ή μακριά από ελεύθερες επιφάνειες, η αντίσταση οφείλεται αποκλειστικά στην τύρβη (ιξώδες) του ρευστού και τις υδροδυναμικές τάσεις που δημιουργούνται στην επιφάνεια του κινούμενου σώματος, ή τις δημιουργούμενες δίνες εντός του ρευστού.

Η υδροδυναμική αντίσταση είναι ανάλογη με το τετράγωνο της ταχύτητας του σώματος, την βρεχόμενη επιφάνεια, τη μορφή του σχήματος και την πυκνότητα του ρευστού. Για τα αλιευτικά σκάφη είναι σημαντική η αντίσταση του κινούμενου σκάφους και η αντίσταση του συρόμενου δικτύου (τράτα).

EN drag

Resistance to motion through a fluid.

FR traînée

L'une des deux forces composantes de la résistance hydrodynamique. La traînée est dirigée vers l'arrière et sa direction est parallèle au déplacement de l'engin.

IT resistenza al traino

Resistenza che un attrezzo da traino oppone durante la pesca.

NL weerstand

De component van de hydrodynamische reactiekracht tegengesteld aan de bewegingsrichting van een lichaam in een vloeistof.

De kracht loodrecht op deze richting wordt liftkracht genoemd.

PT resistência ao avanço; resistência à tracção

É uma das duas forças componentes da resistência hidrodinâmica. A resistência ao avanço é uma força dirigida para trás e a respectiva direcção é paralela ao deslocamento da arte de pesca.

3402

ES maquinilla de cerco con jareta

DA notspil; wirespil

Spil på et notfartøj til indhaling af snurpewiren.

DE Schnürleinenwinde; Ringwadenwinde

GR βαρούλκο έλξης κυκλικών δικτύων

Βαρούλκο ανέλκυσης κυκλικών δικτύων (γρι-γρι) αποτελούμενο από δύο ή τρία τύμπανα που χρησιμοποιούνται για την έλξη των σχοινιών, καθώς και του σάκου των δικτύων.

EN purse seine winch

FR treuil de senne coulissante

Treuil à deux ou trois tambours destiné à virer la coulisse sur les senneurs. Chaque tambour doit pouvoir contenir la coulisse entière. Le troisième tambour permet de virer la remorque de senne séparément.

IT verricello per cianciolo

Verricello destinato a calare, salpare e contenere il cavo di chiusura della rete a circuizione.

NL lier; ringzegenlier

Lier met twee of drie trommels voor het halen en vieren van ringzegennetten. Het vistuig kan in zijn geheel op iedere trommel worden opgewonden. De ringzegenlijn kan dan met de derde trommel worden bediend.
De ringzegen wordt in de Nederlandse zeevisserij niet gebruikt.

PT guincho da retenida

Guincho com dois ou três tambores que se destina a virar a retenida das redes de cerco. Cada tambor deve poder armazenar o comprimento total da retenida. O terceiro tambor destina-se a virar a maçarica e, eventualmente, o cerrador.

3403

ES maquinilla para red de cerco danesa

DA vodspil

Spil til indhaling af vodtovene. Spilakslen er forsynet med to spilkopper, hvorover vodtovene føres under indhalingen. Kompletteres af to tovtromler, som kan opbevare de lange vodtove.

DE Schnürleinenholer; Snurrewadenwinde; Snurreleinenholer

GR βαρούλκο έλξης σχοινιών κυκλικών δικτύων (κν. γρι-γρι)

Βαρούλκο έλξης των σχοινιών στον πυθμένα των κυκλικών δικτύων (γρι-γρι), έτσι ώστε να κλείνει ο σάκος προ της ανέλκυσής του στο σκάφος.

EN seine net winch

FR treuil de senne danoise

Treuil à deux poupées servant à virer les longs cordages de rabattement du poisson et du virage du filet dans la pêche à la senne danoise. Il est complété par deux loveurs ou deux tambours pour le stockage de ces cordages.

IT verricello per sciabica danese

Verricello a due campane usato per virare i lunghi cavi nonché la rete nella pesca con sciabica danese. È completato da due contenitori o due tamburi per raccogliere il cordame.

NL lier voor de Deense zegen; «Snurrevaad"-lier

Lier bestaande uit twee onafhankelijke haspels waarop de zegentouwen van de Deense zegen worden opgeslagen, en met een onafhankelijk draaiende verhaalkop. Het voortuig van het net bestaat o.a. uit twee zeer lange zegentouwen, die over de bodem worden gesleept om de vis naar het net te geleiden. De trommels kunnen op verschillende inschakelbare snelheden winden.

PT guincho de rede de cerco dinamarquesa

Guincho equipado com dois cabeços que se destina a virar os longos cabos de arrasto das redes de cerco dinamarquesas. Este guincho é complementado por duas caixas ou por dois tambores para armazenamento dos cabos de arrasto.

3404

ES maquinilla independiente

Maquinilla de un solo carretel dispuesta, según convenga, para que el cable pase directamente a la polea del pescante.

DA splitwinch

Spil med to tromler, der kan køre uafhængigt af hinanden. Anvendes f.eks. som wirespil i trawlfiskeriet, hvor det fra tid til anden kan være nødvendigt at hale på blot den ene af wirerne.

DE geteilte Kurrleinenwinde

GR βαρούλκο ανεξάρτητης κίνησης· βίντσι ανεξάρτητης κίνησης

Βαρούλκο έλξης των σχοινιών μιας τράτας με δυνατότητα ανεξάρτητης κίνησης των κεφαλών και τυμπάνων.

EN split winch

FR treuil scindé

Treuil de chalut comportant deux bobines indépendantes, actionnées séparément. L'emplacement de chaque bobine est déterminé pour permettre un passage direct de la fune vers la poulie de potence.

IT verricello a tamburi indipendenti

Verricello di rete con due tamburi indipendenti azionati separatamente. L'ubicazione del tamburo è determinata in modo tale da consentire il passaggio diretto del cavo verso la pastecca dell'archetto.

NL split winch; split lier; dubbel uitgevoerde lier

Visserijlier met twee onafhankelijk draaibare liertrommels. Deze zijn zonder middenflens tegen elkaar geplaatst.

PT guincho duplo; guincho separado

Guincho de arrasto equipado com dois tambores independentes, accionados separadamente. O posicionamento de cada tambor é determinado de modo a permitir a saída directa do cabo real para a patesga de arrasto.

3405

ES cabo gancho; verina

Cable que lleva en su extremo un gancho especial, normalmente de caracol, y que se emplea para unir los dos cables de un buque de arrastre por el costado y llevarlos al «perro», pieza por la que pasan para remolcar la red.

DA wireoplægger

Wire (eller på mindre fartøjer en stang) med en krog i enden. Tjener til at samle trawlwirerne agter for at lette manøvreringen af fartøjet under slæbning. Bruges på sidetrawlere til at trække den forreste wire ind til siden for at lægge den ind i kasteblokken ved den agterste galge.

DE Haktau; Hakentau

Mit einem Haken versehenes Tau, mit dessen Hilfe auf Seitenfängern die Kurrleinen nach dem Aussetzen in den Sliphaken gehievt werden.

GR βοηθητικό σχοινί τράτας

Βοηθητικό σχοινί για τράτες πλάγιας έλξης, στερεωμένο έτσι στην τράτα ώστε να ελέγχεται η ομαλή ανέλκυση του σάκου της τράτας, ιδιαίτερα όταν αυτός είναι υπερβολικά γεμάτος με ψάρια.

EN messenger

FR vérine

Cordage terminé par un croc, servant à manœuvrer l'engin sur le pont de pêche; les vérines sont virées habituellement sur les poupées du treuil de pêche. *Sur les chalutiers latéraux, la vérine sert à ramener les funes ensemble, au filage, pour les saisir dans le chien.*

3405

IT trinca

Cavo terminato da un gancio e destinato a manovrare l'attrezzo sul ponte di lavoro; le trinche vengono di solito virate sulle campane del verricello.

NL thuishaler

Lijn voorzien van een haak, waarmee het vistuig aan dek naar de nettentrommel wordt getrokken. Op deze wijze komen de vislijnen los te liggen, zodat de voorlopers kunnen worden losgekoppeld; evenzo bij het uitvieren, om de voorlopers met de vislijnen te verbinden. Ze worden aan ieder nok van een trawl bevestigd.

PT estropo; cabo do pau de carga; mensageiro

Cabo terminado por um gato e que se destina à manobra das artes de pesca sobre o convés de pesca. Pode ser manobrado quer pelos cabeços dos guinchos de arrasto quer por um pau de carga.

Nos arrastões laterais (mensageira), destina-se à reunião dos dois cabos reais no final da largada com vista à sua colocação na patesga de arrasto.

3406

ES halador de palangre

Máquina que permite virar el cabo madre de los palangres. Hay diversos tipos dependiendo del diámetro de cabo que hay que virar.

DA linehaler

Spil til langlinefiskeri til indhaling af linen.

DE **Langleinenholer**

GR ολκός σχοινιών

Μηχανικό σύστημα εισολκής (κν. για το βιράρισμα) των σχοινιών ή του κάβου, π.χ. ένας εργάτης ή βαρούλκο.

EN line hauler

FR vire-ligne

Système mécanisé comportant habituellement une ou plusieurs poulies à gorge étroite, utilisé pour le relevage des lignes (palangres ou autres lignes).

IT salpapalangaro

Meccanismo che comporta di solito una o più pulegge a gola stretta ed è utilizzato per salpare le lenze (palangaro o altro).

NL lijnhaler

Mechanisch werktuig met een of meer schijven waarmee lijnen kunnen worden binnengehaald of gevierd. De lijn moet gemakkelijk in te voeren zijn, zonder het begin door de schijven te moeten voeren.

PT alador de linhas; alador de cabos

Equipamento de força normalmente constituído por uma polia com uma ou mais golas estreitas, utilizado para virar linhas e cabos (palangres, armadilhas).

3407

ES vivero

Cuba situada a bordo, con aportación continua de agua de mar, empleada para mantener el cebo vivo empleado en la pesca de túnidos con caña.

DA dam; tank

Stor beholder eller del af lasten på et fartøj indeholdende havvand, der udskiftes regelmæssigt. Anvendes til opbevaring af levende agn eller fangst, der skal sælges levende.

DE Seewassertank für Lebendköder; Bünn

GR ενυδρείο· δεξαμενή ζώντων αλιευμάτων

Δεξαμενή θαλάσσιου νερού για τη συντήρηση υδρόβιων ζώων, ζώντων δολωμάτων ή ειδικών αλιευμάτων (γαριδοειδή, αστακοειδή κλπ.).

EN live tank; fishbin

Container on board frequently replenished with seawater. Used to keep the catch alive.

FR vivier

Sorte de grande cuve ou compartiment dans lequel l'eau de mer est renouvelée régulièrement. Sert à conserver vivants les poissons utilisés comme appât (pêche à la canne à l'appât vivant) ou les espèces capturées (langouste ou crabe, par exemple).

IT vivaio

Contenitore a bordo di un peschereccio nel quale l'acqua di mare viene continuamente rinnovata e che serve per conservare le esche vive.

NL leeftank; bun

Afgesloten compartiment van een schip waarin zeewater kan worden gecirculeerd (d.m.v. een pomp) voor het levend opslaan van vis en andere zeedieren.
Wordt veel op onderzoeksvaartuigen gebruikt.

PT viveiro

Espécie de cuba de grandes dimensões ou tanque em que a água do mar é renovada periodicamente. Destina-se a conservar vivos quer os peixes utilizados como isco na pesca de salto e vara com isco vivo quer as espécies capturadas (lagostas ou caranguejos, por exemplo).

3408

ES pesca con luz; pesca a la mamparra

Variedad de la pesca al cerco, en la que el cardumen es concentrado por medio de la atracción de uno o varios focos, sumergidos o próximos a la superficie, situados en un bote auxiliar.
En el Mediterráneo español se llama pesca a la mamparra. También se pescan los cefalópodos desde buques provistos de máquinas automáticas que manejan grandes aparejos. Dichos buques disponen de una iluminación extremadamente fuerte que atrae la pesca.

DA lysfiskeri

Fiskeri hvor der anvendes kunstigt lys. Visse fiskearter tiltrækkes af lys og kan derefter fanges på krog, i not eller med et andet fiskeredskab.

DE Lichtfischerei

Fangmethode der Hochsee-, Küsten- und Binnenfischerei, bei der die konzentrierende oder scheuchende Wirkung künstlicher Lichtquellen beim Fischfang ausgenutzt wird.

GR πυροφάνι, αλιεία με πυροφάνι

Είδος αλιείας που βασίζεται στη χρησιμοποίηση τεχνητών πηγών φωτισμού για την προσέλκυση ψαριών στο χώρο αλιείας κατά τη διάρκεια της νύχτας.

EN light-fishing; fishing with light

FR pêche à la lumière; pêche au lamparo

Méthode de pêche employant la lumière artificielle, de surface ou sous-marine, pour attirer le poisson.

IT pesca con attrazione luminosa

Pesca che utilizza la luce artificiale, in superificie o subacquea, per attirare il pesce.

NL visserij met licht

Visserijmethode waarbij de vis naar het vistuig wordt gelokt door middel van lampen.
Wordt veel toegepast in de Middellandse Zee.

PT pesca ao candeio

Método de pesca que emprega a luz artificial, à superfície ou submarina, para atrair os peixes.

English index

A

3303 AB direction
3289 AB-cut
3328 acoustic sounder
3008 actual mesh size
3032 aerial trap
3198 allowable, total - catches
3346 amount of twist
3288 AN-cut
3049 anchored gillnet
3262 anchored line
3360 armoring
3360 armouring
3292 assembled yarn
3145 assembly, dan leno -
3287 AT-cut
3335 atmosphere, standard -
3358 automatic longline

B

3284 B-cut
3029 back strap
3155 backstrop
3386 backstrop ring
3107 bag
3065 bag net
3065 bag shaped net
3004 bag, strengthening -
3068 bait
3357 bait, live -
3223 bait, spoon -
3266 baiting
3250 baitings
3052 ballast
3314 bar
3284 bar cut
3397 bar, shooting -
3167 bar, spreader -
3224 barbless double tunny hook
3126 barrel
3126 barrel, winch -
3034 barricade
3036 barrier
3211 basket trap
3230 basket, eel -
3241 basket, ground -
3251 basket, wire -
3279 basnig
3266 bating
3250 batings
3103 beach seine
3082 beam
3203 beam head
3087 beam trawl
3322 becket bend
3030 becket, halving -
3162 belly
3268 belly, false -
3220 bend
3322 bend, becket -
3322 bend, sheet -
3322 bend, signal halliard -
3322 bend, single -

3322 bend, single sheet -
3407 bin, fish-
3063 blanket net
3138 block
3194 block eye
3137 block suspension
3361 block, power -
3192 block, snatch -
3204 block, towing -
3128 block, warp -
3128 block, warping -
3081 board
3100 board the net (verb)
3206 board, dan leno -
3071 board, diving -
3081 board, otter -
3206 board, pony -
3347 board, suberkrub otter -
3081 board, trawl -
3072 boat dredge
3362 boat, second -
3102 boat seine
3231 boat, dory -
3175 boat, trolling -
3066 boat-operated lift net
3149 bobbin
3150 bobbin
3158 bobbin wire
3149 bobbin, dan leno -
3101 body
3274 bolch
3274 bolch line
3274 bolsh
3022 boltrope
3159 boltrope
3094 boom, outrigger -
3252 bottom longline, floating -
3085 bottom otter trawl
3084 bottom pair trawl
3049 bottom-set gillnet
3254 bottom set longline
3241 bottom set pot
3090 bottom trawl
3012 bottom-side chafer
3127 bracket
3127 bracket, triangular -
3377 braid (verb)
3301 braided netting twine
3209 brail net
3209 brailer
3209 brailer, hand -
3216 branch line
3344 break, time-to-
3339 breaking length
3337 breaking load
3338 breaking load, mesh -
3140 bridle
3372 bull
3092 bull trawl
3181 bull trawling
3197 bulldog grip
3107 bunt
3371 buoy
3039 buoy line
3039 buoy rope
3053 buoyancy
3167 butterfly
3165 by-catch

C

3390 C-hook
3129 cable chock
3273 cable laid rope
3317 cabled netting twine
3293 casing
3270 cast
3059 cast net
3265 cast net, hand -
3264 cast net, supported -
3215 catch
3200 catch quota
3165 catch, by-
3026 catch, discarded -
3349 catch, handing over of -
3349 catch, transhipping of -
3257 catcher, pig -
3198 catches, total allowable -
3280 catching efficiency
3268 chafer
3012 chafer, bottom-side -
3011 chafer, top-side -
3010 chafing
3268 chafing gear
3380 chain, tickler -
3132 cheek
3129 chock
3129 chock, cable -
3179 coast fishery
3009 cod line
3016 codend
3018 codend
3379 codend guide
3394 codend lift
3395 codend lift
3005 codend, double -
3394 codend, lift of -
3015 codend, median lacing of a
        trouser -
3005 codend, trouser -
3005 codends, double -
3009 codline
3294 combination rope
3044 combined gillnet-trammel net
3294 combined rope
3033 corral
3033 corral, fish -
3058 oovor pot
3040 creel
3229 creel, prawn -
3309 cross-over point
3161 crowfoot
3176 cruiser, fishery -
3246 curved hook
3289 cut, AB-
3288 cut, AN-
3287 cut, AT-
3284 cut, B-
3284 cut, bar -
3285 cut, horizontal -
3282 cut, K-
3282 cut, knot -
3311 cut, length of -
3286 cut, N-
3285 cut, T-
3291 cut, type of -
3286 cut, vertical -

3283 cutting
3311 cutting length
3290 cutting rate

D

3145 dan leno assembly
3206 dan leno board
3149 dan leno bobbin
3125 dan leno stick
3272 dan leno triangle
3225 dandy
3003 Danish seine
3396 davit, purse line -
3160 deckie, lazy -
3160 decky, lazy -
3090 demersal trawl
3340 density, linear -
3343 density, resultant linear -
3071 depressor
3258 depressor, rake trawl with -
3222 devon
3222 devonspinner
3209 dip net
3305 direction of netting stretch
3305 direction of stretch
3341 direction of twist
3303 direction, AB-
3307 direction, N-
3306 direction, T-
3341 direction, twist -
3026 discarded catch
3071 diving board
3081 door
3081 door, trawl -
3231 dory
3231 dory boat
3005 double codend
3005 double codends
3091 double rig
3091 double rigging
3224 double tunny hook, barbless -
3316 double weaver's knot
3401 drag
3256 dragging, fly -
3109 draw string
3074 dredge
3072 dredge, boat -
3070 dredge, hand -
3263 dredge, mechanized -
3255 drift line
3169 drift net fishing
3169 drifting
3048 drifting gillnet
3048 driftnet
3126 drum
3188 drum, net -
3191 drum, split -
3126 drum, warp -
3134 drum, warping -
3134 drum, whipping -

E

3328 echo sounder
3238 echo, horizontal - ranging
3230 eel basket
3230 eel pot
3230 eel trap
3280 efficiency, catching -
3148 effort, fishing -

3334 elongation percent
3047 encircling gillnet
3114 encircling net
3134 end, warping -
3226 endless trolling line
3195 engine, towing -
3055 entangling net
3364 entrance
3024 extension piece
3217 eye
3194 eye, block -
3124 eye, kelly's -

F

3385 factor, selectivity -
3129 fairlead
3060 falling gear
3268 false belly
3141 false headline
3389 false link
3375 fast, get - (verb)
3374 fastener
3035 fence
3300 fibre, staple -
3319 filament
3298 filament yarn
3308 final treatment
3237 finder, fish-
3308 finish of netting
3033 fish corral
3269 fish hook
3237 fish loop
3040 fish pot
3370 fish pump
3105 fish school
3227 fish spear
3354 fish tackle
3043 fish trap
3034 fish weir
3196 fish, undersized -
3237 fish-finder
3407 fishbin
3174 fisheries protection
3176 fisheries protection vessel
3176 fishery cruiser
3176 fishery protection vessel
3179 fishery, coast -
3179 fishery, inshore -
3148 fishing effort
3193 fishing gear
3180 fishing ground
3177 fishing guard
3239 fishing intensity
3001 fishing net
3408 fishing with light
3154 fishing net, knotless -
3277 fishing net, knotted -
3104 fishing net, seine -
3257 fishing net, tunny -
3177 fishing patrol
3169 fishing, drift net -
3179 fishing, inshore -
3408 fishing, light--
3046 fixed gillnet
3035 fixed net
3014 flapper
3118 flat link
3051 fleet
3023 float
3112 float line
3112 float rope
3053 floatability
3252 floated line

3252 floating bottom longline
3212 floating gillnet
3252 floating line
3083 floating trawl
3256 fly dragging
3321 folded yarn
3147 foot rope
3038 fyke net

G

3119 G-hook
3119 G-link
3133 gallows, trawl -
3216 ganging
3216 gangion
3171 gantry, stern -
3172 gauge, mesh -
3268 gear, chafing -
3060 gear, falling -
3193 gear, fishing -
3387 gear, guiding-on -
3164 gear, trawl -
3375 get fast (verb)
3056 gillnet
3248 gillnet, monofilament -
3056 gillnet
3046 gillnet on stakes
3049 gillnet, anchored -
3049 gillnet, bottom set -
3044 gillnet, combined - trammel net
3048 gillnet, drifting -
3047 gillnet, encircling -
3046 gillnet, fixed -
3212 gillnet, floating -
3049 gillnet, set -
3354 gilson
3353 gilson winch
3354 gilson wire
3134 gipsy
3134 gipsy head
3240 gorge
3197 grip, bulldog -
3241 ground basket
3180 ground, fishing -
3147 groundrope
3177 guard, fishing -
3379 guide, codend -
3387 guiding-on gear
3245 gun-harpoon

H

3322 halliard, signal - bend
3030 halving becket
3209 hand brailer
3265 hand cast net
3070 hand dredge
3067 hand lift net
3243 hand-harpoon
3349 handing over of catch
3259 handline
3329 hanging
3333 hanging ratio
3332 hanging rope
3245 harpoon, gun -
3243 harpoon, hand-
3244 harpoon, rifle-
3215 haul
3267 haul
3100 haul in (verb)

276

O

P

Q

R

S

T

Índice español

Dansk indeks

DA

Deutscher Index

DE

DE

DE

Ελληνικό ευρετήριο

3240 ίσιο αγκίστρι
3237 ιχθυοανιχνευτής
3034 ιχθυοφραγμός

K

3282 K, κοπή -
3282 K, κόψιμο -
3388 κάβος έλξης
3152 καζίλι, επάνω -
3259 καθετή
3286 κάθετη κοπή
3286 κάθετο κόψιμο
3267 καλάδα
3251 καλάθι, συρμάτινο -
3221 καλαμαριέρα
3261 καλάμι
3260 καλάμι με μηχανισμό
3031 καλαμωτή
3093 καλάρισμα
3181 καλάρισμα τράτας με
     ζευγαρωτά σκάφη
3201 καλάρω
3058 κάλυψης, δοχείο -
3243 καμάκι χειρός
3245 καμάκι, πυροβόλο με -
3082 καμάρι
3335 κανονική ατμόσφαιρα
3286 κανονικό κόψιμο
3133 καπόνι
3396 - δικτύων γρι-γρι
3171 καπόνι, πρυμναίο -
3273 καρλίνο
3149 καρούλι
3150 καρούλι
3149 καρούλι dan leno
3158 καρούλια, κάτω γραντί με -
3035 καρτέρι, δίχτυ -
3071 κατάδυσης, σανίδα -
3205 κατασκευαστής δικτυών
3147 κάτω γραντί
3158 κάτω γραντί με καρούλια
3012 κάτω ποδιά
3368 κατώτερο φτερό
3234 κατώτερο φύλλο δικτυώματος
3219 κεντρίδα αγκιστριού
3094 κέρκος
3134 κεφαλάρι βαρούλκου
3214 κεφαλή αγκιστριού χωρίς μάτι
3134 κεφαλή αλυσέλικτρου
3241 κιούρτοι βυθού
3040 κιούρτος
3229 κιούρτος για γαρίδες
3123 κλειδί
3373 κλείσιμο
3117 κλώνος
3232 κλώνος, συρμάτινος -
3116 κλώσμα σχοινιού
3320 κλωσμένο νήμα
3162 κοιλιά
3313 κόμβος
3316   διπλός -
3313   - διχτυού
3017 κομμάτι δικτυώματος
3024 κομμάτι επιμήκυνσης
3322 κόμπος, απλός -
3277 κόμπους, πλεγμένο δίχτυ με -
3105 κοπάδι ψαριών
3283 κοπή
3284   - Β
3284   διαγώνια -
3282   - K
3286   κάθετη -
3290   μέθοδος -ς

3289   μέθοδος -ς AB
3288   μέθοδος -ς AN
3287   μέθοδος -ς AT
3311   μήκος -ς
3286   - N
3285   οριζόντια -
3285   - T
3291   τύπος -ς
3131 κόρακας εκφυγής
3233 κορυφαίο φύλλο δικτυώματος
3007 κοσκινίσματος, δίχτυ -
3073 κόσκινο
3063 κουβέρτα, δίχτυ -
3223 κουταλάκι
3223 κουτάλι
3150 κουτρουμπούκι
3209 κόφα
3241 κοφινέλα βυθού
3040 κοφινέλο
3211 κοφινέλο
3030 κοψαδούρος
3027 κοψαδούρος, οπίσθιος -
3024 κόψες
3024 κόψη
3283 κόψιμο
3284   - Β
3284   διαγώνιο -
3282   - K
3286   κάθετο -
3286   κανονικό -
3311   μήκος -ατος
3286   - N
3285   οριζόντιο -
3285   - T
3329 κρέμασμα διχτυού
3217 κρίκος αγκιστριού
3390 κρίκος ανοικτός
3118 κρίκος εσοχής
3389 κρίκος λυόμενος
3119 κρίκος-G
3228 κρίκος στίγγου
3028 κυκλική ζώνη
3028 κυκλική νεύρωση
3226 κυκλική συρτή
3114 κυκλωτικό δίχτυ
3168 κυκλωτικό δίχτυ χωρίς στίγγο
3126 κύλινδρος βαρούλκου
3126 κύλινδρος βιντσιού
3101 κύριο σώμα
3101 κύριο τμήμα
3018 κυρίως σάκος
3372 κυρίως σκάφος
3246 κυρτωμένο αγκίστρι
3399 κωδικός μεγέθους κλώσματος

Λ

3218 λαβή αγκιστριού
3108 lampara, δίχτυ -
3235 Larsen, τράτα -
3295 λατζάνα
3146 ληγαδούρα
3177 Λιμενικού Σώματος,αξιωματικοί-
3333 λόγος ανάρτησης
3192 λυκίσκος

M

3132 μάγουλο
3266 μάζεμα
3185 μαϊνάρω

3128 μακαράς
3138 μακαράς
3204 μακαράς ρυμούλκησης
3359 μάνα
3348 μαντάρι
3355 μάντος
3045 μανωμένο δίχτυ
3356 μανωμένων διχτυών, μεσαίο
     δίχτυ των -
3044 μανωμένων διχτυών,
     συνδυασμός απλαδιών και -
3161 μαραφούντι
3143 μάτι
3217   - αγκιστριού
3142   άνοιγμα -ού
3315   διαστάσεις -ού
3013   ελάχιστο μέγεθος -ού
3313   κόμβος -ού
3315   μέγεθος -ού
3172   μετρητής -ού
3173   μήκος -ού
3310   μήκος πλευράς -ού
3314   πλευρά -ού
3008   πραγματικό μέγεθος -ού
3338   φορτίο ρήξης του -ού
3120 μάτιση
3192 ματσαπλί σχιστό
3272 ματσέτα
3125 ματσόξυλο
3304 μέγεθος διχτυού
3013 μέγεθος, ελάχιστο - ματιού
3315 μέγεθος ματιού
3008 μέγεθος, πραγματικό - ματιού
3196 μέγεθος, ψάρι με - κατώτερο του
     ζητούμενου
3290 μέθοδος κοπής
3289   - AB
3288   - AN
3287   - AT
3356 μεσαίο δίχτυ των μανωμένων
     διχτυών
3054 μεσόνερα
3083 μεσοπελαγική τράτα
3077 μεσοπελαγική τράτα με
     ζευγαρωτά σκάφη
3079 μεσοπελαγική τράτα με πόρτες
3349 μεταφόρτωση
3172 μετρητής ματιού
3391 μέτρο, τρέχον -
3330 μήκος δικτυώματος
3339 μήκος θραύσης
3311 μήκος κοπής
3311 μήκος κοψίματος
3173 μήκος ματιού
3310 μήκος πλευράς ματιού
3339 μήκος ρήξης
3331 μήκος σχοινιού ανάρτησης
     διιπύου
3361 μηχανική τροχαλία
3361 μηχανικό παλάγκο
3263 μηχανισμό, δράγα με -
3260 μηχανισμό, καλάμι με -
3097 μηχανότρατα
3096   - οπίσθιας έλξης
3096   - οπίσθιας σύρσης
3095   - πλάγιας σύρσης
3096   - πρυμναίας σύρσης
3052 μολύβι
3248 μονονηματικό απλάδι
3105 μπάγκος ψαριών
3271 μπάλωμα διχτυών
3106 μπάντα
3156 μπάντα, άνω -
3128 μπαστέκα
3392 μπαστέκα
3069 μπέντουλας
3062   - ακτής

GR

303

GR

Index français

FR

FR

V

Y

Z

FR

Indice italiano

313

IT

U

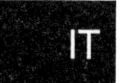

V

Y

Z

Nederlandse index

NL

NL

Z

NL

Índice Português

PT

L

3314 lado da malha
3310 lado da malha, comprimento do -
3314 lado de malha
3108 lâmpara
3060 lançada, arte -
3244 lançado, arpão mecanizado - por espingarda
3245 lançado, arpão mecanizado - por canhão
3215 lance
3267 lance
3176 lancha de fiscalização de pesca
3215 lanço
3267 lanço
3030 laracho
3160 laracho
3027 laracho, estropo do -
3201 largar
3306 largura
3235 Larsen, rede de arrasto -
3052 lastragem
3052 lastro
3095 lateral, arrastão -
3369 lateral, asa -
3159 lateral, cabo de porfio -
3095 lateral, navio de arrasto -
3069 leva, rede de -
3118 ligação, elo de -
3014 língua
3259 linha de mão
3261 linha de vara
3260 linha de vara mecanizada
3262 linha fundeada
3406 linha, alador de -s
3121 livre, pano de rede -
3026 lixo
3286 lombo
3228 lombo, corte ao -

M

3391 m/kg
3359 madre
3143 malha
3285 malha
3142 malha, abertura da -
3338 malha, carga de ruptura da -
3310 malha, comprimento do lado da -
3315 malha, dimensões da -
3314 malha, lado da -
3314 malha, meia -
3338 malha, tensão de ruptura da -
3142 malha, vazio da -
3173 malhagem
3008 malhagem efectiva
3013 malhagem mínima
3312 malheiro
3140 malheta
3139 malheta, brinco da -
3123 manilha
3243 manual, arpão -
3334 mão, ancinho de -
3070 mão, draga de -
3259 mão, linha de -
3059 mão, tarrafa de -
3265 mão, tarrafa de -
3363 maré
3340 massa por unidade de comprimento

3343 massa por unidade de comprimento do produto acabado
3361 mecânico, alador de redes -
3263 mecanizada, draga -
3260 mecanizada, linha de vara -
3244 mecanizado, arpão - lançado por espingarda
3245 mecanizado, arpão - lançado por canhão
3015 mediana, costura - de um saco duplo
3015 mediano, porfio - do saco duplo geminado
3314 meia malha
3252 meia-água, palangre fundeado de -
3150 meia-esfera
3150 meio-melão
3150 -melão, meio- -
3405 mensageiro
3388 mensageiro com pinha
3208 mestre de pesca
3205 mestre de redes
3251 metálica, nassa -
3251 metálica, nassa forrada com rede -
3290 método de corte
3391 metragem por quilo
3196 mínimo, pescado que não apresente o tamanho - legal
3294 misto, cabo -
3356 miúdo
3138 moitão
3128 moitão do cabo real
3194 moitão, alça de -
3137 moitão, alça do -
3293 moitão, caixa do -
3132 moitão, face do -
3136 moitão, roda de -
3392 molinete
3202 molinete da amarra
3195 molinete de reboque
3236 monofilamento
3248 monofilamento, rede de emalhar de -
3333 montagem, coeficiente de -
3138 moutão
3102 mugiganga
3292 multifilar, fio -
3040 murejona
3251 murejona

N

3286 N, corte -
3307 N, direcção -
3040 nassa
3211 nassa
3241 nassa
3251 nassa forrada com rede metálica
3251 nassa metálica
3229 nassa para camarões
3230 nassa para enguias
3097 navio de arrasto
3095 navio de arrasto lateral
3096 navio de arrasto pela popa
3176 navio de fiscalização de pesca
3313 nó
3322 nó de escota
3318 nó direito
3316 nó duplo
3322 nó simples
3324 nó, pano de rede com -s

3325 nó, pano de rede com -s com dois sistemas de fornecimento de fio
3326 nó, pano de rede sem -s
3277 nó, rede de pesca com -s
3323 nó, rede de pesca de um só fio com -s
3154 nó, rede sem -
3335 normal, atmosfera -
3307 normal, direcção -

O

3124 oito
3249 oito
3249 oito, elo em -
3217 olhal de anzol

P

3184 palangre
3358 palangre automático
3254 palangre de fundo
3255 palangre derivante
3252 palangre fundeado de meia-água
3252 palangre fundeado de superfície
3254 palangre fundeado de fundo
3226 palangre sem fim
3183 palangre, pesca com -
3115 palangreiro
3152 pana, cabo da -
3141 pana, cabo da - auxiliar
3141 pana, falso cabo da -
3017 pano de rede
3324 pano de rede com nós
3325 pano de rede com nós com dois sistemas de fornecimento de fio
3121 pano de rede livre
3007 pano de rede selectivo
3326 pano de rede sem nós
3330 pano de rede, comprimento de um -
3304 pano de rede, dimensões do -
3305 pano de rede, direcção de estiramento de um -
3278 pano de rede, estiramento de um -
3025 papagaio
3092 parelha
3181 parelha, pesca de -
3253 parelha, rede envolvente-arrastante de -
3084 parelha, rede de arrasto pelo fundo de -
3092 parelha, rede de arrasto de -
3031 parreira
3214 pata de um anzol
3192 patesca
3204 patesca de arrasto
3128 patesca do cabo real
3192 patesga
3405 pau de carga, cabo do -
3203 patim
3161 pé-de-galinha
3161 pé-de-galinha dos cabos reais
3017 peça de rede
3050 peça de rede
3366 pegamento
3146 pegamento, fio para -
3375 pegar
3146 pegar, fio para -

Bibliography

BEYLEN, J. VAN, ET AL.
Maritieme Encyclopedie
Uitgeverij C. de Boer, Bussum, 1970

BOER, E.J. DE
Visserijmethoden
het Visserijschap, Rijswijk, 1984

V. BRANDT, A.
Fish-catching methods of the world
Fishing News Books Ltd., Farnham, Surrey, England, 1984

BRIDGER, J. FOSTER, J.J., MARGETTS, A.R., STRANGE, E.S.
Glossary of United Kingdom fishing gear terms
Fishing News Books Ltd., Farnham, Surrey, England, 1981

BRONSVELD, J.J.
Visserijkunde
Het Visserijschap, 's-Gravenhage, 1983

CASTRO E SILVA, R.
Arte naval moderna. Aparelho e manobra de navios
Edição do autor
Editorial da Marinha, 9.ª edição
Damaia, Lisboa, 1979

DE LA CUEVA SANZ, M.S., SUBSECRETARÍA DE LA MARINA MERCANTE
Artes y aparejos, tecnología pesquera
Litografía EGRAF, Madrid, 1974

DLUHY, R.
Schiffstechnisches Wörterbuch, Band 1, Deutsch-Englisch
Vincentz Verlag, Hannover, 1983

DLUHY, R.
Schiffstechnisches Wörterbuch, Band 2, Englisch-Deutsch
Vincentz Verlag, Hannover, 1975

DOELMAN, H.D.
Aan de Haak - Spectrum Sportvissers Handboek
Het Spectrum, Utrecht/Antwerpen, 1977

ESPARTEIRO, A. MARQUES
Dicionário ilustrado da marinha
Livraria Clássica Editora
Lisboa, 1970

GRUSS, R
Dictionnaire Gruss de marine
Éditions maritimes et d'outre-mer, 1978

KERCHOVE, R. DE
International maritime dictionary
Van Nostrand Reinhold Company, New York, 1961

LAYTON, C.W.T.
Dictionary of nautical words and terms
Brown, Son & Ferguson Ltd., Glasgow, 1978

LEGA NAVALE ITALIANA
Dizionario enciclopedico marinaresco
U. Mursia & C., Milano, 1972

LEGARRA, J., Y OTROS
Inventario de artes de pesca en Euskadi
Editorial Itxaropena, SA, Zarautz, 1984

LEITÃO, HUMBERTO, LOPES, J. VICENTE
Dicionário da linguagem da marinha antiga e actual
Centro de Estudos Históricos Ultramarinos
Lisboa, 1963

LOZANO CABO, F.
Oceanografía, biología marina y pesca
Paraninfo, SA, Madrid, 1983

MARTÍNEZ-HIDALGO, J.M.
Diccionario náutico
Ediciones Garriga, Barcelona, 1977

MASTROPASQUA, V.
Dizionario tecnico nautico - Italiano-Inglese, Inglese-Italiano
Libreria Editrice Mario Bozzi, Genova, 1967

KRISTJONSON, H.
Modern Fishing Gear of the World
Food and Agriculture Organization of the United Nations, Rome
Fishing News Books Ltd., Farnham, Surrey, England, 1981

NEDELEC, C.
Definition and classification of fishing gear categories
Food and Agriculture Organization of the United Nations, Rome, 1982

POST, G.F., DUYNDAM, D.
Samenstellen en repareren van netwerk
het Visserijschap, Rijswijk, 1984

SCHARNOW, U. ET AL.
Lexikon der Seefahrt
VEB Verlag, Berlin, 1976

SCHOENMAKER, W.
*Nieuw Nautisch-Technisch Woordenboek, Engels-Nederlands,
     Nederlands-Engels*
Educaboek BV, Culemborg, Nederland, 1980

SILVA, A. A. BALDAQUE DA
Estado actual das pescas em Portugal
Imprensa Nacional
Lisboa, 1981

SILVA LOPES, ANA M. S.
O vocabulário marítimo português e o problema do mediterraneísmo
Revista Portuguesa de Filologia, XVI e XVII
Faculdade de Letras da Universidade de Coimbra
Instituto de Letras da Universidade de Coimbra
Instituto de Estudos Românticos
Coimbra, 1975

SUÁREZ GIL, L.
Diccionario técnico marítimo, inglés-español, español-inglés
Editorial Alhambra, SA, Madrid, 1983

VANDENBERGHE, J.-P., CHABALLE, L.Y.
  Elsevier's nautical dictionary
  Elsevier Scientific Publishing Company, Amsterdam, 1978

OFFICIAL JOURNAL OF THE EUROPEAN COMMUNITIES
  L 24, 1983
  L 194, 1984
  L 318, 1984

NEDERLANDSE NORM:
  NEN-ISO 858, 1976
  NEN-ISO 1530, 1976
  NEN-ISO 1532, 1976
  NEN-ISO 1805, 1976
  NEN-ISO 1806, 1976
  NEN-ISO 5307, 1990

INTERNATIONAL STANDARD:
  ISO 139
  ISO 858, 1973
  ISO 1107, 1974
  ISO 1139
  ISO 1140, 1975
  ISO 1141, 1975
  ISO 1142, 1973
  ISO 1181, 1973
  ISO 1346, 1975
  ISO 1530, 1973
  ISO 1531, 1973
  ISO 1532, 1973
  ISO 1805, 1973
  ISO 1806, 1973
  ISO 1968, 1973
  ISO 1969, 1976
  ISO 1970, 1973
  ISO 2062
  ISO 2075, 1972
  ISO 2307, 1972
  ISO 3090, 1974
  ISO/DIS 3169

INTERNATIONAL STANDARD:
  ISO 3660

NORMA ESPAÑOLA:
  UNE 40-212-74
  UNE 40-213-75
  UNE 40-219-73
  UNE 40-220-73
  UNE 40-221-73
  UNE 40-273-76
  UNE 40-278-76
  UNE 40-295-76
  UNE 40 296-75
  UNE 40-297-75
  UNE 40-300-75
  UNE 40-301-76
  UNE 40-302-75
  UNE 40-305-75
  UNE 40-317-76, parte I
  UNE 40-317-76, parte II
  UNE 40-401-79

DEUTSCHE NORMEN:
DIN 53842, Teil 1, 1976
DIN 53842, Teil 2, 1976
DIN 53844, 1976
DIN 53845, 1974
DIN 53846, 1982
DIN 60900, Blatt 1, 1975
DIN 60900, Blatt 3, 1975
DIN 60905, Blatt 1, 1970
DIN 61250, 1967
DIN 61251, 1966
DIN 61252, 1967
DIN 61253, Blatt 1, 1971
DIN 61253, Blatt 2, 1971
DIN 61254, 1967
DIN 61255, 1976
DIN 61256, 1976
DIN 83305, Teil 1, 1984
DIN 83305, Teil 2, 1984
DIN 83305, Teil 3, 1984
DIN 83305, Teil 4, 1984

NORME ITALIANE:
UNI 8100, 1980
UNI 8266, 1981
UNI 8680, 1984
UNI 8286, 1981
UNI 8287, 1981

NORMES FRANÇAISES:
G 36-101, 1969
G 36-102, 1969
G 36-103, 1970
G 36-104, 1972
G 36-105, 1974
G 36-106, 1974
G 36-150, 1970
G 36-151, 1970
G 36-152, 1972
G 36-153, 1972

Illustrations
Principal sources used

BEYLEN, J. VAN, ET AL.
Maritieme Encyclopedie
Uitgeverij C. de Boer, Bussum, 1970

V. BRANDT, A.
Fish-catching methods of the world
Fishing News Books Ltd., Farnham, Surrey, England, 1984

BRIDGER, J., FOSTER, J.J., MARGETTS, A.R., STRANGE, E.S.
Glossary of United Kingdom fishing gear terms
Fishing News Books Ltd., Farnham, Surrey, England, 1981

DE LA CUEVA SANZ, M.S., SUBSECRETARÍA DE LA MARINA MERCANTE
Artes y aparejos, tecnología pesquera
Litografía EGRAF, Madrid, 1974

LEGARRA, J., Y OTROS
Inventario de artes de pesca en Euskadi
Editorial Itxaropena, SA, Zarautz, 1984

LOZANO CABO, F.
Oceanografía, biología marina y pesca
Paraninfo, SA, Madrid, 1983

SCHARNOW, U., ET AL.
Lexikon der Seefahrt
Transpress Verlagsgesellschaft, Berlin, 1976

IRPEM
Inventario degli attrezzi da pesca

ESCOLA PROFISSIONAL DE PESCA (LISBOA)
Tecnologia da Pesca

ESCOLA PROFISSIONAL DE PESCA (LISBOA)
Remo, Vela e Motor

FAO
Vessel types

FAO
Training series 1, Pair trawling with small boats

FAO
Training series 3, Fishing with bottom gillnets

FAO, DOC. 222
Definición y clasificación de las diversas categorías de artes de pesca

AENOR
Normas UNE sobre redes de pesca